PHYSICS RESEARCH AND TECHNOLOGY

ADVANCES IN SEMICONDUCTOR RESEARCH

PHYSICS OF NANOSYSTEMS, SPINTRONICS AND TECHNOLOGICAL APPLICATIONS

PHYSICS RESEARCH AND TECHNOLOGY

Additional books in this series can be found on Nova's website
under the Series tab.

Additional e-books in this series can be found on Nova's website
under the e-book tab.

PHYSICS RESEARCH AND TECHNOLOGY

ADVANCES IN SEMICONDUCTOR RESEARCH

PHYSICS OF NANOSYSTEMS, SPINTRONICS AND TECHNOLOGICAL APPLICATIONS

DOMINIQUE PERSANO ADORNO
AND
SERGEY POKUTNYI
EDITORS

nova publishers
New York

LIBRARY OF CONGRESS CATALOGING-IN-PUBLICATION DATA

ISBN: 978-1-63321-755-3

Published by Nova Science Publishers, Inc. † New York

CONTENTS

PREFACE

The world of semiconductor research is continuously expanding our knowledge of the physics governing phenomena at micro and nano scales, driving the development of new technologies and rapidly enhancing the quality of our everyday life. The huge amount of scientific papers published up today in this field of research confirms the great interest of the scientific community for semiconductor science and its future applications. However, this enormous growth of available scientific information sometimes makes the familiar channels of communication considerable less effective, because of the difficulties for experts in a given field to keep up with the current literature. Advances in Semiconductor Physics Series has been conceived mainly to improve this situation.

This monograph presents a collection of selected contributions reporting some of the most stimulating and challenging results obtained by recent researches in the field of semiconductor physics. About the same number among theoretical, experimental and simulative studies have been included in this book, driven by the basic idea that all these different types of investigations are equally important in increasing our understanding of the physics of semiconductors. All contributions are designed and written with the aim of reporting research findings having a wide interest for an audience as broad as possible, including both graduate and master students, as well as specialists in experimental, theoretical or applied research in the physics of semiconductors.

In order to increase its readability, the book is structured into separate sections, describing recent results from researches on the following topics: (i) semiconductor nanophysics, optics and spectroscopy of nanostructures, (ii) spintronics and (iii) technological applications.

In the twenty-first century a large part of advances in semiconductor research is related to the physics of nanosystems, including zero-dimensional (quantum dots), one-dimensional (quantum wire) and two-dimensional (quantum layer or quantum wells) structures. This research area investigates a new class of nanomaterials, nanoparticles, layered nanostructures, molecular nanofilms, with sizes in the range 1÷10 nanometers, in which several optical, electrical, magnetic, thermal and superconducting properties, not available in bulk materials, can be explored. The radical differences between the physical properties of such nanosystems and those of bulk materials are due to the fact that spatial limitations may cause unforeseen nanodimensional effects, in which the spatial quantization of the energy spectra of quasi-particles (electrons, holes, excitons, phonons, plasmons) plays a crucial role. Intensive experimental, simulative and theoretical research in such nanostructures can guide scientists to solve the general problem of controlling the physical fundamental parameters in crystalline nanostructures and devices, such as the band gap, effective masses of charge carriers and their mobility, absorption and emission energy spectra, opening the way to large possibilities of technological applications. In this framework, nanomaterials, in particular, are promising structures for creating elements to control optical signals in semiconductor nanolasers and optical bistable elements and transistors, technology photosynthetic nanostructures, etc. In the first section of this monograph a general study on quantum size effects in semiconductor nanocrystals is presented. This section also highlights interesting results on noise induced critical phenomena in Si nanocrystals in the presence of metastable states and bi-functionality of semiconductor nanoparticles with a switching mechanism. The last two chapter of this section are entirely devoted to the interesting topic of spectroscopy of exciton states in quasi - zero - dimensional semiconductor nanostructures.

Relevant contributions to semiconductor research have been collected also in the field of spintronics, or spin electronics, regarding the study of active control and manipulation of spin degrees of freedom in solid-state systems. It offers a possible direction towards the development of hybrid devices that could perform logic operations, communication and storage, within the same material technology: electron spin could be used to store information, which could be transferred as attached to mobile carriers and, finally, detected. Despite these advantages, for the operability of prospective spintronic devices, the features of spin relaxation at relatively high temperatures, jointly with the influence of transport conditions, should be firstly well understood. Within the second section, a chapter is specifically devoted to the physics of spin dynamics in semiconductors working under streaming regime, where each free charge carrier, an electron or a hole, accelerates quasi-ballistically until reaching the optical-phonon energy, then emits an optical phonon and starts the next period of acceleration. The inclusion of spin degree of freedom into the streaming-regime kinetics gives rise to rich and interesting spin-related phenomena. Unusual relaxation phenomena and resonances are also presented in a general overview of spin dynamics in one-dimensional semiconductors. Finally for this section, specific simulative investigations of electron spin relaxation processes in Silicon crystals are presented.

The last part of the book provides selected insights on the fundamental properties of semiconductor structures having important technological applications, in terms of improving both optical absorption, oxidative kinetics and terahertz electric signals in high electron mobility transistors. The synthesis and properties of magnetosensitive nanocomposites with functions of medical-biological nanorobots are discussed.

In summary, this monograph contains a rich selection of important results of the current research in semiconductor physics, highlighting the great fertility and liveliness of scientific research in the various fields of this topic. Many questions raised in this book have practically never been addressed in the monograph literature. In this respect, the Editors are confident that this book will be useful as a reference and also contribute to further progress in the study of semiconductors.

The Editors:

Dominique Persano Adorno
Ph.D. in Applied Physics
Assistant Professor of Solid State Physics

Department of Physics and Chemistry
University of Palermo
Viale delle Scienze, Ed.18
90128 Palermo (Italy)
phone: +39 091 23899086
E-mail: dominique.persanoadorno@unipa.it
web site: http://www.unipa.it/dominique.persanoadorno/

Sergey I. Pokutnyi
Doctor of Sciences (Physics, Mathematics)
Professor of Theoretical Physics
Chuiko Institute of Surface Chemistry of National Academy of Sciences of Ukraine
17 General Naumov Str., 03164
Kyiv, Ukraine
E-mail: Pokutnyi_Sergey@inbox.ru

ACKNOWLEDGMENTS

It is a pleasure to be able to express our appreciation to:

– NOVA Publishing for giving us this great opportunity to highlight and disseminate such important results in the field of semiconductor research, leaving us free to opportunely select the scientific contents, and always providing a high level professional collaboration.

– All authors, who enthusiastically choose to join us in this research enterprise and hardly worked to prepare their contributions. We would like to thank twice most of them because of their participation both as author and reviewer of other chapters. We also appreciated the assistance of Dr. Nicola Pizzolato for his technical and scientific support in the preparation of the book.

– Our families, for their encouragement, support and kindness.

SECTION 1.
PHYSICS OF NANOSYSTEMS

In: Advances in Semiconductor Research
Editors: D. Persano Adorno and S. Pokutnyi
ISBN: 978-1-63321-755-3

Chapter 1

QUANTUM SIZE EFFECTS IN SEMICONDUCTOR NANOCRYSTALS

Faming Gao ***and Lihua Gao***
Key Laboratory of Applied Chemistry,
Yanshan University, Qinhuangdao, China

Abstract

Most semiconducting clusters show size-dependence effects in the nanometer region. Along with the energy gap of the cluster, the photoluminescence can even be varied through the red to blue region of the visible spectrum. Quantum size effects in semiconductor clusters have been a subject of extensive studies for the last two decades. We review recent developments of theoretical studies on the quantum size effects. The useful theoretical models based on the exciton confinement such as Effective-mass approximation (EMA), Empirical tight-binding method (ETBM) and Empirical pseudopotential method (EPM) have been established. In particular, a chemical bond method of quantum confinement is introduced in detail. In order to calculate readily the energy-gap shifts for clusters of all size, this method express the effects through algebraic relations rather than through variational solutions of the wave equation. It shows that a competition between the positive Kubo energy-gap shift and the negative surface energy shift plays the crucial role in the optical gaps of the nanosystems.

Keywords: quantum size effects, nanocrystals, semiconductor, chemical bond

Introduction

For the last two decades, semiconductor nanocrystals have attracted considerable interest due to their unique properties and potential use in the fields of nonlinear optics, luminescence, electronics, catalysis, solar energy conversion, and optoelectronics[1-3]. Current researches in semiconductor nanocrystals are mainly focused on the improved and controlled clusters growth, suitable process technologies, unique materials characterization, and better

* E-mail address: fmgao@ysu.edu.cn. Corresponding Author Address: No. 438 Hebei Street, Qinhuangdao, 066004, Hebei Province, China.

understanding of the structure-property relationships. Further research of nanotechnology will certainly lead to significant breakthrough in semiconductor industry. The development of new semiconductor clusters depends not only on materials engineering at a practical level, but also on a comprehensive understanding of the properties of materials, as well as the fundamental science behind these properties. Structure-property relationships have always been the intrinsically and most important aspect to materials science and engineering.

Reducing the size of semiconductor to nanometer range leads to the appearance of quantum size effect, surface effect, small-size effect, macroscopic quantum tunneling effect, dielectric confinement effect, etc., in which quantum size effect has been one of the most studied subjects, both in practice and theory. Quantum size effect gives rise to the discrete energy levels depending on the size of the structure as it is known from the simple potential well treated in introductory quantum mechanics. Among many properties that exhibit a dependence upon size in semiconductor cluster, blue shift (increase) of band gap energy is of particular importance. Along with the energy-gap of the cluster, the photoluminescence can even be varied through the red to the blue region of the visible spectrum. Thus, this allows scientists the unique opportunity to tune the electronic and chemical properties of a material simply by controlling its particle size. Theoretically, predicting accurately the spectra of clusters as a function of their size is one of the fundamental nanoscientific issues, and of utmost importance.

The pioneering work to allow for the influence of size quantization on interband absorption in a semiconductor sphereis the Effective Mass Approximation (EMA) proposed by Al. L. Efros and A. L. Efros [1]. It assumes a particle in a potential well with an infinite potential barrier at the particle boundary. Three size regimes on the basis of quantum scales are identified:

Firstly, when the radius R of a semiconductor is less than the Bohr radii of electrons a_e, and holes a_h, the blue shift can be expressed as [1]:

$$\Delta E = \frac{\hbar^2 \pi^2}{2\mu R^2} \tag{1}$$

where μ is the reduced mass of an electron-hole pair. The magnitude of Coulomb attraction can be neglected due to that the kinetic energy of localization is much larger, and this case is termed Strong Confinement.

Secondly,when a_h<R<a_e,the shift is expressed as:

$$\Delta E = \frac{\hbar^2 \pi^2}{2m_e R^2} \tag{2}$$

here m_e is the effective electron mass, and this is the heavy hole case.

Thirdly, when R >{ a_e, a_h },

$$\Delta E = \frac{\hbar^2 \pi^2}{2MR^2} - E^*_{Rg} \tag{3}$$

M is the total mass of electron m_e and hole m_h, and E^*_{Rg} is the effectiveRydberg energy:

$$E^*_{Rg} = \frac{e^4}{2\varepsilon^2\hbar^2(m_e^{-1} + m_h^{-1})} \tag{4}$$

This case is termed Weak Confinement. In the EMA model, the energy band is parabolic near the band-edge.

Later Brus[5-7] modified the EMA model and proposed the following expression for the band gap shift of the finite sized system:

$$\Delta E_g = \frac{\hbar^2\pi^2}{2R^2}\left(\frac{1}{m_e} + \frac{1}{m_h}\right) - 1.786\frac{e^2}{\varepsilon R} - 0.248E^*_{Rg} \tag{5}$$

The first term is the kinetic energy of both the electron and the hole, representing a relationship between quantum localization energy or confinement energy and the radius R. The second term shows the Columbic interaction energy between the electron and hole. ε is the dielectric constant. The third term is size independent and is generally small compared to the other two terms, which can be negligible, except for the semiconductors with small dielectric constant [2]. According to Eq. (5), the band gap increases as the quantum dot radius decreases, since that the quantum localization term shifts to higher energy with decreased R, meanwhile the Columbic term shifts excited electronic state to lower value. The EMA model apparently provides a good understanding for the blue shift of the optical absorption threshold. However, this model is suitable only for very large nanocrystals. When R decreases to a small regime, the eigenvalues of the lowest excited states are located in a region of the energy band that is no longer be approximated as parabolic, and the EMA model breaks down. This model gives quantitatively accurate results only for very large clusters [1]. Hence, there is a need for microscopic calculations geared towards mapping out the energy-level structure of these clusters and explain the salient features of their absorption spectra. Krishna et al., [3, 4] carried out a systematic investigation of quantum confinement effects in clusters using pseudopotentials. Their study provides a global understanding of the changes taking place in the electronic structure of clusters as the cluster dimensions, crystal structure, physical shape, and the lattice constant are varied. Pokutnyi[5-7] proposed a modified effective mass approximation to describe the exciton energy spectrum in semiconductor quantum dots with the radii a comparable to the exciton Bohr radius a_{ex}^0 .It is shown that, when the reduced effective mass of the excitons is assumed as a function of the radius of the quantum dot, the modified EMA method could be used to describe the exciton for quantum dots with radii $a \approx a_{ex}^0$.

Öğüt et al. [8] have implemented *ab* initio realspace calculations of quasiparticle and local density approximation (LDA) band gaps, self-energy corrections, exciton Coulomb energies, and optical gaps for Si quantum dots with diameters up to 2.72 nm using the higher-order finite pseudopotential method. Their calculated optical gaps are in very good agreement with the optical absorption data from Si nanocrystallites. Although *ab* initio calculations appear to be a more desirable way to investigate the optical gaps of clusters [8],for most nanosystems containing from a few thousands to 10^6 atoms, *ab* initio calculations are still too laborious to be practical.Wang [9] proposed an empirical tight-binding model (ETBM) to calculate the band gap of PbS with a sp^3pseudofuctions centered on Pb and S atoms, which can give a better description of the band structure and is more adequate for small crystallites. Then Lippens and Lannoo[10] developed the ETBM model to study the size dependence of the band gap for CdS, ZnS and CdSe, using a 13 parameters sp^3s^* model without the spin-orbit interaction. The crystallites observed had a symmetric shape and dangling orbitals were neglected. The recursion method was adopted for the reason that it is an iterative algorithm and does not require the storage of all the elements of Hamiltonian matrix. Ren et al. [11] used the ETBM model to study the electronic structures of surface-hydrogenated Si cluster, by using the group theory to simplify the diagonalization of their matrices. Similar researches demonstrated that the ETBM can give useful information about semiconductor nanocrystals.

There are also other approaches to investigate the size-dependent band gaps of semiconductor clusters, such as the effective bond order method (EBOM)[15, 16], the empirical pseudopotential method (EPM) [12], perturbation theory [13], etc. However, major computational efforts and difficulties do not allow one to calculate the properties of large sized nanocrystals, hence they can only treat thousand-atom electronic structure calculations. Furthermore, almost all existing theoretical calculations for band gap on semiconductor clusters to date assume that interior-to-interior optical transitions dominate the optical properties, neglect the surface effects, and effectively treat the crystal surface as a perfectly reflecting wall [19].Nevertheless, the optical properties of clusters can be significantly affected by the surface states due to the high surface-to-volume ration. A 5nm CdS semiconductor dot has roughly 15% atoms at the surface. Such a high surface-to-volume ratio may induce an enhanced or reduced transfer rate of photogenerated charge carries because of the high density of surface sites which arise from unsatisfied bonds at the reconstructed surface. The energies of these surface states generally lie in the bandgap of the semiconductor dots[20]. Hence, theycan trap charge carriers (electron or hole) and behave as reducing (electron) or oxidizing (hole) agents [14].This behavior at the surface can affect the overall conductivityand optical properties. As a result, surface states have significant effects on the optical andoptoelectronic properties of the semiconductor dots. It would lead to insufficient prediction for the spectra of clusters, and hinder the understanding of novel properties if we neglect the surface effect of semiconductor clusters, especially for those with the dimensions of only a few nanometers.

Methods

Chemical Bond Theory for Quantum Size Effects

Recently, a chemical bond method of quantum confinement effects is proposed to study the band gap of quantum clusters with different sizes and configurations [22, 23].In their opinions, the optical gap shift may be basically decomposed into contributions due to the surface effects and quantum effects, both of which are size dependent. The surface effects have a high correlation with the bonding and fraction of atoms at the surface, and cannot be neglected when the size of semiconductors decreases to nanoscale. The quantum effects are related to the reduction of the band structure into discrete quantum levels because of the limited atoms of the nanocrystals. Consequently, the energy gap shift ΔE, which may be measured by the shift of absorption threshold of the nanocrystal relative to that of the bulk, can be written as the sum of the quantum effect shift part ΔE^{kubo} and the surface effect shift part $\Delta E^{\text{surface}}$[15]:

$$\Delta E_g(D) = \Delta E^{\text{kubo}} + \Delta E^{\text{surface}} \tag{6}$$

For a bulk crystal, the interaction of an infinite number of atomic orbitals results in a continuum of levels, i.e. band. The numbers of atomic orbitals which take part in combination will become less along with the decrease of crystal size, causing that the band continuum becomes broken down to quantized energy levels. The gap between the highest occupied molecular (HOMO) and the lowest unoccupied molecular orbital (LUMO) energy level may be called the Kubo gap [24]. Accordingly, Kubo energy shiftΔE^{kubo}stems from that the electrons are confined inside a finite volume V. An empirical expression for the Kubo energy shift was proposed as follow [15]:

$$\Delta E^{\text{kubo}}=(m_e e^4/2\hbar^2)(VN_e)^{-1/3} \tag{7}$$

where N_e is the valence electron density of the bonds:

$$N_e = (n_e)^* N_b \tag{8}$$

$$n_e = (Z^*_{\text{cation}} / N_{\text{C cation}} + Z^*_{\text{anion}} / N_{\text{C anion}}) \tag{9}$$

N_b is the density of chemical bond (*i.e.*the number of chemical bonds per cubic centimeter) which can be obtained from the crystal structure parameters. $n_e$is the number of effective valence electrons per chemical bond. Z^*_{cation} and Z^*_{anion} are the numbers of effective valence of cation and anion, respectively. $N_{\text{C cation}}$ and $N_{\text{C anion}}$ represents the nearest coordination number of cation and anion, respectively. In nanoscale systems, the average surface atom coordination number equals to the total bond number at the surface divided by the number of the surface atoms in the nanocrystal, which were employed to study the electronic structure of nanocrystal by Mason [16].

On the other hand, the termination of the lattice periodicity in the surface of nanoclusters not only creates a potential boundary but also reduces the coordination number of the surface atoms [17] and increases the ratio of surface-to-volume. For the surface region of semiconductor clusters, the bond length contracts due to the surface coordination number imperfection. It is believed that the bond contraction at the surface and the rise in the surface-to-volume ratio of nanoscaled materials are responsible for some derivatives of properties from the bulk crystals. Here, the structure of a nanoparticle may be classified into two regions: the surface and the inside. The former has large dangling bonds and atoms with imperfect coordination number, while the latter is assumed the same as the bulk. The difference between the surface and the inside of nanoparticles is mainly the coordination number loss in surface, which is just the origin of surface effect. Therefore, from the chemical bond point of view, a nanocluster contains two different kinds of bonds: surface bonds and inside bonds. Since the electron feels the effects of the nanocrystal boundaries in addition to the periodic potential, the surface energy shift part should be resolved into contributions from the two types of bonds, which may be expressed as [15]:

$$\Delta E^{\text{surface}} = F^{\text{surface}}(E_{\text{g}}^{\text{surface}} - E_{\text{g}}^{\text{bulk}}) \tag{10}$$

where F^{surface} is the fraction of surface bonds composing the nanocluster. $E_{\text{g}}^{\text{surface}}$ and $E_{\text{g}}^{\text{bulk}}$ represent the average gap of the surface bond of the nanocluster and the bulk phase, respectively. These average gaps may be obtained using Phillips' chemical bond theory [18, 19].

Phillips' chemical bond theory is an simple means of obtaining the energy-gaps of semiconductor crystals with reasonable accuracy[27, 28],and hasbeen successfully applied to many aspects, especially those that are difficult to address by first principles technique[20]. For a binary compound, the Hamiltonian and potential may be expressed as [21]:

$$\hat{H} = -\frac{\hbar^2}{2m_e}\nabla^2 + V(\mathbf{r}) \tag{11}$$

$$V(\mathbf{r}) = \sum_G [V_s(G)S_s(\mathbf{G}) + iV_A(G)S_A(\mathbf{G})]\,\mathbf{exp}(i\mathbf{G}\cdot\mathbf{r}) \tag{12}$$

Here **G** is a reciprocal lattice vector, $S_s(\mathbf{G})$ and $S_A(\mathbf{G})$ are the symmetric and antisymmetric structure factors, and $V_s(G)$ and $V_A(G)$ are the symmetric and antisymmetric form factors. The matrix elements of the Hamiltonian are expressed as [18]:

$$H_{11}=H_{22}=0 \tag{13}$$

$$H_{12}=H_{21}{}^{*}=E_{\text{h}}+iC, \tag{14}$$

To first order in $V(\mathbf{r})$, the symmetric, homopolar part $E_{\text{h}}$and the antisymmetric, heteropolar part C arise from the real partand the imaginary partof Eq. (12), respectively. The decomposition of average energy gap E_{g} between the highest occupied and the lowest unoccupied energy level may be:

$$E_{\mathbf{g}}^2 = E_{\mathbf{h}}^2 + C^2 \tag{15}$$

In the case of a pure covalent Group IV clusters, E_{g} just equal to E_{h} due to C=0 by symmetry, and E_{h} is a function of nearest-neighbor distance. Thus E_{g}can be expressed as a function of the bond length d:

$$E_{\mathbf{h}} = 39.74 / d^{2.48} \tag{16}$$

where d and E_{h} are in Åand eV, respectively.

$$C = 14.4(\beta \cdot N_C^2) \cdot (Z^*_{\text{cation}} - Z^*_{\text{anion}}) \exp(-k_s \cdot d/2)/(d/2) \quad (17)$$

$$k_s = 2(3\pi^2 N_e)^{1/6} / (\pi a_B)^{1/2} \quad (18)$$

here βdepends on the crystal structure and can be determined via the long-wave dielectricconstant $\varepsilon(\infty)$ of the crystal. a_B is Bohr radius.N_C is the averagecoordination number.

It is shown from the formulae above that bond length plays an important role in calculating the energygaps and the bond length is different for the surface and inside ones.Here, the bond length data of bulk crystal is taken as the inside bond length, while the bond length data of the crystal-cage with only surface atoms is taken as the outside bond length. Figure 1 is the schematic of a nanocrystal sphere and a crystal-cage with only surface atoms.The geometry optimization by first principles for bulk crystal and the crystal-cage was carried out to determine the bond length for surface and inside of semiconductor clusters.

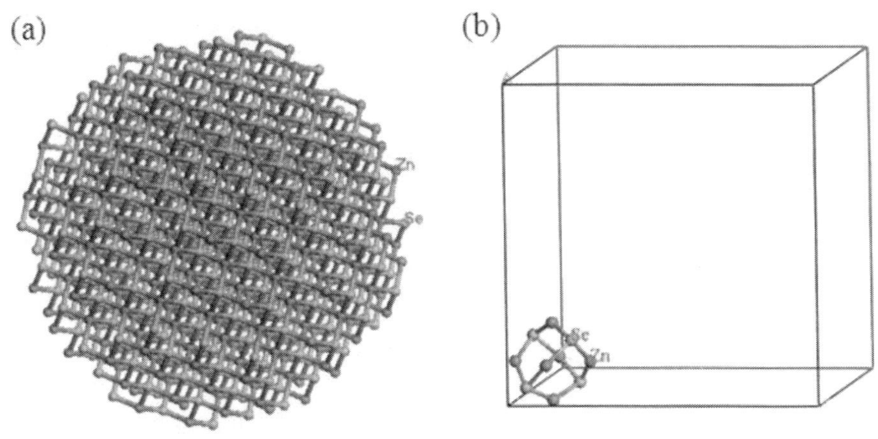

Figure 1. Nanocrystal structure of semiconductor: (a) schematic of a nanocrystal sphere of ZnSe; (b) the crystal-cage of ZnSe with only surface atoms.

The Band Gap Shifts of II-VI, III-V Compound Nanocrystals

CdS, CdSe, ZnS,InP, InAs and GaAs crystals are typical binary semiconductors, whose clusters have been studied using above theory. Figure 2a-f gives the energy gap shifts calculated for CdS, ZnS, InP, InAs, GaAs clusters with zinc blende structure and CdSe clusters with wurtzite sturcute, respectively, as well as the experimental data from other models [15].

Since the change in energy gaps is expressed as a function of the size with respect to that of the bulk, the zero of the energy axis corresponds to the bulk values. The EMA results are shown by the dashed line, which is found to indeed overestimate the bandgaps for InP clusters. While the values calculated from the ETBM model underestimate the experimental energy shifts for the ZnS clusters. What's more, the curve of energy shifts with decreasing cluster size shows a zigzag shape, which is different from traditional results of smooth curves, indicating that the energies do not always shift to blue with decreasing cluster size.

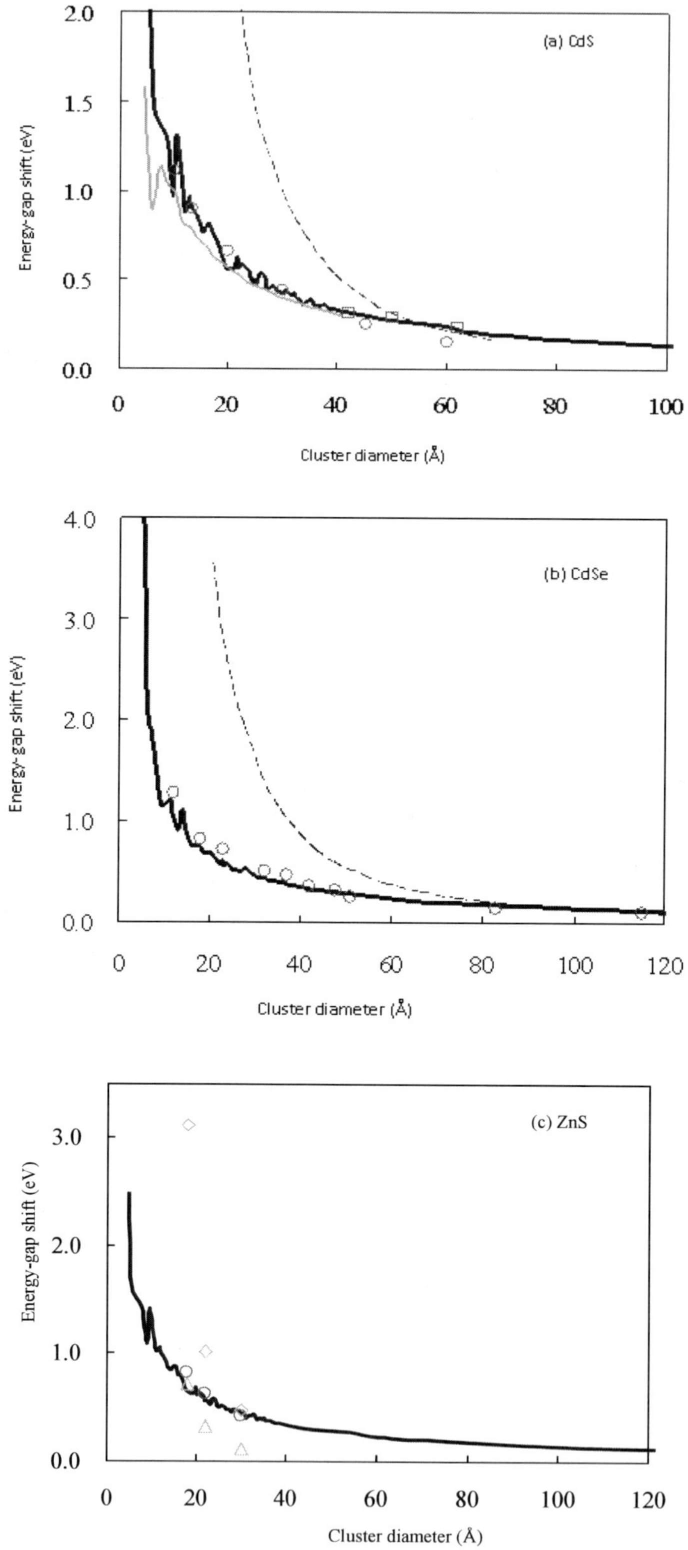

Figure 2. Continued on next page.

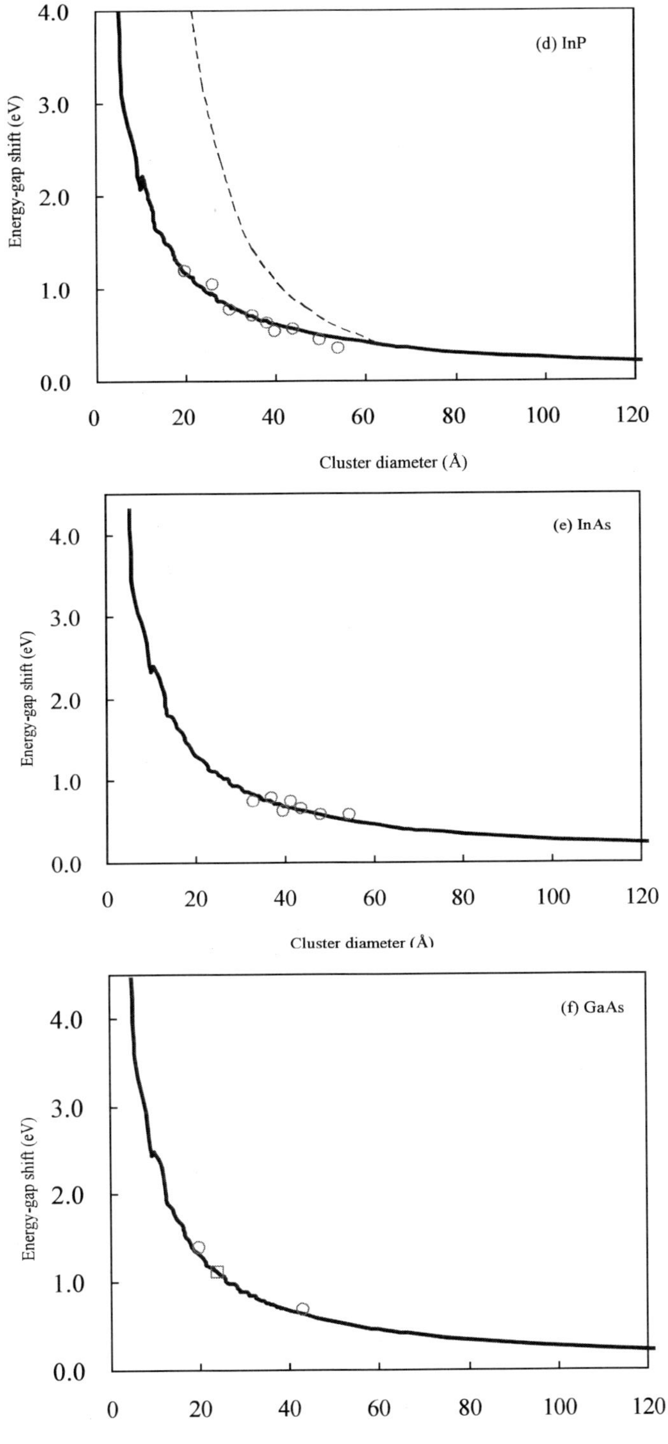

Figure2. Experimental and calculated energy-gap shifts of zinc-blende (a) CdS, (c) ZnS, (d) InP, (e) InAs, (f) GaAs and (b) wurtzite CdSe. The blue solid lines are from the calculations of the spherical clusters using the chemical bond theory. The curves obtained fromEMA are shown by the dashed line. The experimental data are denoted by: (a) ○[2] and □[22]; (b) ○[23]; (c) ○, ◊ from EMA results and △ from ETBM results [24]; (d) ○[25]; (e) ○[26]; (f) ○[27] and □[28].

The Band Gap Shifts of Diamond and Si Nanocrystals

The diamond-cages with the structure of diamondoidsare a unique form of diamond clusters, which attractive unprecedented investigations about the solid state properties as a function of size and shape. The smallest diamondoid is adamantane, and other further members, such as diamantane, triamantane, tetramantane, tetramantane and pentamantane, are derived by adding one face fused cage at a time to the preceding diamondoid structure[38, 39]. The calculated band gap shift ΔE for dimond cages are shown in Table 1. It can be seen that the surface energy shifts for the diamond-cage with the structure of diamondoids are negative values, which slightly reduce the Kubo energy shift. And there is a bit difference in surface energy shifts between two isomers of tetramantane because of different atomic number at the surface [15].

Table 1. Theoretical and experimental energy gap shifts for the carbon cages with the diamondoid structure and the spherical sillicon clusters of different sizes. Energy-gap shift ΔE, kubo energy shift ΔE^{kubo} and surface energy shift $\Delta E^{surface}$ are in eV. The diameter of silicon clusters is D in Å. ΔE_{cal} and ΔE_{exp} are the calculated and experimental results, respectively [15]

diamond cage	ΔE^{kubo}	$\Delta E^{surface}$	ΔE_{cal}	ΔE_{exp}
C_{10}(adamantine)	3.98	-0.31	3.67	3.71
C_{14}(diamantane)	3.55	-0.31	3.25	3.28
C_{16}(triamantane)	3.27	-0.28	2.99	3.01
C_{22}(1(2)3-tetramantane)	3.06	0.25	2.81	2.87
C_{22}(123-tetramantane)	3.06	-0.27	2.79	—
C_{26}(1(2, 3)4-pentamantane)	2.89	-0.21	2.69	2.66
Si (D = 23.1 Å)	1.24	0.02	1.26	1.30
Si (D = 23.9 Å)	1.20	0.01	1.21	1.25
Si (D = 24.3 Å)	1.18	0.02	1.20	1.15
Si (D = 25.0 Å)	1.12	0.01	1.13	1.12
Si (D = 26.8 Å)	1.09	0.01	1.10	1.00
Si (D = 30.4 Å)	0.95	0.01	0.96	0.94

The results of the nano Si with spherical shapes are also shown in Table 1. The surface energy shifts of the nano Si can be neglected as they are very small. Thus, their energy-gap shifts rest with kubo energy shifts. There is a good agreement between the calculated and the experimental results, demonstrating the predictive power of chemical bond method.

The Band Gap Shifts of Oxides Nanocrystals

Titanium oxide (TiO_2) is a well-known photocatalysis materialwhich has been extensively studied with regard to synthesis, properties and applications. Since the catalytic properties may be affected by energy gap shift, the investigation of quantum size effect on band gap is of intense interest. Serpone [29] synthesized anatase TiO_2 nano clusters with average sizes 2.1, 13.3 and 26.7nm, and characterized the clusters by absorption and photoluminescence spectroscopies. Under their conditions, no significant size quantization (such as blue shifts in UV-vis spectra) was observed. Recently, TiO_2 clusters were synthesized using the DPA G4 template, and showed the quantum size effects [30]. A blue shift with decreasing particle size was observed, and the energy gap was dependent on the crystal form of the material. Using the proposed chemical bond method [31], the energy gap shift of spherical rutile and anatase cluster may be predicted, as shown in Figure 3 [15]. Consistency between the predictions and Satoh's experimental results [30] confirmed the accuracy of the chemical bond theory.

It can be seen from Figure 3(b)' inset that the energy gap shift ΔE approaches zero of the energy axis when $D>$ 6nm, which may explain why no blue shifts in UV-vis spectra were observed in Serpone's work [29]. Another interesting phenomena is that some anatase TiO_2 clusters show considerable redshift. For example, the energy shift ΔE is -0.31eV for anatase TiO_2 clusters with the size 0.95nm, causing the reduction of the energy gaps down to visible light range. It is known that the photoactivity of anatase TiO_2 is limited to the UV region, and how to extend the energy gap of TiO_2 into the visible range has attracted great interest. Doping seems an effective approach [42, 43], however, doped materials suffer from a thermal instability [32]. According to the calculated results, controlling the size of the semiconductor cluster may be a possible method to extend the energy gap of TiO_2 into visible region.

For rutile TiO_2 clusters with the size 1.0±0.2nm, 1.23±0.2nm and 1.49±0.2nm, the calculated average energy gap shift, 0.83eV, 0.55eV and 0.42eV, are very close to the corresponding experimental ones, 1.0eV, 0.6eV and 0.47eV, respectively. And for anatase TiO_2 clusters with size 1.09±0.2nm, 1.23±0.2nm and 1.57±0.2nm, the calculated average energy-gap shifts, 0.53eV, 0.52eV and 0.32eV, are also close to the corresponding experimental ones[41], 0.68eV, 0.47eV and 0.24eV, respectively. The calculated energy gap shift is the sum of the kubo energy shift and the surface energy shift, therefore, we can evaluate the variations of kubo and surface energy shift as a function of size for TiO_2 clusters separately.

As shown in Figure 3 [15], the kubo energy shift ΔE^{kubo} smoothly approaches to the bulk values with increasing size of the cluster. However, the curve of surface energy shift $\Delta E^{surface}$ shows an intense zigzag shape, which is almost the same with that of ΔE, suggesting that the shape of the energy gap shift curve is dominated by that of the surface energy gap shift.

ZnO is a versatile material that is bio-safe and biodegradable, and has been widely used in medical applications and environmental science.Like TiO_2 clusters, ZnO also shows quantum confinement effects in the experimentally accessible rage of sizes (<7 nm). The calculated energy gap shift of spherical ZnO nanocrystals with different diameters are listed in Table 2[44, 45]. It can be seen from the calculated results that the bond length of surface and inside atoms remain constant respectively during the variation of D for a given semiconductor, while the surface bond lengths of the nanocrystal are a little shorter than the inside ones because of surface reconstruction. Beside the effect of bond length, the effect of change in coordination number of surface atoms on band gap is also important based on the formulae. The valence electron densities of the surface bonds N_e are much higher that that of the inside ones due to the lower average coordination number N_c of the surface atoms. In addition, the heteropolar coulomb gap C of the surface bonds decrease by almost 3 times.

That's to say, the change of bond length, the coordination number of surface atoms as well as the heteropolar coulomb gap play important roles in band gap shift.

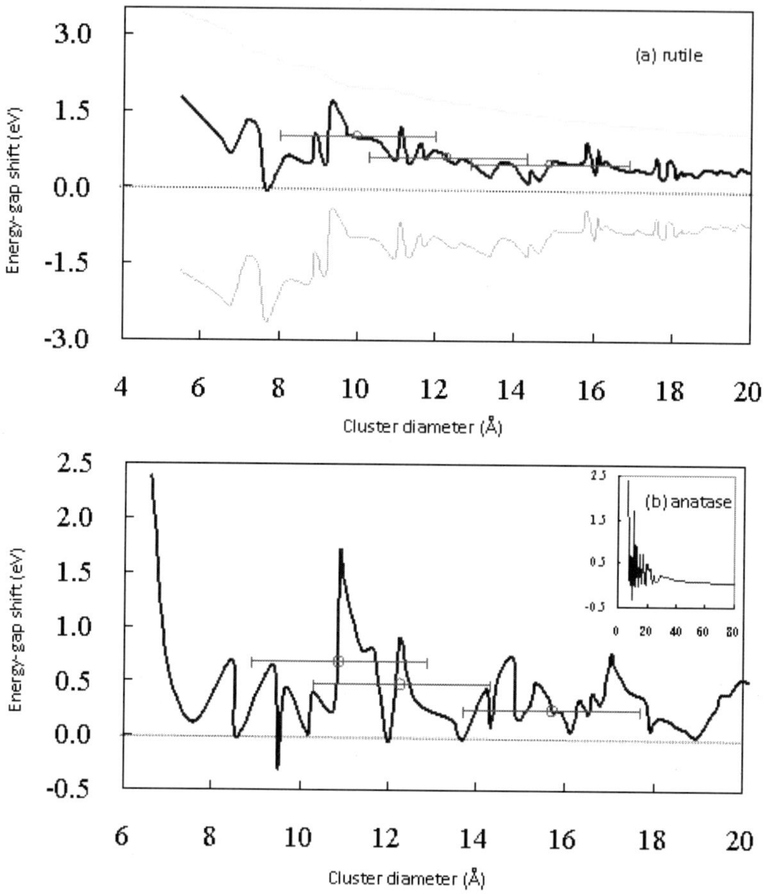

Figure 3. Dependence of energy-gap shifts of TiO_2 on cluster sizes and sturctures. Solid lines are the calculations of spherical clusters using the chemical bond theory. Open circles represent the experimental data[22]. (a) Rutile TiO_2 clusters:the blue, yellow and green lines represent energy-gap shift ΔE, kubo energy shift ΔE^{kubo} and surface energy shift $\Delta E^{surface}$, respectively. (b) Anatase TiO_2 clusters:the blue line represent energy-gap shift ΔE. Inset shows the trend of ΔE.

Although the surface atoms decrease with the reduced diameters, both the proportion of surface atoms to bulk atoms and the proportion of surface bonds $F^{surface}$ increase more obviously, which lead to a gradually enhanced surface effect and a higher energetic state of surface atoms. The transition energy from the valence to the conduction band increase, effectively increasing the band gap energy. According to the calculated results, $\Delta E^{surface}$ is more sensitive to $F^{surface}$ compared with$E_g^{surface}$ which remainsapproximatelyconstant during the variationof D. The relationship between $\Delta E^{surface}$ and $F^{surface}$ is almost linear for a given nanocrystal alloy.

Table 2. Theoretical data for the Spherical Nano ZnO clusters. N_c is average coordination number; energy-gap shift ΔE, kubo energy shift ΔE^{kubo} and surface energy shift $\Delta E^{surface}$ are in eV

D(Å)		d(Å)	N_e(Å^{-3})	N_c	C(eV)	E_h(eV)	E_g(eV)	ΔE^{kubo}	$\Delta E^{surface}$	ΔE
57. 5	inside	1.988	0.331	4.000	9.145	7.230	11.658	0.422	-0.316	0.106
	surface	1.930	0.544	2.659	3.678	7.781	8.607			
67. 5	inside	1.988	0.331	4.000	9.145	7.230	11.658	0.359	-0.268	0.092
	surface	1.930	0.547	2.641	3.622	7.781	8.583			
105	inside	1.988	0.331	4.000	9.145	7.230	11.658	0.231	-0.172	0.059
	surface	1.930	0.542	2.666	3.701	7.781	8.617			
120	inside	1.988	0.331	4.000	9.145	7.230	11.658	0.202	-0.151	0.051
	surface	1.930	0.543	2.663	3.693	7.781	8.613			

The Band Gap Shifts of Semiconductor Nanowires

Nanotechnology based on nanowires has opened up new directions in band gap engineering of optoelectronic devices. Among the III-V semiconductor nanowires, InP nanowires are one the most extensively investigated targets in recent years due to their potential applications in future nanoelectronic and nano-optoelectronic devices[46, 47]. Using the chemical bond theory, the energy-gap shifts of InP as well as Si nanowires are calculated and plotted against the number of stoms shown in Figure 4[31].

In order to study the quantum confinement effects of nanowires, three possible structural models of tapered and twinned InP nanowires were constructed and expressed as ABCABC, ABCACBA and ABCAACBA, as shown in Figure 4(a)-(c). The first structural model is a normal stacking sequence of zinc blende crystal where each letter represents a bilayer composed of a pair of one group III and one group V atom. The second model ABCACBA (here A is the twin plane) may lead to a straight-edged nanowire. And the third model ABCAACBA(AA are two sequential twin planes) may lead to a tapered-twinned nanowire. The wires grow in <111> direction. L and D are the length and diameter of nanowires, respectively.For nanowires, the large aspect ratio L/D may also play an important role in the shift of energy gaps besides the structural morphology [33]. For non-polar Si semiconductor nanowires, the straight-edge nanowires (red curve) have the biggest energy gap shift, while the tapered nanowires (blue curve) have the smallest ones. For polar InP nanowires, however, a totally reversed sequence of energy shift appears, i.e., the blue curve lies on top of one another. In addition, the difference between the curves of InP nanowires are significant greater than that of Si nanowires. The energy gap shifts in the straight and twinned InP nanowires tend to negative values, instead of decreasing toward zero like that of the tapered nanowires according to the calculated results, which is in agreement with Weert' experimental results[49, 50].

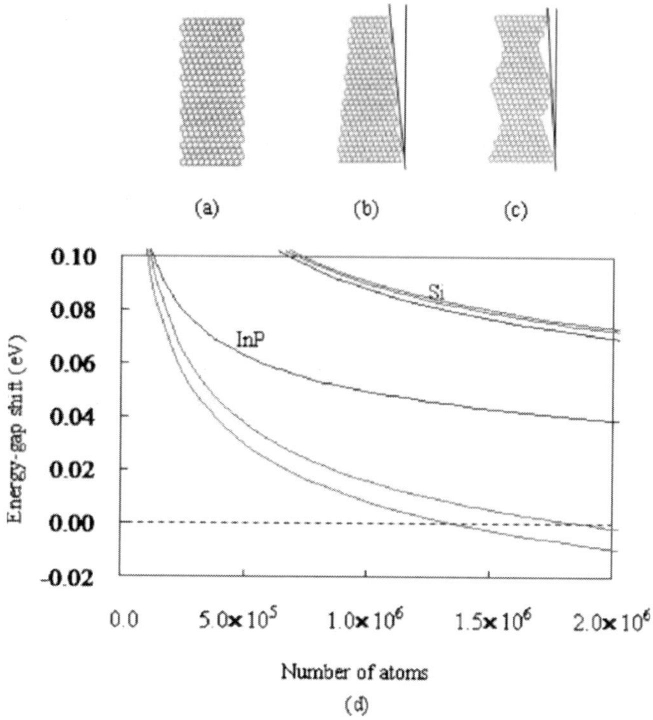

Figure 4. (a), (b) and (c) are the structure models of the straight-edge, tapered and the tapered-twinned InP nanowires along <111>, respectively. (d) The energy-gap shifts of the Si nanowires with diameter of 7nm and InP nanowires with diameter of 7.6nm. The red, green and blue curves represent the energy-gap shifts of the straight-edge, tapered and twinned nanowires, respectively [31].

Many other nanowires show similar large redshifts relative to energy gap of the bulk. According to Eq. (6), the energy gap shift is the result of a competition between the positive Kubo energy gap shift and thenegative surface energy gap shift. When aspect ratios $L/D \rightarrow \infty$, Kubo energy gap shift $\Delta E^{kubo} \rightarrow 0$, while the surface energy shift $\Delta E^{surface}$turn to negative values for both the straight and twinned InP nanowires with D = 7.6nm, which make the energies shift to red.

Further studies about the InP cubic nanowires of 5*5*L ,10*10*L , 20*20*L , 50*50*L and 100*100*L nm also show negative energy gap shift.

The Band Gap Shifts of Core/Shell Semiconductor Dots

For nanocrystals composed of a single material, the electronic properties can be tuned by the choice of material, size and shape. Additional control may be obtained using a core/shell structure where the type or thickness of the shell and core can tailor the optical, electronic, electrical and chemical properties of semiconductor nanocrystals. These core/shell quantum dots and nanowires allow a wide range of problems to be addressed, and have attracted much interest due to the notable potential applications in lasers and biomedicine. Some of the most interesting systems, for example ZnS, CdS shell and CdSe, CdTe core systems are widely studied in experiments. For core/shell quantum dots, the atoms of the core material have normal coordination. The kubo energy shift and surface energy shift can be written as[23]:

$$\Delta E^{kubo}(\text{core/shell}) = (m_e e^4/2\hbar^2)(V^{core}N_e^{\ core} + V^{shell}N_e^{\ shell})^{-1/3} \quad (19)$$

$$\Delta E^{surface}(\text{core/shell}) = F^{surface}\ [E_g^{\ surface}(\text{shell}) - E_g^{\ bulk}(\text{shell})] \quad (20)$$

The calculated energy gaps for the InP/ZnS, CdSe/CdS, CdSe/ZnS and CdTe/ZnS core/shell quantum dots are listed in Table 3, which are in agreement with the experimental values[51-54].

The Band Gap Shifts of Semiconductor Homogeneous Nanoalloys

Besides the size control, alloying could also tune the band gap by choosing the appropriate composition the materials and/or changing the ratio x of one component in nanocrystal alloys, since the interatomic interaction among different elements or compounds are different [55-61].

Table 3. Theoretical data for the core/shell nanocrystals. n_b is the number of bonds; calculated energy-gap $E_{g\ calc}$, experimental energy-gap $E_{g\ exp}$, kubo energy shift ΔE^{kubo} and surface energy shift $\Delta E^{surface}$ are in eV. D,Dc andDs are diameter of nanocrystal, diameter of core and shell thickness in Å, respectively

Core/shell semiconductors	D (Dc/Ds) (Å)		n_b	ΔE^{kubo}	$\Delta E^{surface}$	$E_{g\ calc}$	$E_{g\ exp}$
InP/ZnS	45 (28/17)	inside	3518	0.649	-0.361	1.637	1.687
		surface	732				
CdSe/CdS	44 (37/7)	inside	2470	0.721	-0.438	2.014	2.049
		surface	583				
	51 (37/14)	inside	3903	0.629	-0.371	1.991	1.992
		surface	743				
	58 (37/21)	inside	6388	0.541	-0.313	1.960	1.966
		surface	1006				
	65 (37/28)	inside	9336	0.481	-0.279	1.934	1.946
		surface	1305				
	72 (37/35)	inside	13041	0.433	-0.248	1.918	1.927
		surface	1587				
CdSe/ZnS	50 (40/10)	inside	4196	0.583	-0.367	1.949	2.011
		surface	862				
	64 (40/24)	inside	10639	0.452	-0.257	1.927	1.989
		surface	1422				
CdTe/ZnS	45 (25/20)	inside	3526	0.649	-0.361	1.718	1.76
		surface	732				

Some alloy nanocrystals show higher luminescence efficiency and stability due to the larger particle size, higher crystallinity, highercovalency, lower inter-diffusion, and spatial composition fluctuation. These indicate that the band gap can be tuned and even be "designed" as hoped in applications by changing the nanocrystal diameters and the types or fractions of the components, while the stability of composite structures comprising different semiconductor materials can be greatly improved if compared with the plain nanocrystals. The chemical bond theory may be employed to study the band gap of semiconductor homogeneous alloyswith various size and compositions.

For semiconductor nanocrystal alloys, expressed as A_xB_{1-x} (A and B represent the two semiconductor crystals composing the alloy), the band gap $E_g(x, D)$ function is similar to that of single semiconductor and can be expressed as the sum of the band gap for bulk alloys and the band gap shift induced by the ratio x and the size D[34]:

$$E_g(x,D) = E_g(x,\infty) + \Delta E_g(x,D) \tag{21}$$

$E_g(x, \infty)$is a composition-dependent function for bulk nanocrystal alloys and usually can be expressed as the empirical formula[35] as follow based on the experimental data:

$$E_g(x,\infty) = E_g(0,\infty) + [E_g(1,\infty) - E_g(0,\infty) - b]x + bx^2 \tag{22}$$

where b, named the bowing parameter, is also a composition-dependent function when the alloy components have different lattice constants [36].

$$\Delta E_g(x,D) = \Delta E_g^{\text{kubo}}(x,D) + \Delta E_g^{\text{surface}}(x,D) \tag{23}$$

$$\Delta E^{\text{kubo}}(x,D) = (m_e e^4 / 2\hbar^2)[V(x,D)N_e(x,D)]^{-1/3} \tag{24}$$

$$\Delta E_g^{\text{surface}}(x,D) = \frac{n_A(x,D)}{n_A(x,D)+n_B(x,D)}\Delta E_{gA}^{\text{surface}}(D) + \frac{n_B(x,D)}{n_A(x,D)+n_B(x,D)}\Delta E_{gB}^{\text{surface}}(D) \tag{25}$$

here$n_A(x, D)$ and $n_B(x, D)$ are the bond numbers of semicondutor A and B in nanocrystal alloys, respectively. $\Delta E_{gA}^{\text{surface}}(D)$ and $\Delta E_{gB}^{\text{surface}}(D)$ are the surface effect shift part of nanocrystal A and B with diameter D, respectively, which can be calculated by formulae (6)-(18).

Since different nanocrystal structures and diameters of particle size will surely lead to different atom numbers of surface and inside nanocrystals, as well as the proportion F^{surface} of surface bonds, sphere models with different diameters and structures were built to determine the numbers of atoms and bonds. For the same crystal with different diameters, we suppose the bond lengths of surface and inside the crystal remain constant respectively. Here D values are selected base on the known experimental results of the nanocrystal alloys.

Table 4 listed the available experimental band gap functions of bulk alloys and the calculated band gap shifts and band gap energy in the range of x =0~1 with corresponding structure and diameter D. It can be seen thatE_g (x, D) increases with decreasing D for a given nanocrystal alloy, similar to the cases of single semiconductor compounds composing the alloy. Owing to that most of the E_g (x, ∞) for bulk nanosemiconductors alloy are definite, E_g (x, D) of nanosemiconductor alloys with different D are mainly determined by ΔE_g (x, D). For example, the ΔE_g (x, D) of zb-$(ZnS)_x(CdS)_{1-x}$varies from about 0.303 to 0.132eV when D increases from 4.4 to 9.0nm. The biggest and smallest ΔE_g (x, D) are 0.499eV for zb-$(CdS)_x(CdSe)_{1-x}$ nanocrystal alloy with D = 2.8nm and 0.121eV for zb-$(ZnSe)_x(CdSe)_{1-x}$ with D = 10.0nm, respectively. The energy gap shift show obvious size-dependent properties, especially for the nanosemiconductor alloys with the size of several nanometers.

Table 4. The experimental band gap functions of bulk alloys and the calculated band gap shift and band gap with corresponding *D*[34]

Alloy	$E_g(x,\infty)$ (eV)	*D*(nm)	*x*	$\Delta E_g(x,D)$ (eV)	$E_g(x,D)$ (eV)
zb-$(ZnS)_x(CdS)_{1-x}$	$2.460+0.439x+0.827x^2$ [37]	4.4	0~1	0.259~0.303	2.719~4.029
		9.0		0.132~0.150	2.592~3.876
w-$(ZnS)_x(CdS)_{1-x}$	$2.501+0.328x+0.921x^2$ [37]	4.8		0.250~0.280	2.751~4.030
		5.8		0.200~0.230	2.701~3.980
		6.4		0.179~0.211	2.680~3.961
		8.0		0.131~0.169	2.632~3.919
zb-$(CdS)_x(CdSe)_{1-x}$	$1.74+0.49x+0.3x^2$ [38]	2.8		0.499~0.395	2.239~2.925
		3.1		0.480~0.378	2.220~2.908
		3.6		0.417~0.295	2.157~2.825
		4.3		0.343~0.260	2.083~2.790
		5.8		0.256~0.203	1.996~2.733
zb-$(ZnSe)_x(CdSe)_{1-x}$	$1.675+0.659+0.387x^2$ [37]	10.0		0.150~0.124	1.825~2.845
				0.136~0.121	1.811~2.842
w-$(ZnSe)_x(CdSe)_{1-x}$	$1.74+0.51x+0.35x^2$ [39]	5.2		0.265~0.246	2.005~2.846
		5.8		0.241~0.222	1.981~2.822
		6.3		0.228~0.202	1.968~2.802
		6.8		0.209~0.190	1.949~2.790
		7.5		0.192~0.172	1.932~2.772
w-$(CdSe)_x(CdTe)_{1-x}$	$1.43-0.553x+0.862x^2$ [40]	3.5		0.433~0.406	1.863~2.145
		5.0		0.316~0.292	1.746~2.031
		6.5		0.241~0.224	1.671~1.963

Beside the size effects, the compositions of the alloys also perform an important function in the shift of the energy gap. Taking zb-$(CdS)_x(CdSe)_{1-x}$ with D=2.8nm for example, the energy gap shift changes from 0.499 to 0.395 eV when x varies from 0 to 1. When D increases to 5.8nm, the energy gap shift varies from 0.256 to 0.203 in the range of x=0~1. With the increased D, the energy shift changes less along with x. Generally, the effect of composition on energy gap shift is not as significant as the size effect, however, the composition-dependent energy shift is considerable large for both the bulk and nanoscaled semiconductors. With the change of x, for instance, the energy gap shift expand 1.266, 1.310, and 1.284 eV for bulk zb-$(ZnS)_x(CdS)_{1-x}$ and nanoscaled alloys with D = 4.4 and 9nm, respectively. Therefore, in some cases, if adjusting size may result in system instability or hard fabrication, modulating composition can avoid these drawbacks and meantime satisfy the same desirable band gap energy of nanocompounds.

Figures 5-9 show comparisons of the theoretical $E_g(x,D)$ functions (solid lines) for nanosemiconductor alloys and available experimental results (symbols)[34]. The x-axis represents composition change from pure nanocrystal with smaller energy gap to another pure nanocrystal with bigger one.

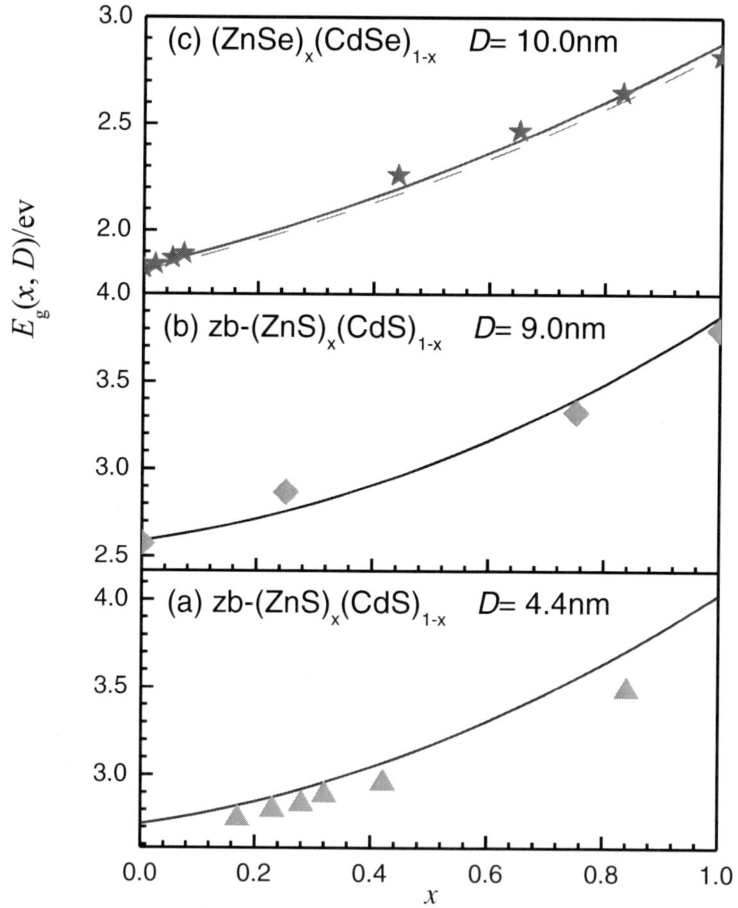

Figure 5. $E_g(x,D)$ functions for zb-$(ZnS)_x(CdS)_{1-x}$ nanocrystal alloys with the diameters of (a) 4.4 and (b) 9.0nm, and $(ZnSe)_x(CdSe)_{1-x}$ with the diameter of (c)10.0nm, respectively. The solid lines represent the calculated band gap energies of zinc blende structure.The red dashed line in (c) denotes the calculated band gap energy of $(ZnSe)_x(CdSe)_{1-x}$ with wurtzite structure. The experimental results are denoted by ▲[43], ◆[64] and★[44], respectively[34].

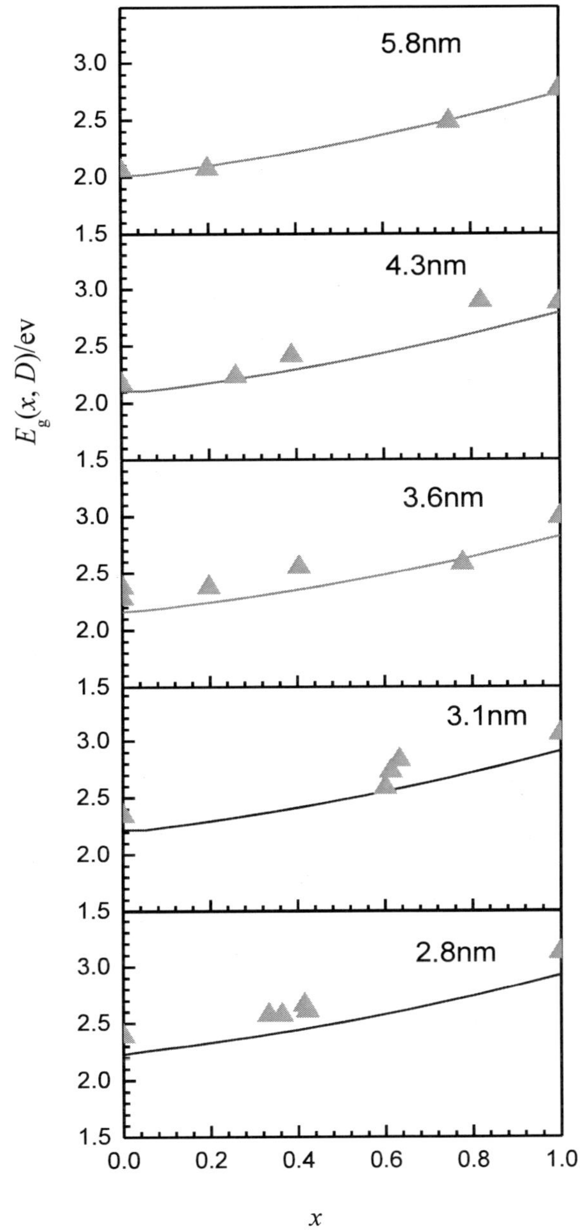

Figure 6. $E_g(x, D)$ functions for zb-$(CdS)_x(CdSe)_{1-x}$ nanocrystal alloy with the diameters of 2.8, 3.1, 4.3, 3.6 and 5.8nm, respectively. The solid lines denote the calculated band gaps, and the symbols represent the experimental results [38].

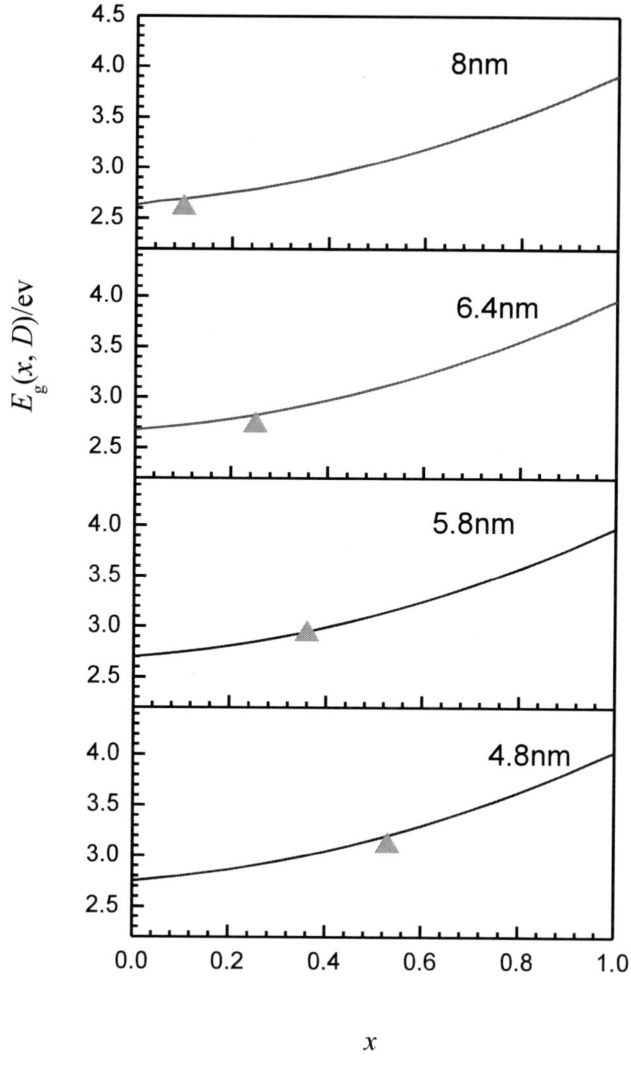

Figure 7. $E_g(x, D)$ functions for w-$(ZnS)_x(CdS)_{1-x}$ nanocrystal alloy with the diameters of 4.8, 5.8, 6.4 and 8.0nm, respectively. The solid lines denote the calculated band gaps and the symbols represent the experimental results [47].

(ZnSe)x(CdSe)1-x alloys show higher luminescence and stability compared to the case for the components with narrower energy gaps. For (ZnSe)x(CdSe)1-x bulk alloy, it is believed that there is a structural phase transition from wurtzite to zinc-blende structure approximately in the composition region $0.3<x<0.7$[45]. While it is also reported that MBE-grown (ZnSe)x(CdSe)1-x layers on GaAs(100) result in films of the zinc-blende structure over the whole composition range[46]. However, the differences between the calculated results for nanosemiconductors with zinc-blende and wurtize structures are comparable small that can be neglected. For example, the discrepancies of energy shift are 0.003eV for zb-ZnSe and w-ZnSe and 0.022eV for zb-CdSe and w-CdSe semiconductor particles with D=10nm, respectively. In addition, the experimental values of the bowing factors and energy gap shift are almost the same for nanosemiconductors with these two different structures. The comparison between the available experimental and calculated energy gaps of (ZnSe)x(CdSe)1-x with zinc blende structure (solid line) and wurtzite structure (red dashed line) are plotted in Fig. 2, where it can seen that the discrepancy of energy gap is much small, indicating that it is it is acceptable to neglect the phase transition and take zine blende or wurtzite structure as model for calculation yet without great error.

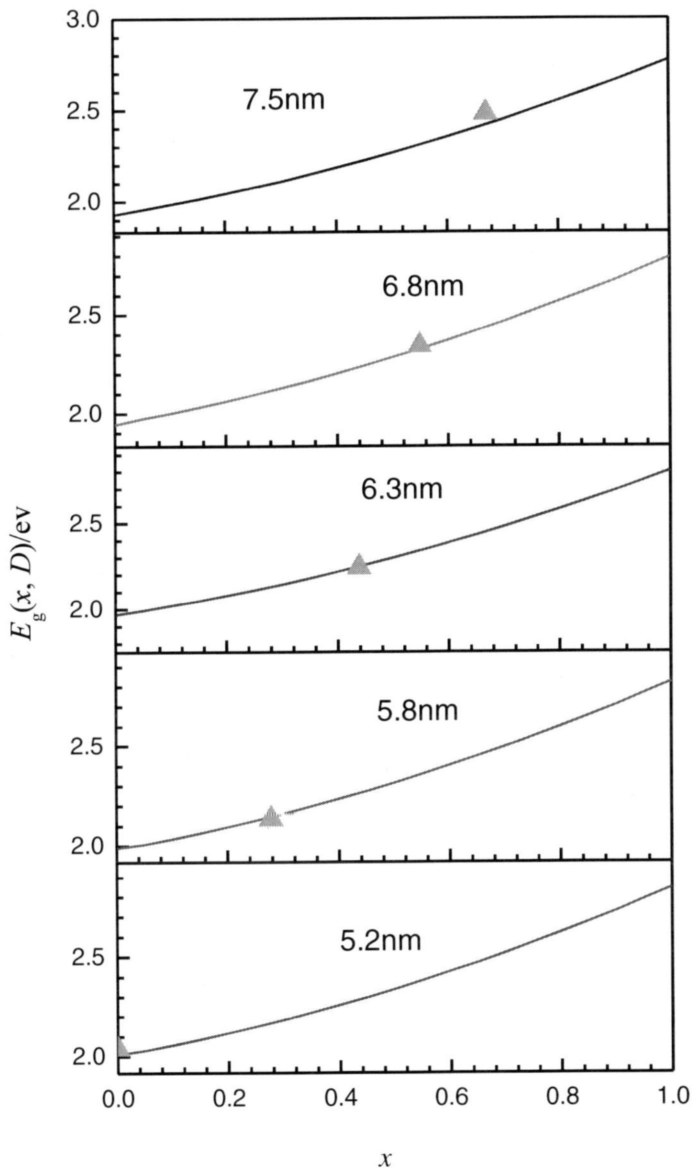

Figure 8. $E_g(x, D)$ functions for w-$(ZnSe)_x(CdSe)_{1-x}$ nanocrystal alloy with the diameters of 5.2, 5.8, 6.3, 6.8 and 7.5nm, respectively. The symbols represent the experimental results [48].

Figures 6-9 show the energy gap functions for homogeneous zb-$(CdS)_x(CdSe)_{1-x}$, w-$(ZnS)_x(CdS)_{1-x}$, w-$(ZnSe)_x(CdSe)_{1-x}$ and w-$(CdSe)_x(CdTe)_{1-x}$ alloys with different diameters. There are considerable good agreements between the available experimental and calculated results. The surface effect will become weaker with increasing D, which may induce large discrepancy in calculation due to the close relationship between the surface properties and the band gaps. The nanocrystal alloys with high quality and small diameters are difficult to synthesize, while the chemical bond model can provide prediction and theoretical guidance before the synthesis.

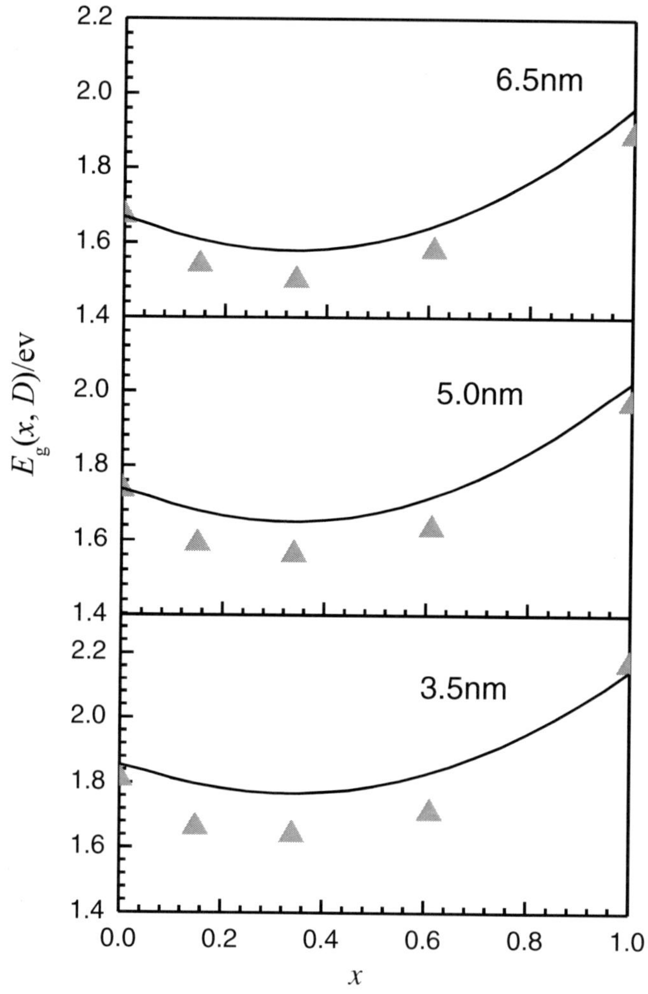

Figure 9. $E_g(x, D)$ functions for w-$(CdSe)_x(CdTe)_{1-x}$ with D =3.5, 5.0, and 6.5nm, respectively. The symbols represent the experimental results [49].

Conclusion

In summary, a few theories for quantum size effect are reviewed. In particular, a chemical bond method of quantum size effect is introduced in detail.It can be used to calculate the energy gap shift and further the energy gap of single nanocrystals particles, nanowires, core/shell semiconductors, semiconductor homogeneousalloys. The calculation may be carried out on a pocket calculator. Its simplicity stems from incorporating the quantum mechanical effects and surface effects into chemical bond properties via algebraic relationships rather than through variationalsolutions of the wave equation. It will provide guide to design the materials with expected properties.

This work is supported by the National Natural Science Foundation of China (Grant 21371149).

References

[1] A. L. Efros, *Soviet Physics Semiconductors Ussr* 16 (7), 772-775 (1982).
[2] Y. Wang and N. Herron, *The Journal of Physical Chemistry* 95 (2), 525-532 (1991).
[3] M. V. Rama Krishna and R. A. Friesner, *Physical Review Letters* 67 (5), 629-632 (1991).
[4] M. V. Rama Krishna and R. A. Friesner, *The Journal of Chemical Physics* 96 (2), 873-877 (1992).
[5] S. I. Pokutnyĭ, *Semiconductors* 41 (11), 1323-1328 (2007).
[6] S. I. Pokutnyi, *Semiconductors* 47 (6), 791-798 (2013).
[7] S. I. Pokutnyi, *Semiconductors* 46 (2), 165-170 (2012).
[8] S. Öğüt, J. R. Chelikowsky and S. G. Louie, *Physical Review Letters* 79 (9), 1770-1773 (1997).
[9] Y. Wang, A. Suna, W. Mahler and R. Kasowski, *The Journal of Chemical Physics* 87 (12), 7315-7322 (1987).
[10] P. E. Lippens and M. Lannoo, *Physical Review* B 39 (15), 10935-10942 (1989).
[11] S. Y. Ren and J. D. Dow, *Physical Review* B 45 (12), 6492-6496 (1992).
[12] L. W. Wang and A. Zunger, *Physical Review* B 51 (24), 17398-17416 (1995).
[13] G. Y. Wu, T. C. McGill, C. Mailhiot and D. L. Smith, *Physical Review* B 39 (9), 6060-6070 (1989).
[14] D. Bera, L. Qian, T.-K. Tseng and P. H. Holloway, *Materials* 3 (4), 2260-2345 (2010).
[15] F. Gao, *Inorganic Chemistry* 49 (22), 10409-10414 (2010).
[16] M. G. Mason, *Physical Review* B 27 (2), 748-762 (1983).
[17] C. Q. Sun, *Physical Review* B 69 (4), 045105 (2004).
[18] J. C. Phillips, *Reviews of Modern Physics* 42 (3), 317-356 (1970).
[19] J. A. Van Vechten, *Physical Review* 187 (3), 1007-1020 (1969).
[20] F. Gao, J. He, E. Wu, S. Liu, D. Yu, D. Li, S. Zhang and Y. Tian, *Physical ReviewLetters* 91 (1), 015502 (2003).
[21] T. K. Bergstresser and M. L. Cohen, *Physical Review* 164 (3), 1069-1080 (1967).
[22] T. Trindade, P. O'Brien and X.-m. Zhang, *Chemistry of Materials* 9 (2), 523-530 (1997).

[23] C. B. Murray, D. J. Norris and M. G. Bawendi, *Journal of the American ChemicalSociety* 115 (19), 8706-8715 (1993).
[24] J. Nanda, S. Sapra, D. D. Sarma, N. Chandrasekharan and G. Hodes, *Chemistry ofMaterials* 12 (4), 1018-1024 (2000).
[25] H. Fu and A. Zunger, Physical Review B 56 (3), 1496-1508 (1997).
[26] A. A. Guzelian, U. Banin, A. V. Kadavanich, X. Peng and A. P. Alivisatos, *Applied Physics Letters* 69 (10), 1432-1434 (1996).
[27] M. A. Malik, M. Afzaal, P. O'Brien, U. Bangert and B. Hamilton, *Materials Scienceand Technology* 20 (8), 959-963 (2004).
[28] M. A. Olshavsky, A. N. Goldstein and A. P. Alivisatos, *Journal of the AmericanChemical Society* 112 (25), 9438-9439 (1990).
[29] N. Serpone, D. Lawless and R. Khairutdinov, *The Journal of Physical Chemistry* 99 (45), 16646-16654 (1995).
[30] N. Satoh, T. Nakashima, K. Kamikura and K. Yamamoto, *Nat Nano* 3 (2), 106-111 (2008).
[31] F. Gao, *Applied Physics Letters* 98 (19), - (2011).
[32] R. Asahi, T. Morikawa, T. Ohwaki, K. Aoki and Y. Taga, *Science* 293 (5528), 269-271 (2001).
[33] C. N. R. Rao, G. U. Kulkarni, A. Govindaraj, B. C. Satishkumar and P. J. Thomas, *Pure and Applied Chemistry* 72 (1-2), 21-33 (2000).
[34] L. Gao and F. Gao, *Applied Physics Letters* 103 (5), 0531011-0531014 (2013).
[35] A. G. Thompson and J. C. Woolley, *Canadian Journal of Physics* 45 (2), 255-261 (1967).
[36] R. Hill and D. Richardson, Journal of Physics C: *Solid State Physics* 6 (5), L115 (1973).
[37] S. Adachi, *Properties of Semiconductor Alloys: Group-IV, III-V, II-VI, semiconductors.* (John Wiley & Sons, Ltd, West Sussex, 2009).
[38] L. A. Swafford, L. A. Weigand, M. J. Bowers, J. R. McBride, J. L. Rapaport, T. L. Watt, S. K. Dixit, L. C. Feldman and S. J. Rosenthal, *Journal of the American Chemical Society* 128 (37), 12299-12306 (2006).
[39] J. U. Kim, J. J. Lee, H. S. Jang, D. Y. Jeon and H. Yang, *Journal of Nanoscience andNanotechnology* 11 (1), 725-729 (2011).
[40] J. B. Mullin, edited by R. K. Willardson and H. L. *Goering* (Reinhold, New York, 1962), Vol. 1.
[41] S. Guha, B. J. Wu, H. Cheng and J. M. DePuydt, *Applied Physics Letters* 63 (15), 2129-2131 (1993).
[42] Y. F. Zhu, X. Y. Lang and Q. Jiang, *Advanced Functional Materials* 18 (9), 1422-1429 (2008).
[43] X. Zhong, S. Liu, Z. Zhang, L. Li, Z. Wei and W. Knoll, *Journal of MaterialsChemistry* 14 (18), 2790-2794 (2004).
[44] J.-P. Ge, S. Xu, J. Zhuang, X. Wang, Q. Peng and Y.-D. Li, *Inorganic Chemistry* 45 (13), 4922-4927 (2006).
[45] A. S. Nasibov, Y. V. Korostelin, P. V. Shapkin, L. G. Suslina, D. L. Fedorov and L. S. Markov, *Solid State Communications* 71 (10), 867-869 (1989).
[46] N. Samarth, H. Luo, J. K. Furdyna, R. G. Alonso, Y. R. Lee, A. K. Ramdas, S. B. Qadri and N. Otsuka, *Applied Physics Letters* 56 (12), 1163-1165 (1990).

[47] X. Zhong, Y. Feng, W. Knoll and M. Han, *Journal of the American Chemical Society* 125 (44), 13559-13563 (2003).
[48] X. Zhong, M. Han, Z. Dong, T. J. White and W. Knoll, *Journal of the AmericanChemical Society* 125 (28), 8589-8594 (2003).
[49] R. E. Bailey and S. Nie, *Journal of the American Chemical Society* 125 (23), 7100-7106 (2003).

In: Advances in Semiconductor Research
Editors: D. Persano Adorno and S. Pokutnyi
ISBN: 978-1-63321-755-3

Chapter 2

BI-FUNCTIONALITY OF SEMICONDUCTOR NANOPARTICLES WITH A SWITCHING MECHANISM

Rui Qing and Wolfgang M. Sigmund
Rhines Hall, Department of Materials Science and Engineering
University of Florida, Florida, US

Abstract

This chapter is dedicated to the introduction of a unique phenomenon in semiconductor nanoparticles, i.e., bi-functionality in ceria based and ceria-zirconia based solid solution materials. Governed by a switching mechanism, the nanoparticles can act reversibly either as a photocatalyst or a free radical scavenger. The property depends on both the existence of a bandgap and the facilitated off-stoichiometric defects generation/exchange within these nanoparticles. In the example presented in this chapter the switch is realized by applying UV light radiation onto the target materials in aqueous suspension form. UV light with different wavelengths can serve to enable the different functionalities of the nanoparticles. The selection wavelength is largely determined by the bandgap of the synthesized material. Thus quantum confinement effect also plays a role in defining the use of light for the bi-functionality. With further tuning of the bandgap by forming a solid solution of the material and aliovalent doping, combined with an *in vivo* screening mechanism, such bi-functionality is a promising property in potential biomedical applications..

Keywords: ceria, zirconia, nanoparticles, photocatalyst, free radical scavenger, reactive oxygen species

Introduction

Traditionally used for mechanical properties in ceramic polishing and glass sensitizing applications, ceria [cerium (IV or III) dioxide] is now being exploited more for their electronic properties. The unique capability of cerium ions to switch between [Ce^{3+}] and [Ce^{4+}] ionic states enables it to be widely used in various catalytic and functional systems.

Examples include histochemistry, three-way catalytic converters, electrolytes in solid oxide fuel cells (SOFC), gas sensors, oxygen storage materials, etc. [1-7].

Not until recently did researchers start to relate human disorders and living body malfunctions to the detrimental effects induced by reactive oxygen species (ROS) [8, 9]. The accreting impact of ROS can lead to the damage of living cells and cause various kinds of human diseases such as cancers, obesity, cardiovascular dysfunction, Parkinson's disease, etc.

In order to effectively prohibit the detrimental effect from these ROS and attempt to cure the human diseases, treatments on human body with free radical scavengers is becoming popular. Besides the traditionally used organic enzymes, ceria nanoparticles, despite being a semiconductor material, are found to possess this capability, as confirmed by many groups [10-12]. The lifespan of the living cells were extended significantly by ceria. It is believed that ceria could serve as an alternative of commercialized enzyme free radical scavengers and protect the living cells. Details of these free radical scavengers will be introduced in later sections.

Considering that ceria based materials are semiconductors, such compounds should also behave as photocatalysts and create ROS species when agitated by light with photonic energy high enough to overcome their bandgap. The fast transition between [Ce^{3+}] and [Ce^{4+}] valence states and their capability of reacting with environmental oxygen species could actually provide a natural sink for electrons generated by photonic excitation processes and retard the recombination rate, as will be introduced in the next section. One early paper did demonstrated that cerium dioxide was a potential photocatalyst that could be used to split water with the presence of electron acceptors, i.e., [Fe^{3+}], providing a possibility of using free radical scavengers as free radical formers [3].

However, the concept of bi-functional, or reversely performing nanoparticles with a switching mechanism, was not brought out until Qing et al. in Sigmund's lab did a systematic study using UV radiation and dye degradation method. Here in this chapter we would like to present this exciting new aspect of ceria based semiconductor nanoparticle materials with controlled functionality.

Photocatalysis of Semiconductors

One major application of semiconductor materials is photocatalysis, which is defined to be the chemical reaction induced by photo-radiation with the presence of a catalyst [13]. The catalytic materials that facilitate the reactions without being consumed are named photocatalysts. A common example for such a type of reaction is photosynthesis by plants, in which chlorophyll serves as the catalyst. Photocatalysts are widely used in a broad range of fields, primarily in environmental and catalytic applications [14-17]. Compared to the other type of photocatalyst, i.e., organometallic complexes, semiconductors have attracted greater interests for their optimal catalytic performance and non-pollutive nature. There have been many reports of using these materials in applications such as the removal of pollutants in water, self-cleaning glasses, etc. [14-17].

Due to the presence of a bandgap, semiconductors are all theoretically photocatalytically active. Yet previous research indicates that metal oxides and compounds exhibit better catalytic activity compared to other complexes [18, 19]. Yet the basic working principles are common and can be illustrated as follows.

To initiate a photocatalytic reaction, a photon with energy higher than the bandgap of the material impinges on the sample so as to excite an electron in the valence band (VB) to the conduction band (CB), leaving a hole in the VB with positive charge. The electron-hole pair, named as exciton, either recombines and releases the absorbed photonic energy or reacts with environmental chemical species, primarily those adjacent to or on the surface of the photcatalysts and generates free radicals. During the reaction, photonic energy and the excitons are consumed, leaving the lattice structure of the catalyst unchanged. Several parameters in the reaction , which are the light absorbing property of the semiconductor, recombination rate of agitated electrons and holes as well as the rate for redox reactions that transform electrons and holes to free radicals contribute to the overall efficiency in this process. The schematic presentation for photocatalysis is illustrated in Figure 1 [20].

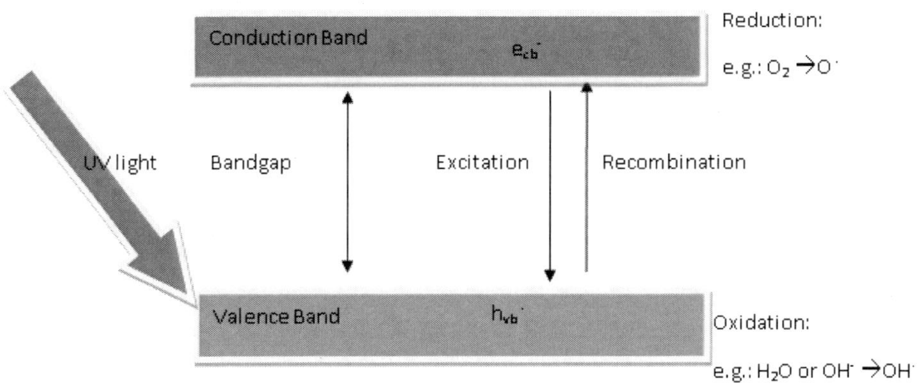

Figure 1. Schematic representation of the reactions taking place in a semiconductor photocatalyst (Reproduced with permission from ref. [20]).

To date, titania (titanium dioxide, TiO_2) is still the most used semiconductor photocatalyst. So here we use as an example to elucidate the typical redox reaction during photocatalysis. The main reactions and the time needed are summarized as follows [21-23]:

Exciton generation

$$TiO_2 + hv \rightarrow h_{vb}^+ + e_{cb}^- \quad 10^{-15}\text{s} \tag{1}$$

Electron-hole trapping

$$h_{vb}^+ + Ti^{IV}OH \rightarrow (Ti^{IV}OH)^· \quad 10^{-8}\text{s} \tag{2}$$

$$e_{cb}^- + Ti^{IV}OH \rightarrow (Ti^{III}OH) \quad 10^{-9}\text{s} \tag{3}$$

$$h_{vb}^+ + Ti^{IV} \rightarrow Ti^{III} \quad 10^{-9}\text{s} \tag{4}$$

Electron-hole recombination:

$$h_{vb}^+ + Ti^{III}OH \rightarrow (Ti^{IV}OH) \quad 10^{-7}\text{s} \tag{5}$$

$$e_{cb}^{-} + (Ti^{IV}OH)^{\cdot} \rightarrow (Ti^{IV}OH) \quad 10^{-8}\text{s} \tag{6}$$

Oxidation and reduction:

$$(Ti^{IV}OH)^{\cdot} + Red^{0} \rightarrow (Ti^{IV}OH) \quad + Red^{\cdot+}\ 10^{-7}\text{s} \tag{7}$$

$$e_{cb}^{-} + Ox \rightarrow (Ti^{IV}OH) \quad + Ox^{\cdot+}\ 10^{-3}\text{s} \tag{8}$$

From these reactions we can see that, in order to enhance the photocatalytic efficiency of a catalyst, the recombination rate should be retarded so as to increase the chance of generating free radicals. Researchers have successfully increased the activity of titania based materials by a variety of means, such as doping them with transition metals (e.g., Cr, Fe), coupling them with metals (e.g., Pt) and coupling them with a semiconductor (e.g., Aeroxide® TiO_2 P25: anatase phase titania with rutile phase titania). The photocatalytic reaction rate can also be controlled by regulating the recombination rate of excitons.

Photocatalysis is highly surface sensitive, as shown in the processes above. Thus nanosize particles are preferred and exhibit superior efficiency compared to their bulk counterparts. There are many research reports focusing on the size impact and use of nanoparticles as photocatalysts [24-26]. Nanoparticles inhabit the region between single molecules and larger colloids and show unique behaviors during applications, several of which are worthy to be noticed. First, as nanosize particles often have a high specific surface area, materials exhibit an enhanced photocatalytic efficiency when the solid loading is the same with the bulk material. Second, quantum confinement, first discussed by Frohlich in 1937 and reviewed by many others, has a wide range of impact on the electrical, optical and photocatalytic properties of these nanosize semiconductors [27-31]. It refers to the bandgap change or absorbance onset of a semiconductor due to the decrease of particle size. When the size is comparable to the de Broglie wavelength of electronic carriers or excitonic Bohr radius, there will be spatial confinement on electronic carrier movements. Quantized energy levels and discrete energy states take the place of the continuous band structure. The effect is displayed in the absorption spectrum by an observable blue shift. As the electronic carriers are "trapped" in discrete states, the redox potentials for chemical reactions are subjected to change. Yet whether this impact is beneficial or detrimental to photocatalytic processes is dependent on actual existing species.

Semiconductor Free Radical Scavengers

Before introducing the semiconductor free radical scavengers, i.e., ceria, a few concepts in terminology need to be clarified. First, the term "free radical" refers to a molecule that has one or more unpaired electrons. The loose electrons make the molecule highly unstable and have the tendency to react with other species. In the human body, the free radicals, typically reactive oxygen species (ROS), move through the bloodstream and take electrons from the organs or give away their unpaired electrons to reach a stable state. These reactions often cause cell damage, and consequently body malfunction and diseases. Second, in order to prevent or alleviate the consequences rendered from these reactions, "free radical scavengers", typically a vitamin, mineral or enzyme referred to as "antioxidants", are introduced into the body so as to hunt down and eliminate the unstable molecules. The free radical scavengers prevent the redox process that is required for electrons to pass between cells. Unlike catalysts, traditional free radical scavengers consume themselves in the process and cannot be restored automatically.

As expected, the role of free radical scavengers are extensively tested in the biomedical field. Various antioxidants are researched. Currently, the strategy is to insert these chemicals into the affected area and hope these substances will target diseased cells directly. The intake of free radical scavengers can be realized by repeatedly eating antioxidant-rich food or direct injection when in emergency.

The biomedical use of ceria nanoparticles was discovered by Rzigalinski et al. with other researchers also confirming in different cultures [10-12]. It was found that the lifespan of the living cells tested were extended significantly by ceria treatment. Thus ceria serves as an alternative to commercialized enzyme free radical scavengers and protect the living cells.

However, the performance of pure ceria in these applications is limited by its oxygen exchange capability, as governed by the equilibrium ratio between [Ce^{3+}] and [Ce^{4+}] and, mostly importantly, the transition rate between them. In 1950, Duwez and Odell established the first phase diagram of ceria-zirconia quasi-binary oxide system. Proven to be effective in catalytic applications, ceria-zirconia solid solutions are also studied and reported as an effective reactive oxygen species scavenger. Recently, Tsai et al. found that the free radical scavenging efficiency of these nanoparticles was markedly enhanced by this zirconium substitution of cerium. The rate constant against superoxide and peroxide free radicals of the synthesized compounds in colloidal suspensions even surpass that of superoxide dismutase (SOD), the best commercial radical scavenging enzyme to date [32].

As mentioned above, the protective effect on living cells from these semiconductors comes from their capability of scavenging ROS. Due to the unique redox property of this material, as will be mentioned in later sections, ceria can efficiently accommodate environmental oxygen species in response to external oxidative stress. Transition between [Ce^{3+}] and [Ce^{4+}] ionic states renders the material capable of eliminating surrounding ROS, which in turn protect the cells from being oxidized.

The introduction of zirconium into the lattice further facilitates this process and enhances the protective power of the solid solution material by enhancing the transition rate of cerium between its two valence states. Moreover, these nanoparticles are more advantageous over the traditional enzyme radical scavengers in the biomedical field because of their self-regeneration capability. Das et al. proposed a reaction pathway through which this property is realized [8, 33].

The ceria based materials release O_2 back into the environment after the radical scavenging process is completed. The auto-restoration process keeps the reactivity of the semiconductor free radical scavengers and offers a continuous protection to the living body [8], [34]. Until now, there is not yet any report concerning the time limit of ceria as a free radical scavengers.

Crystal Structure and Redox Behaviors of Ceria Based Nanoparticles

Ceria (CeO_2) has a fluorite crystal structure with Fm3m space group. The crystals are comprised of face-centered cubic (FCC) unit cells of cerium cations with oxygen anions residing in the octahedral sites as shown in Figure 2. Each cerium ion has eight nearest-neighbor (NN) oxygen ions. The oxygen is tetrahedral connected to NN cations. Ceria responds to the environmental oxidative condition and exhibits non-stoichiometric defects. It has another stoichiometry, i.e., Ce_2O_3, which has a space group of P3m1 and is pale yellow in color [35]. The color of CeO_2 is bright yellow.

Due to the fast response of the material to the environment, ceria possesses an excellent oxygen buffering property due to its high tolerance of oxygen ion incorporation/removal in the lattice. The crystal structure is largely retained during this process [36]. The materials exhibit ionic-electronic mixed conductivity. The electronic conductivity is based on a polaron hopping mechanism due to the large bandgap of the material. The electrons involved in this process are highly localized and convert the valence state of the adjacent cerium ions.

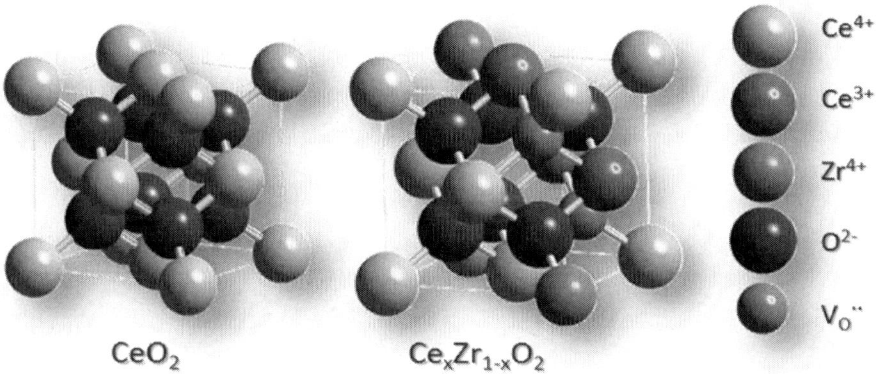

Figure 2. Schematic Structure for CeO_2 and $Ce_xZr_{1-x}O_2$ crystals. (Reproduced with permission from ref. [20]).

Oxygen deficiencies are critical in determining the electronic and chemical properties of ceria. The concentration of the defects can be modified by several methods. One common way of introducing oxygen vacancies into the lattice is via intrinsic defects, which is dependent on the partial pressure of environmental oxygen and temperature. Also introducing the aliovalent doping can alter the equilibrium defects concentration. Ceria is able to accommodate a large number of oxygen vacancies by substituting cerium with acceptor dopants. Also by isovalent substitution of cerium, i.e., Zr, the oxygen defect concentration at stabilized state is increased. The phenomenon is more obvious at high temperatures. This is attributed to the band structure modification that helps the electron association/dissociation process.

The ease of change in cerium's valence state is explained by the similar energy of its 4f and 5d electronic states. The localization and delocalization of electrons in the 4f state over its low energy barrier to 5d state leads to the reversible transition of cerium's two valence states.

During the removal of an oxygen atom, two electrons are left behind to form reduced [Ce^{3+}] ions. The electrons then occupy states from split-offs of cerium 4f band, which is initially empty. They then lie in the O_{2p}-Ce_{5d} bandgap. In the meantime, oxygen vacancies migrate in clusters based on a hopping mechanism that follows the Arrhenius equation. The state of oxygen deficiency in ceria is often defined by the chemical ratio between [Ce^{3+}] and cerium. Due to the difficulty in forming oxygen interstitials, this ratio is directly used to determine the concentration of oxygen defects in the material.

Besides the change in static equilibrium concentration of oxygen defects, the change can also happen dynamically in ceria. In fact, the off-stoichiometry spontaneously responds to external initiatives such as electric field, surface stresses and oxidative stress from ROS. The material is also considered an excellent oxygen buffer that is widely used in the solid oxide fuel cells (SOFC) applications.

However, while the buffering of oxygen is common for all type of ceria based materials, nanosize ceria possesses properties that are different from the bulk materials. Increased numbers of oxygen vacancies and defect formation rate are typically observed for ceria with decreased particle size. The enhanced specific surface area (SSA) for the nanoparticles creates more interfaces which exhibit different properties compared to the bulk. Facilitated transition in the two valence states implies lower activation energy for the change in oxygen vacancies. The higher reactivity significantly decrease the operation temperature of the ceria based materials in applications such as SOFC.

One other phenomenon that governs the electrical, optical and photocatalytic property of the nanosize particles is the quantum confinement effect. Details about this phenomenon can be found in early sections and won't be repeated here.

Bi-Functionality of $Ce_xZr_{1-x}O_2$ Nanoparticles in Suspension Form

Ceria can either perform as photocatalysts and free radical scavenger. It is quite interesting how these two almost opposite reactions can occur and be regulated, especially in the same system. Qing et al. studied the photocatalytic performance of $Ce_xZr_{1-x}O_2$ nanoparticles in colloidal suspensions and compared them to the commercialized Aeroxide TiO_2 P25. The nanoparticles were fully dispersed in suspension form and allowed to decompose the azo dye, i.e., Procion Red MX-5B, that was added into the solution under the radiation of UV light. Different wavelengths were used to regulate the reaction direction and evaluate the rate.

Details of the experimental design, synthesis and the characterization for those nanoparticles won't be covered here, but some basic information is provided. Synthesized via reverse micelle method, the $Ce_xZr_{1-x}O_2$ nanoparticles typically exhibited a hydrodynamic diameter of around 3-4 nm as determined by the dynamic light scattering method in suspension form. The measured diameters were then used to calculate the specific surface area (SSA) for determining the rate constants in photocatalysis experiments, and was found to be approximately 109±20 m^2/g, which is two times of the SSA for Aeroxide® TiO_2 P25 particles. A sample image of the nanoparticle characterized by TEM technique is shown in Figure 3. Red circles are used to identify the nanoparticles.

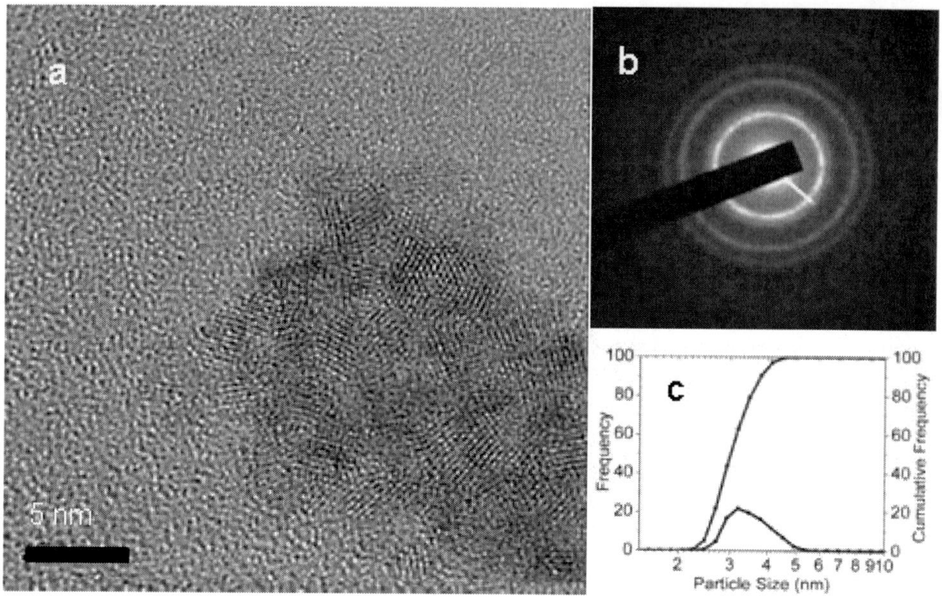

Figure 3. Size and size distribution for ceria nanoparticles. (a): TEM image of CeO_2 nanoparticles, scale bar 5 nm; (b): electron diffraction pattern of CeO_2 nanoparticles; (c): size distribution for CeO_2 nanoparticles. (Reproduced with permission from ref. [37]).

As mentioned in the introduction, bandgaps for nanosize $Ce_xZr_{1-x}O_2$ particles are determined to be larger than their bulk counterparts due to the quantum confinement effect. By using the plot from UV/Visible light absorption spectra (shown in Figure 4) and applying the Tauc relation, as shown below, to transform the data, the bandgap values are obtained.

$$a(hv) = A(hv - E_g)^{m/2}$$

Here, A is constant, α is absorption coefficient, (hv) is the photonic energy, m=1 for direct band transitions, and m = 4 for indirect band transitions [48]. A direct band transition was presumed for ceria and m = 1.

E_g values for synthesized samples are summarized in Table 1, where the impact from quantum confinement effects is clearly shown. Also, depsite the fact that the zirconium facilitates the valence state transition in fluorite ceria lattice, it increases the bandgap of the solid solution material. The bandgap profile follows the Vegards' law and the non-linear increase is explained by the bowing effect in higher orders. The bandgap for the nanoparticles plays a deterministic role in selecting the wavelength of UV light that can be used as the "switch" for enabling the different function in these nanoparticles, as will be elucidated in later sections.

Details of the experiment design and setup can be found in the original paper and so are not described here [37]. Shown in Figure 5 are the dye degradation results for Procion MX-5B azo dye in the solution with $Ce_xZr_{1-x}O_2$ nanoparticles or Aeroxide P25 TiO_2 photocatalysts under the UV radiation at wavelength of 302 nm and 365 nm. It is clear that 302 nm UV radiation allows $Ce_xZr_{1-x}O_2$ nanoparticles to act as a photocatalyst against the dye selected. Also a noticeable trend of better photocatalytic activity from high zirconium content solid solution nanoparticles is observed. The mechanism for this phenomenon will be illustrated in the next section. Within all the experimental sets, commercialized Aeroxide TiO_2 P25 exhibits the best catalytic activity and is the only photocatalyst that degrades all the dye. The overall reaction follows the Langmuir-Hinshelwood theory. Yet two types of rate-determining reactions are observed.

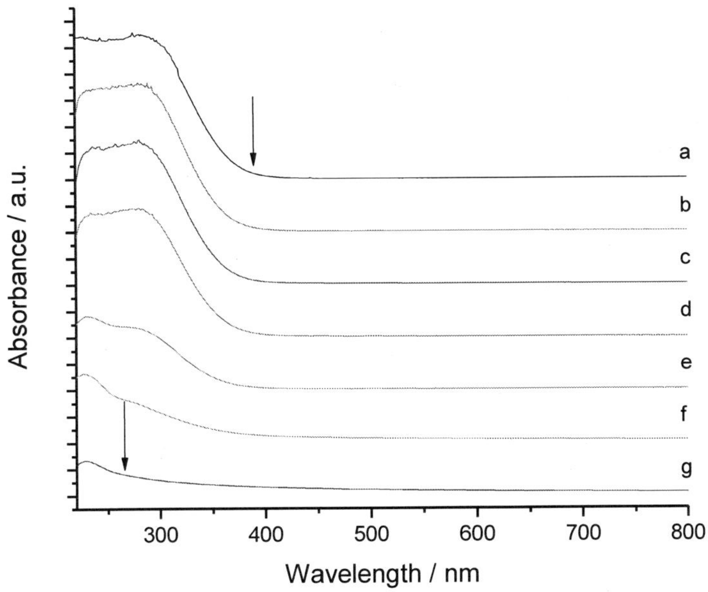

Figure 4. UV/VIS absorption spectra for synthesized $Ce_xZr_{1-x}O_2$ nanoparticles in aqueous suspension: (a): CeO_2; (b): $Ce_{0.8}Zr_{0.2}O_2$; (c): $Ce_{0.7}Zr_{0.3}O_2$; (d): $Ce_{0.6}Zr_{0.4}O_2$; (e): $Ce_{0.4}Zr_{0.6}O_2$; (f): $Ce_{0.2}Zr_{0.8}O_2$; (g): ZrO_2. (Reproduced with permission from ref. [37]).

Table 1. The calculated bandgap for direct band transition of $Ce_xZr_{1-x}O_2$ nanoparticles

Sample	Bandgap (eV)
CeO_2	3.58±0.03
$Ce_{0.8}Zr_{0.2}O_2$	3.69±0.02
$Ce_{0.6}Zr_{0.4}O_2$	3.70±0.04
$Ce_{0.4}Zr_{0.6}O_2$	3.73±0.05
$Ce_{0.2}Zr_{0.8}O_2$	3.83±0.01
ZrO_2	4.55±0.09

For cerium rich nanoparticles until $Ce_{0.6}Zr_{0.4}O_2$, a pseudo zero-order reaction is found. Dye concentration changes linearly as a function of time. It is expected that the limitation of dye surface adsorption and slow degradation are the rate determining factors for the reaction. However, for zirconium rich nanoparticles, faster reactions rendered the change in residual dye concentration to be the determining factor for reaction rate, which shows a pseudo-1st order reaction behavior. For experiments under 365 nm UV radiation, none of the ceria-zirconia nanoparticles exhibit any photocatalytic activity. This is explained by their large bandgap. 365 nm UV light does not contain enough photonic energy for the generation of excitons in the crystal lattice. UV light with a wavelength no larger than 348 nm is necessary for activating the photocatalytic reactions.

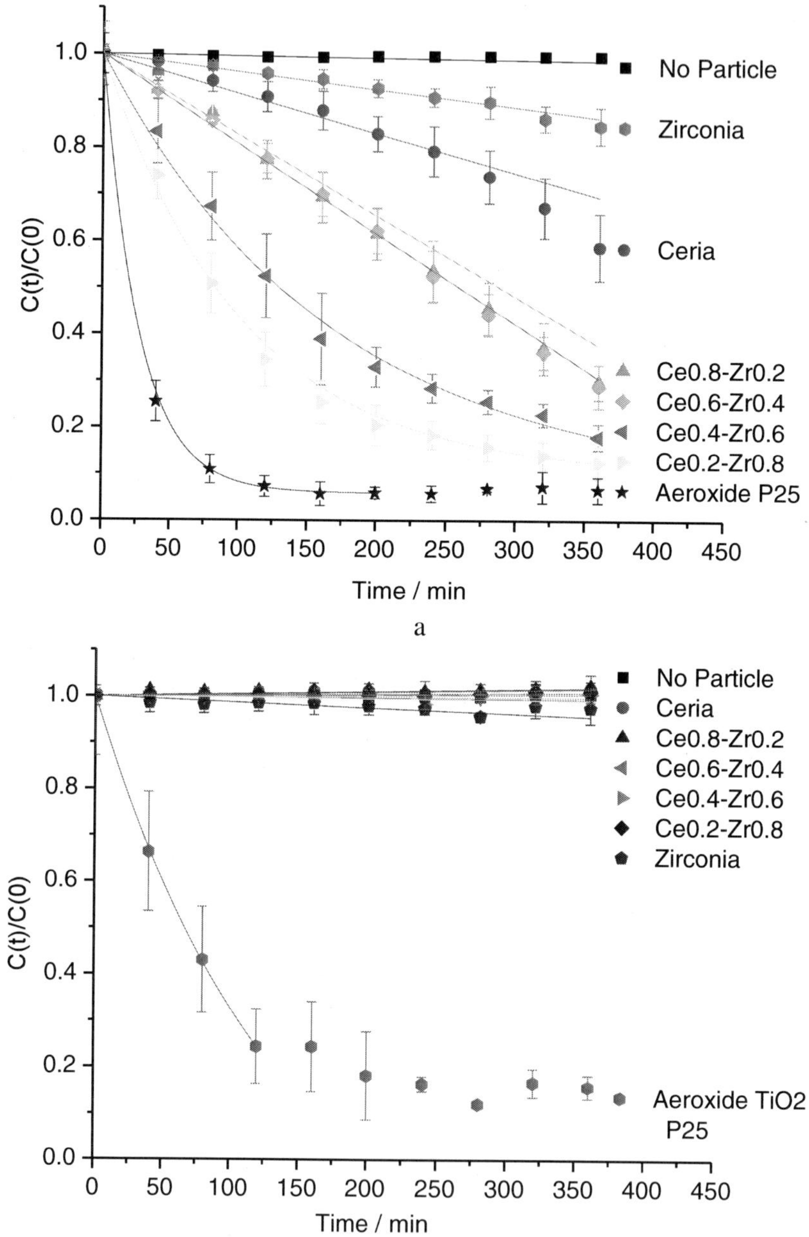

b

Figure 5. (a) Time dependence of normalized residual dye concentration for samples under 302 nm UV light irradiation with $Ce_xZr_{1-x}O_2$ nanoparticles aqueous suspension in static state. (b) Time dependence of normalized residual dye concentration for samples under 365 nm UV light irradiation with $Ce_xZr_{1-x}O_2$ nanoparticles aqueous suspension in static state. Fitted lines are drawn to guide the viewer. (Reproduced with permission from ref. [37]).

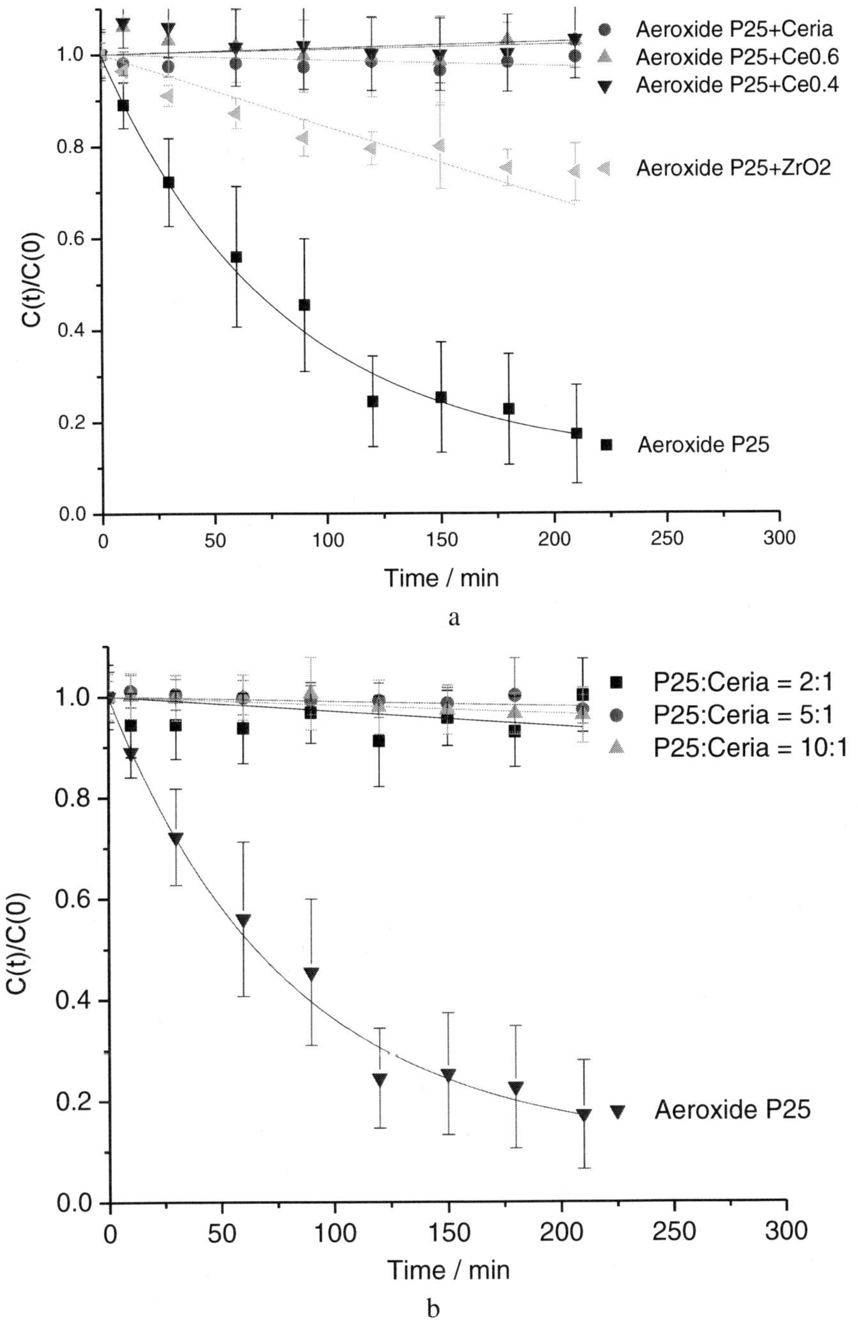

Figure 6. (a) Time dependence of normalized residual dye concentration for mixed samples in static aqueous suspension with different composition under 365 nm UV light irradiation. (b) Time dependence of normalized residual dye concentration for mixed samples in static aqueous suspension with different SSA ratio under 365 nm UV light irradiation. Fitted lines are drawn to guide the viewer. (Reproduced with permission from ref. [37]).

In Figure 6 the photocatalysis results using a $Ce_xZr_{1-x}O_2$ and Aeroxide TiO_2 P25 mixture particles are presented. 365 nm UV light, which can agitate electrons in TiO_2 but not in the

$Ce_xZr_{1-x}O_2$, is chosen for the experiments. The amount of the two types of particles is normalized to their SSA value accordingly. From Figure 6(a) we can see that the addition of $Ce_xZr_{1-x}O_2$ nanoparticles can effectively inhibit the photocatalytic activity of the TiO_2 material, regardless of the composition. All the experimental sets retained the dye concentration even after 6 hours of UV radiation, indicating that the free radical scavenging capability from $Ce_xZr_{1-x}O_2$ nanoparticles surpassed the photocatalytic activity of commercial TiO_2. In Figure 6(b) different amounts of ceria nanoparticles are used to eliminate the impact of surface absorption of ceria on TiO_2 particles, which might also account for the decrease in photocatalytic activity of the latter. It clearly shows that even with a ratio of 1:10, $Ce_xZr_{1-x}O_2$ nanoparticles still can completely stop the catalytic reaction of the TiO_2 under 365 nm light radiation. This again verifies the free radical scavenging capability of ceria-based nanoparticles against photocatalysis from the commercial Aeroxide P25.

The results also indicate that $Ce_xZr_{1-x}O_2$ nanoparticles might actually be a better photocatalyst than Aeroxide TiO_2 P25, as they can surpass their own free radical scavenging power when the impinging light has sufficient photonic energy to initiate exciton pair in the semiconductor. Yet the validity of this statement and the full potential of ceria in photocatalysis still require further research effort to verify and explore.

Mechanism and Equilibrium of Ceria

While the measurement of the bandgap explains the activity of $Ce_xZr_{1-x}O_2$ nanoparticles under UV radiation of different wavelengths, the inner factors that contribute to either the photocatalysis or the free radical scavenging capability of ceria based nanomaterials still need to be revealed. From XPS spectra, as shown in Figure 7, of unirradiated and irradiated nanoparticles, it is shown that substitution of cerium by zirconium in ceria's fluorite lattice can promote the generation of the $[Ce^{3+}]$ valence state, increasing the concentration of oxygen vacancies in the material. However, UV radiation can induce the incorporation of environmental oxygen species into the lattice and eliminate its intrinsic non-stoichiometric defects. Considering that the introduction of zirconium facilitates the transition of cerium's two valence states, the rapid change in stoichiometry of the oxide helps to enhance both the photocatalysis and free radical scavenging capability of the nanoparticles.

Even though the actual reaction pathway for photocatalysis induced by $Ce_xZr_{1-x}O_2$ nanoparticles remains largely unknown, mechanism has been proposed by borrowing the concept of TiO_2-based photocatalysts and is presented below:

$$hv \rightarrow h_{vb}^{+} + e_{cb}^{-}$$
$$h_{vb}^{+} + OH \rightarrow OH^{\cdot}$$
$$Ce^{4+} + e^{-} \rightarrow Ce^{3+}$$
$$Ce^{3+} + O_2 \rightarrow Ce^{4+} + O_2^{\cdot -}$$
$$O_2^{\cdot -} + H^{+} \rightarrow HO_2^{\cdot}$$
$$2HO_2^{\cdot} \rightarrow O_2 + H_2O_2$$

Comparing this mechanism for photocatalysis with the reaction pathway for free radical scavenging of ceria-zirconia nanoparticles gives the following equations.

$$O_2^{2-} + Ce_xZr_{1-x}O_2 + V_O^{\bullet\bullet} + 2e^- \rightleftarrows O_2^{2-} \cdots Ce_xZr_{1-x}O_2 + V_O^{\bullet\bullet} + 2e^- \longrightarrow$$

$$2O^- \cdots Ce_xZr_{1-x}O_2 + V_O^{\bullet\bullet} + 2e^- \longrightarrow 2O^{2-} \cdots Ce_xZr_{1-x}O_2 + V_O^{\bullet\bullet} \longrightarrow$$

$$O^{2-} + Ce_xZr_{1-x}O_2 + O_O^X \longrightarrow \frac{1}{2}O_2 + O^{2-} + Ce_xZr_{1-x}O_2 + V_O^{\bullet\bullet} + 2e^-$$

It is clear that similar oxygen species (free radicals) are involved as intermediates for both processes. Thus in the as-discussed system, the proceeding of one reaction shall compete with the other. The equilibrium can be easily traced back to Le Chatelier's principle, where the concentration of the oxygen species and external initiatives, i.e., UV radiation in this experiment, are the factors that governs the direction of the reaction.

Based on the reported experimental results, it is assured that 302 nm UV illumination can continuously supply the energy needed for generation of the exciton pair out of the $Ce_xZr_{1-x}O_2$ nanoparticles, including the formation of free radicals for decomposition of the dye. In comparison, 365 nm UV light radiation cannot agitate the nanoparticle to generate sufficient exciton pairs to act as a photocatalyst.

So, in this case, the solid solution behaves solely as a free radical scavenger against the ROSs. This free radical scavenging was proved by titania - $Ce_xZr_{1-x}O_2$ mixture experimental sets. It is also noteworthy that free radical scavenging activity from the $Ce_xZr_{1-x}O_2$ nanoparticles surpassed the photocatalytic activity of titania, hindering the overall photocatalysis.

Referring to both of the reaction pathways stated above, we see that the generation of extra electrons from the stoichiometric defects in $Ce_xZr_{1-x}O_2$ lattice can be released into the solution as well as the surface of those TiO_2 particles. These electrons increase the electron - hole recombination rate, hindering the overall photocatalytic reactions. Here, UV light with different wavelengths acted as a switch for the synthesized $Ce_xZr_{1-x}O_2$ nanoparticles. By choosing the external radiation sources wisely, the solid solution materials perform in opposite directions.

Another trend worth discussing for the two opposite properties from these nanoparticles is enhancement in performance by introducing zirconium into ceria lattice. A six fold increase was reported for the superoxide radical scavenging capability from these $Ce_xZr_{1-x}O_2$ nanoparticles with 60% zirconium as compared to pure ceria [32]. This corresponds well with the photocatalytic activity enhancement from the solid solution materials.

Publications on the use of ceria-based material as a three-way catalyst revealed that the introduction of zirconium content into $Ce_xZr_{1-x}O_2$ lattice facilitates the transition between [Ce^{3+}] and [Ce^{4+}] valence state, through which environmental electrons could be accommodated/released. The fast change between the two chemical states is beneficial to both photocatalysis and free radical scavenging reactions.

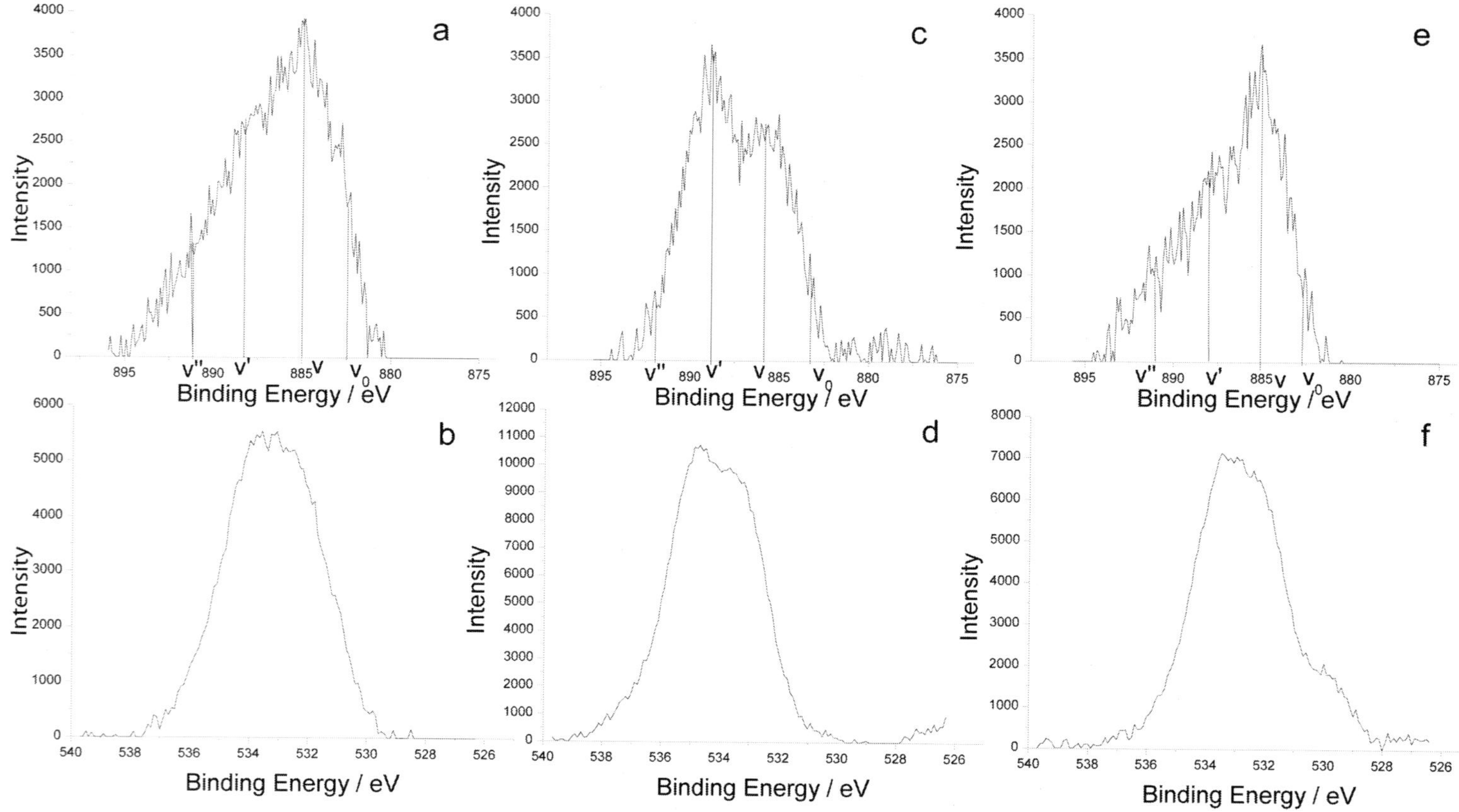

Figure 7. X-ray photoelectron spectroscopy (XPS) multiplex intentsity for oxygen 1s peak and cerium 3d peak in $Ce_xZr_{1-x}O_2$nanoparticles: (a): Ce 3d peak for CeO_2 nanoparticles before UV irradiation; (b): O 1s peak for CeO_2 nanoparticles before UV irradiation; (c): Ce 3d peak for $Ce_{0.4}Zr_{0.6}O_2$ nanoparticles before UV irradiation; (d): O 1s peak for $Ce_{0.4}Zr_{0.6}O_2$ nanoparticles before UV irradiation; (e): Ce 3d peak for CeO_2 nanoparticles after UV irradiation; (f): O 1s peak for CeO_2 nanoparticles after UV irradiation. (Reproduced with permission from ref. [37]).

Conclusion

Here, $Ce_xZr_{1-x}O_2$ nanoparticles are found to be capable of performing adverse reactions in the same colloidal system using UV light as a switch, different from the previous reports of multifunctional semiconductors [38-40] in which the multi-functionality was primarily based on the diverse applications of the same materials in different fields,. Such a concept is not complicated but as of yet not fully understood. Distinguishable bi-functionality was found and reported in these semiconductor materials, i.e., $Ce_xZr_{1-x}O_2$ solid solution nanoparticles. It is believed that this versatility of the semiconductor material will be beneficial in various applications, especially in biomedical fields. Furthermore, introducing dopants into the semiconductor could provide accepter/donor states in the forbidden band, modifying the band structure. The catalytic activity in either direction can then be regulated. Appropriate bandgap values will allow the use of different light wavelengths as a functional switch of the nanoparticles with controlled reaction rates. With a proper screening mechanism, such bi-functionality could be found useful, especially in killing diseased cells while protecting the normal ones. Besides the biomedical field, these versatile materials can also be used in selective decomposition of pollutants in wastewater, etc.Even though more detailed modification and experimental design is needed to fully utilize such semiconductor materials in these applications, ceria-based solid solutions are already considered a promising candidate.

References

[1] K. Eguchi, *Journal of Alloys and Compounds* 250 (1-2), 486-491 (1997).

[2] G. Telek, J. Y. Scoazec, J. Chariot, R. Ducroc, G. Feldmann and C. Roze, *Journal of Histochemistry & Cytochemistry* 47 (9), 1201-1212 (1999).

[3] G. R. Bamwenda and H. Arakawa, *Journal of Molecular Catalysis a-Chemical* 161 (1-2), 105-113 (2000).

[4] C. Descorme, Y. Madier and D. Duprez, *Journal of Catalysis* 196 (1), 167-173 (2000).

[5] P. Jasinski, T. Suzuki and H. U. Anderson, *Sensors and Actuators B-Chemical* 95 (1-3), 73-77 (2003).

[6] Trovarelli, *Catalysis by Ceria and Related Materials*. (Imperial College Press, London, 2005).

[7] Dalslet, P. Blennow, P. V. Hendriksen, N. Bonanos, D. Lybye and M. Mogensen, *Journal of Solid State Electrochemistry* 10 (8), 547-561 (2006).

[8] M. Das, S. Patil, N. Bhargava, J. F. Kang, L. M. Riedel, S. Seal and J. J. Hickman, *Biomaterials* 28 (10), 1918-1925 (2007).

[9] Y. Y. Tsai, J. Oca-Cossio, K. Agering, N. E. Simpson, M. A. Atkinson, C. H. Wasserfall, I. Constantinidis and W. Sigmund, *Nanomedicine-Uk* 2 (3), 325-332 (2007).

[10] B. A. Rzigalinski, *Technol Cancer Res* T 4 (6), 651-659 (2005).

[11] R. W. Tarnuzzer, J. Colon, S. Patil and S. Seal, *Nano Letters* 5 (12), 2573-2577 (2005).

[12] D. Schubert, R. Dargusch, J. Raitano and S. W. Chan, *Biochemical and Biophysical Research Communications* 342 (1), 86-91 (2006).

[13] M. Kaneko and I. Okura, *Photocatalysis: Science and Technology*. (Springer, 2003).

[14] Fujishim and K. I. Honda, *Kog Kagaku Zasshi* 74 (3), 355-& (1971).

[15] Fujishima and K. Honda, *Nature* 238 (5358), 37-38 (1972).

[16] Fujishima, K. Kohayakawa and K. Honda, *Journal of the Electrochemical Society* 122 (11), 1487-1489 (1975).

[17] W. Jaegermann, in *Modern Aspects of Electrochemistry*, edited by B. E. C. R.E. White, J. O'M. Bockris (Plenum Press, New York, 1996), Vol. 30, pp. 1–185.

[18] M. A. Fox and M. T. Dulay, *J. Photoch. Photobio A* 98 (1-2), 91-101 (1996).

[19] Mills and S. LeHunte, *J. Photoch. Photobio A* 108 (1), 1-35 (1997).

[20] R. Qing, University of Florida, 2009.

[21] S. T. Martin, H. Herrmann, W. Y. Choi and M. R. Hoffmann, *Journal of the Chemical Society-Faraday Transactions* 90 (21), 3315-3322 (1994).

[22] S. T. Martin, H. Herrmann and M. R. Hoffmann, *Journal of the Chemical Society-Faraday Transactions* 90 (21), 3323-3330 (1994).

[23] M. R. Hoffmann, S. T. Martin, W. Y. Choi and D. W. Bahnemann, Chemical Reviews 95 (1), 69-96 (1995).

[24] Levy, Journal of Electroceramics 1 (3), 239-272 (1997).

[25] Beydoun, R. Amal, G. Low and S. Mcevoy, *Journal of Nanoparticle Research* 1 (4), 439-458 (1999).

[26] P. V. Kamat and D. Meisel, *Comptes Rendus Chimie* 6 (8-10), 999-1007 (2003).

[27] H. Frohlich, *Proceedings of the Royal Society of London Series a-Mathematical and Physical Sciences* 160 (A901), 0230-0241 (1937).

[28] H. Frohlich, *Proceedings of the Royal Society of London Series a-Mathematical and Physical Sciences* 172 (A948), 0094-0106 (1939).

[29] H. Frohlich, *Phys. Rev.* 56 (4), 349-352 (1939).

[30] M. Gratzel, *Heterogeneous Photochemical Electron Transfer*. (CRC Press Inc., Boca Raton, 1989).

[31] D. Yoffe, *Adv Phys* 42 (2), 173-266 (1993).

[32] Y. Y. Tsai, J. Oca-Cossio, S. M. Lin, K. Woan, P. C. Yu and W. Sigmund, *Nanomedicine-Uk* 3 (5), 637-645 (2008).

[33] S. Das, S. Singh, J. M. Dowding, S. Oommen, A. Kumar, T. X. T. Sayle, S. Saraf, C. R. Patra, N. E. Vlahakis, D. C. Sayle, W. T. Self and S. Seal, *Biomaterials* 33 (31), 7746-7755 (2012).

[34] J. M. Dowding, T. Dosani, A. Kumar, S. Seal and W. T. Self, *Chem. Commun.* 48 (40), 4896-4898 (2012).

[35] M. Mogensen, N. M. Sammes and G. A. Tompsett, *Solid State Ionics* 129 (1-4), 63-94 (2000).

[36] F. Zhang, Q. Jin and S. W. Chan, *J. Appl. Phys.* 95 (8), 4319-4326 (2004).

[37] R. Qing and W. Sigmund, *J. Photoch. Photobio A* 266, 55-63 (2013).

[38] F. Niu, J. Zhai, L. Jiang and W. G. Song, *Chemical Communications* (31), 4738-4740 (2009).

[39] F. Rupp, M. Haupt, H. Klostermann, H. S. Kim, M. Eichler, A. Peetsch, L. Scheideler, C. Doering, C. Oehr, H. P. Wendel, S. Sinn, E. Decker, C. von Ohle and J. Geis-Gerstorfer, *Acta Biomaterialia* 6 (12), 4566-4577 (2010).

[40] H. Q. Tang, J. Li, Y. Q. Bie, L. H. Zhu and J. Zou, *Journal of Hazardous Materials* 175 (1-3), 977-984 (2010).

In: Advances in Semiconductor Research
Editors: D. Persano Adorno and S. Pokutnyi ISBN: 978-1-63321-755-3

Chapter 3

ENHANCEMENT OF THE LIFETIME OF METASTABLE STATES IN Er-DOPED Si NANOCRYSTALS BY EXTERNAL COLORED NOISE

S. Spezia*, D. Valenti, D. Persano Adorno and B. Spagnolo
Dipartimento di Fisica e Chimica, Group of Interdisciplinary Physics
Università di Palermo and CNISM-INFM,
Unità di Palermo, Palermo, Italy

Abstract

The changes in the lifetime of a metastable energy level in Er-doped Si nanocrystals in the presence of an external source of colored noise are analyzed for different values of noise intensity and correlation time. Exciton dynamics is simulated by a set of phenomenological rate equations which take into account all the possible phenomena inherent to the energy states of Si nanocrystals and Er^{3+} ions in the host material of Si oxide. The electronic deexcitation is studied by examining the decay of the initial population of the Er atoms in the first excitation level $^4I_{13/2}$ through the fluorescence and the cooperative upconversion by energy transfer. Our results show that the deexcitation process of the level $^4I_{13/2}$ is slowed down within wide ranges of noise intensity and correlation time. Moreover, a nonmonotonic behavior of the lifetime with the amplitude of the fluctuations is found, characterized by a maximum variation for values of the noise correlation time comparable to the deexcitation time. The indirect influence of the colored noise on the efficiency of the energy transfer upconversion activated from the level $^4I_{13/2}$ is also discussed.

Keywords: Photoluminescence, Stochastic modeling and numerical simulations, Cooperative energy transfer upconversion, Fluctuation phenomena and random processes, Silicon nanocrystals, Erbium

---

* E-mail address: stefano.spezia@unipa.it. (Corresponding Author)

Introduction

The maximum efficiency of the photovoltaic cells, consisting of single-junctions, is limited to about 30% (for one sun illumination) by intrinsic losses such as the inability to absorb below-band gap photons. This intrinsic characteristic, called the Shockley-Queisser limit, is due to the discrete nature of the band structure of semiconductors. [1-7] Only photons of energy at least equal to the band gap are absorbed and can contribute to the photogeneration of electrons in a photovoltaic device. The photons with energy lower than the energy band gap are transmitted through the solar cell and do not contribute to the output electric power. [5]

A way to exceed this limit is to utilize the low energy photons via electronic *upconversion* process. [3, 5-6] Upconversion processes of photons of low energy into photons of higher energy offer a promising possibility of increasing the efficiency of a solar cell, allowing to exploit also photons of the solar spectrum with energy lower than the band gap of the material used. In particular, detailed balance calculations have shown that the application of an *upconverter* to a solar cell increases its efficiency limit from 31% to 37.4%. [3] Therefore, a full investigation of the efficiency of an upconverting material becomes an essential point in the design of a solar cell/upconverter integrated system.

An upconverter consists of a material with a band gap almost equal to that of the solar cell and containing intermediate levels. The upconverters usually combine an active ion, whose energy level scheme is employed for absorption of low energy photons, and a host material, in which the active ion is embedded. The most efficient upconversion has been found by using rare earths, i.e. lanthanides elements such as Erbium, Ytterbium, etc. [7-8] Moreover, the silicon nanocrystals (Si nc) in silicon oxide represent a possible candidate to incorporate rare earths, such as erbium atoms (Er), due to their compatibility with the silicon technology.

Among the upconversion mechanisms, a crucial role is played by the energy transfer upconversion (ETU) processes, which are based on cooperative effects. ETU processes have demonstrated the highest efficiency with respect to other mechanisms, i.e. two-photon absorption, which request very high power of the incident radiation. Moreover, Pilla et al. and Jacinto et al. have recently showed that the quantum efficiency of the fluorescence of a given electronic energy level is directly related to both the lifetime of the considered metastable state and the rate of ETU process. [9-10]

Over the last few decades experimental and theoretical studies have more and more stressed the importance of the noise fluctuations in natural and artificial systems. Noise-induced effects have been experimentally observed and theoretically studied in different physical and biological contexts. [11-28] In particular, stochastic resonance, [24-28] resonant activation [29-31] and noise enhanced stability [32-38] phenomena in several systems have been widely discussed. Specifically, the noise enhanced stability (NES) effect implies that a system remains in a metastable state for a time longer than the deterministic counterpart, and the lifetime has a maximum in correspondence with a well determined noise intensity. [34-35, 38]

In a previous work, Pacifici et al. have discussed a possible phenomenological description of the Er-doped Si nanocrystals, based on a set of rate equations which allow to quantitatively justify the results of optical measurements. [39]

The purpose of the present work is to study the effects of an external colored noise on the deexcitation dynamics of the first excited energy level of Er^{3+} ions embedded in Si nc. By superimposing an external noise source to the intrinsic one, it is possible to modify the dynamic response of the system. The intrinsic noise, which in principle could be modeled as a white noise, is already taken into account in the transition rates and represents the effect of the host temperature. In particular, by using a stochastic model based on rate equations, we focus on the influence of a multiplicative colored noise, modeled as an Ornstein-Uhlenbeck process. [40] As we have discussed, the physical condition of maximizing the efficiency of the upconversion phenomenon coincides with that of enhancing the average lifetime of particular energy levels of the Er atoms. The metastable state of interest is the first excited level of Er^{3+} ions, on which the effects of the external colored noise have been investigated by solving numerically the equations of the model. Our findings show that the presence of the external source of colored noise can determine an increase of the deexcitation time. In particular, this increase is maximum at specific values of the noise intensity and dependent on the ratio between the noise correlation time and the characteristic time of the deexcitation process obtained in the absence of fluctuations.

Er-Doped Si Nanocrystals

Si Nanocrystal-Er Atom Interaction

In 2003, photoluminescence measurements have allowed to identify a schematic diagram of the energy levels for the system of Si nanocrystals doped with Er atoms [11] (see Figure 1).

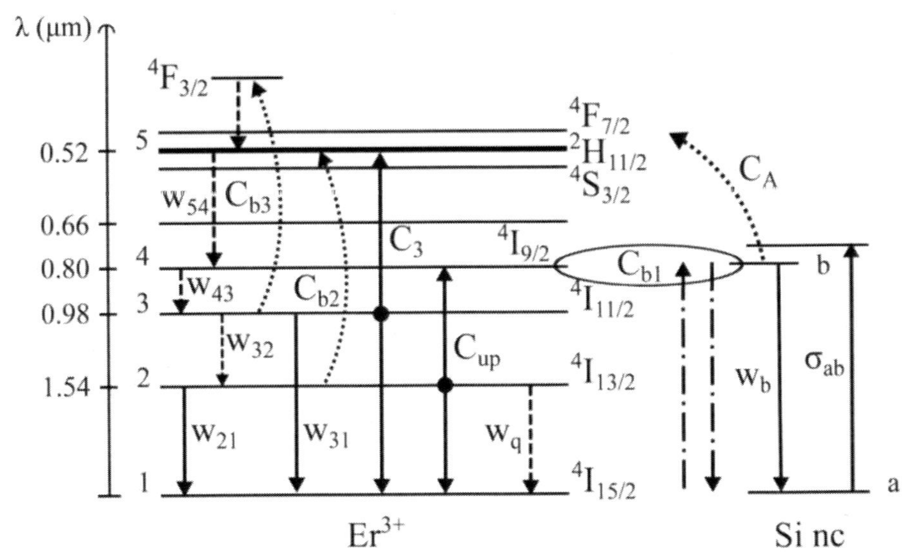

Figure 1. Energy level scheme for the Er-doped Si nanocrystals system. The Si nanocrystals are modeled by a two-effective energy level scheme, while Er^{3+} ions are represented as a five-effective energy level diagram.

In this scheme, a Si nanocrystal (nc) is represented by a three level system, consisting of two band edge levels and an interfacial level, which acts as a trap for the exciton. [41–43] Since we assume the trapping phenomenon to be a very fast process in comparison with the typical transition times, we can consider the Si nc as an effective two-level system, where the ground state is represented by the level *a* and the upper energy state by the level *b*. Within this assumption, σ_{ab} is the effective cross section of excitation, and describes both the creation of an exciton and its subsequent trapping at the interfacial energy level. Both processes occur after the absorption of a 488 nm photon. w_b is the total recombination rate of an exciton for an isolated Si nc, and is given by the sum of both radiative and non-radiative recombination rates. The Er^{3+} ion is described as a five-effective level system. In fact, due to the Stark splitting, the fifth level is composed of $^4S_{3/2}$, $^2H_{11/2}$, and $^4F_{7/2}$, which are strongly overlapping, and can be therefore taken into account as a single level. w_{ij} indicate the total transition rates from level *i* to level *j*, where $i, j = 1,2,...,5$ and $i > j$.

All the parameters of Figure 1 are listed in Table 1. Specifically, we can distinguish two classes of parameters: the transition rates expressed in s^{-1} and the coefficients describing energy transfers expressed in cm^3 s^{-1}. Among the second-class parameters, the coefficient C_{b1} describes the coupling between the excited level of Si nc and the Er ground state, which is responsible for the energy transfer between the Si nc's and Er ions surrounding it. This energy transfer is represented by the two antiparallel arrows, standing for the non-radiative deexcitation of an excited Si-nc (down arrow) and the following Er^{3+} excitation to the $^4I_{9/2}$ level (up arrow). The parameters C_{b2} and C_{b3} describe the excitation of level 5 of Er^{3+} from the levels *2* and *3*, respectively. In other words, these coefficients describe the further excitations of an excited Er^{3+} state through the energy transfer from the excited level *b* of the Si nc. The constant C_A takes into account a phenomenon very similar to the Auger effect occurring in Er-doped crystalline Si (Ref. [44]). In the Er-Si nc system this effect consists in an energy transfer from an excited Er level back to a confined exciton of a Si nc, which is therefore promoted to a higher energy level. This effect is present only at high pump powers and mainly involves the long lived $^4I_{13/2}$ level. The scheme also represents the quenching effect, which is due to the energy migration over the sample due to the energy transfer between two neighbouring Er^{3+} ions, one in the first excited level, the other in the ground state. The value of the total rate of the deexcitation from the first excited Er level is $w_{21} + w_q$, with $w_{21} = 4.2 \times 10^2$ s^{-1}. The rate $w_q = 8\pi C_{Er} N_q N_0$ takes into account the quenching effect, where N_q is the concentration of the quenching centers, C_{Er} is the coupling coefficient describing the interaction between the ground state and the first excited level of Er, and N_0 is the total Er concentration.

The value of the product $C_{Er} N_q$ has been obtained by experiments[39] and it is equal to 3.2×10^{-20} cm^3 s^{-1}. Other two processes are represented by a double arrow (with a single point of origin), and are indicated by the constants C_{up} and C_3. They are the cooperative up-conversion coefficients describing the interaction of two neighbouring Er^{3+} ions which are both in the first or in the second excited state, respectively. In the first case one of the two ions decays to the ground state giving its energy to the other one, which will be excited to the

$^4I_{9/2}$ level. In the second case the interaction will bring an ion to the $^2H_{11/2}$ level and the other one to the ground state.

Table 1. Parameters of the physical model

Symbol	Value	Reference
λ_{pump}	488 nm	Ref. [39]
σ_{ab}	2×10^{-16} cm^2	Ref. [45]
σ	1×10^{-19} cm^2	Ref. [45]
w_b	2×10^4 s^{-1}	Ref. [45]
w_{21}	4.2×10^2 s^{-1}	Ref. [45]
$w_q = 8\pi C_{Er} N_q N_0$	$8.1\times10^{-19} N_0$ s^{-1}	Ref. [45]
w_{32}	4.2×10^5 s^{-1}	Ref. [39]
w_{43}	1×10^7 s^{-1}	Ref. [39]
w_{54}	1×10^7 s^{-1}	Ref. [39]
C_{b1}	3×10^{-15} cm^3 s^{-1}	Ref. [39]
C_{up}	7×10^{-17} cm^3 s^{-1}	Ref. [39]
C_{b2}	3×10^{-19} cm^3 s^{-1}	Ref. [39]
C_{b3}	3×10^{-19} cm^3 s^{-1}	Ref. [39]
C_A	3×10^{-19} cm^3 s^{-1}	Ref. [39]
C_3	7×10^{-17} cm^3 s^{-1}	Ref. [39]

In the model we take into account also the possibility that a direct absorption of a 488 nm photon by a Er atom leads to a transition from the ground state to the $^4F_{7/2}$ level. This process is characterized by an excitation cross section $\sigma = 1\times10^{-19}$ cm^2 (Ref. [39]).

Deterministic Model for the Si Nanocrystal-Er Atom Interaction

By considering all the processes represented in the scheme of Figure 1, it is possible to write down a set of rate equations of first order describing the time evolution of the concentration of Si nc's and Er^{3+} ions in each energy level

$$\frac{dn_b}{dt} = \sigma_{ab}\varphi n_a - w_b n_b - \sum_{i=1}^{3} C_{bi} n_b N_i \tag{1}$$

$$\frac{dn_a}{dt} = -\sigma_{ab}\phi n_a + w_b n_b + \sum_{i=1}^{3} C_{bi} n_b N_i \tag{2}$$

$$\frac{dN_5}{dt} = \sigma\phi N_1 + \sum_{i=2}^{3} C_{bi} n_b N_i + C_3 N_3^2 - (w_{51} + w_{54}) N_5 \tag{3}$$

$$\frac{dN_4}{dt} = C_{b1} n_b N_1 + C_{up} N_2^2 + w_{54} N_5 - w_{43} N_4 \tag{4}$$

$$\frac{dN_3}{dt} = w_{43} N_4 - (w_{32} + w_{31}) N_3 - C_{b3} n_b N_3 - 2C_3 N_3^2 \tag{5}$$

$$\frac{dN_2}{dt} = w_{32} N_3 - (w_{21} + w_q) N_2 - C_{up} N_2^2 - C_{b2} n_b N_2 - C_A n_b N_2 \tag{6}$$

$$\frac{dN_1}{dt} = (w_{21} + w_q) N_2 + C_{up} N_2^2 + C_A n_b N_2 + w_{31} N_3 + C_3 N_3^2 - \sigma\phi N_1 - C_{b1} n_b N_1 \tag{7}$$

where ϕ is the flux of photons incident onto the sample, n_a, n_b and N_i, with $i = 1,2,3,4,5$, are the populations of the different levels of Si nanocrystals and Er^{3+} ions, shown in Figure 1. In particular, $n_0 = n_a + n_b$ and $N_0 = \sum_i N_i$ define the total concentrations of excitable Si nc's and Er atoms, respectively. We note that in Eqs. (1)-(7) the interaction between two energy levels is given by the product of the concentrations of centers whose energy values are those of the levels involved. Therefore the interaction probability will be described by rates which are inversely proportional to the square of the mean interaction volumes. This is due to the fact that a short-range dipole-dipole-like interaction, whose strength depends on the inverse sixth power of the mean distance between the two centers, is introduced.

Moreover, we note that in Er^{3+} the non-radiative deexcitation rates between contiguous levels are larger for transitions from the higher energy levels: $w_{54}, w_{43} >> w_{32} >> w_{21}$. This implies that in Er^{3+},the higher energy states are much less populated than the others.

Stochastic Model for the Si Nanocrystal-Er Atom Interaction

Now we extend the model for the Si nc–Er^{3+} ions interaction [39] above detailed by including a stochastic term that mimics an external source of random fluctuations. In particular this contribution is inserted as a multiplicative noise term, $N_2\mu(t)$, in the rate equation for the energy level 2 of Er^{3+} ions. Eq. (6) is therefore replaced by

$$\frac{dN_2}{dt} = w_{32} N_3 - (w_{21} + w_q) N_2 - C_{up} N_2^2 - C_{b2} n_b N_2 - C_A n_b N_2 + N_2 \mu(t) \tag{8}$$

Here $\mu(t)$ describes a colored noise and is given by the archetypal exponentially correlated process, known as Ornstein-Uhlenbeck (OU) process [40]

$$\frac{d\mu(t)}{dt} = -\frac{dt}{\tau_C}\mu(t) + \frac{\sqrt{D}}{\tau_C}\xi(t), \tag{9}$$

where $\xi(t)$ represents a Gaussian white noise within the Ito's scheme with zero mean, correlation function $\langle \xi(t)\xi(t') \rangle = \delta(t-t'),$ and intensity D, while τ_C is the correlation time of the OU process. The correlation function of the OU process is

$$\langle \mu(t)\,\mu(t') \rangle = \frac{D}{2}\exp\left(-\frac{|t-t'|}{\tau_C}\right) \tag{10}$$

and gives $D\delta(t-t')$ in the limit $\tau_C \to 0$.

Stationary State for the Population N_2 and Decay Time

The lifetime of the metastable level 2 has been investigated as a function of both the noise intensity and the correlation time, integrating Eqs. (1)-(5), (7), (8), and averaging over a certain number of numerical realizations.

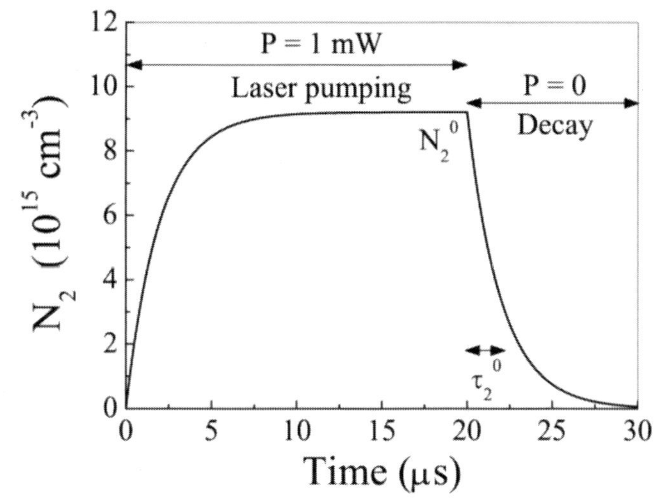

Figure 2. Time evolution of the population of the first excited state of Er^{3+} ions, N_2, as a function of time, in the presence of *laser pumping* of power P □ 1 mW ([0,20] μs) and during the time *decay*, P □ 0, ([20,30] μs). The laser pump has been switched off after the population N_2 has reached the stationary value N_2^0. τ_2^0 is the lifetime of the population of Er^{3+} ions in the level 2, obtained in the absence of noise.

In Figure 2 we show the population of Er atoms in the energy level 2, i.e. N_2, as a function of time, obtained in the absence of external noise ($D = 0$). In particular, we divide the simulation time in two time windows. In the first one, the numerical integration of Eqs. (1)-(5), (7), (8) is performed using a value of the *laser pumping* power, *P*, equal to 1 mW. In the second time interval, after the system has reached the steady state condition i.e., $N_2 \,\square\, N_2^0$, the laser beam is switched off ($P = 0$). This determines a time decay of the population N_2, and the lifetime τ_2 of the population of Er^{3+} ions in the level 2 is calculated as the time for which a reduction of N_2^0 by a factor e^{-1} is observed. For the value of pump power used here (1 mW), this lifetime is approximately equal to *2* ms. This value obtained in the absence of external noise, will be denoted, from now on, by the symbol τ_2^0. For the case shown in Figure 2, $N_2^0 \cong 9.20 \times 10^{15}$ cm^{-3}.

Effects of the External Noise Source

In Figure 3 we show the population of Er^{3+} ions, N_2, in the energy level of the first excited state $^4I_{13/2}$, as a function of the time. The analysis is carried out in the absence of noise ($D \,\square\, 0$), and for different values of the noise intensity *D* (panel (a): $D^{1/2} \,\square\, 10$ s^{-1}; panel (b): $D^{1/2} \,\square\, 10^2$ s^{-1}) and noise correlation time τ_C (= $0.1\tau_2^0$, τ_2^0, $10\tau_2^0$).

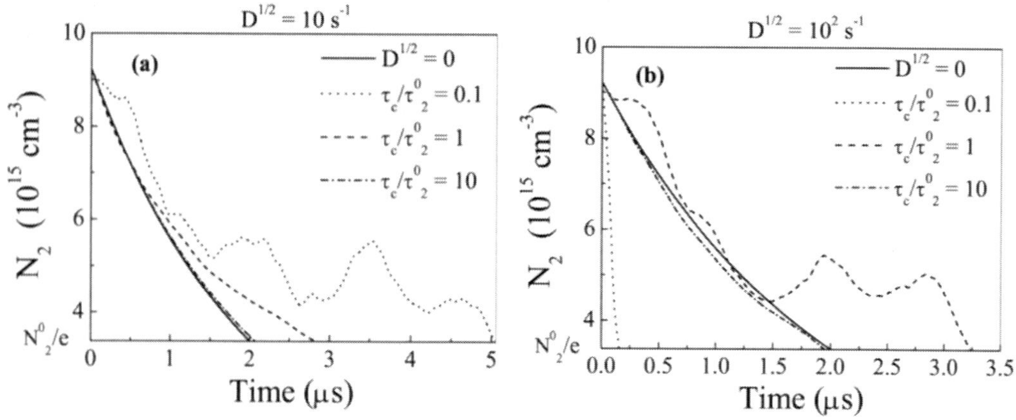

Figure 3. Time evolution of the population of the first excited state of Er^{3+} ions, N_2^0, as a function of the time, obtained after the pumping with a laser beam of power 1 mW has been turned off. The curves have been obtained in the absence of noise ($D \,\square\, 0$), and for $D^{1/2} \,\square\, 10$s^{-1} (panel a), $D^{1/2} \,\square\, 10^2$s^{-1} (panel b), using three different values of the noise correlation time (τ_C = $0.1\tau_2^0$, τ_2^0, $10\tau_2^0$).

For $\tau_C \ll \tau_2^0$ the deexcitation process of $^4I_{13/2}$ level is considerably affected by the external fluctuations, showing different trends which depend on the value of the noise intensity. Specifically, for $D^{1/2}$ □ $10\ s^{-1}$ the external noise causes a slowing down of the decay from the first excited level of Er atoms (dotted-line curve in panel (a) of Figure 3), while determines a faster depletion for $D^{1/2}$ □ $10^2\ s^{-1}$ (see dotted-line curve in panel (b) of Figure 3). This effect is quite counterintuitive since the colored noise has a negligible memory (small value of τ_C) with respect to the characteristic decay time τ_2^0 of the first excited level of Er atoms. A different behaviour should be observed if Er^{3+} ions were isolated. Conversely, in our system the Er atoms interact with the Si nc, exchanging energy, and the transition ${}^4I_{13/2} \rightarrow {}^4I_{15/2}$ competes with the deexcitation processes from higher energy levels, e.g. ${}^4I_{11/2} \rightarrow {}^4I_{13/2}$ characterized by a rate $w_{32} = 4.2\times10^5\ s^{-1}$. Concerning this, one can note that the inverse of the noise correlation time, $\tau_C = 0.1\tau_2^0$, is of the same order of magnitude of the transition rate w_b between the energy levels *b* and *a* of the Si nc. These are indirectly affected by fluctuations through the energy exchanges with the Er levels. As a consequence, for short memory values, the time evolution of the energy level 2 of Er^{3+} ions is influenced in a qualitatively different way by the external noise source, and dependent on the values of *D*.

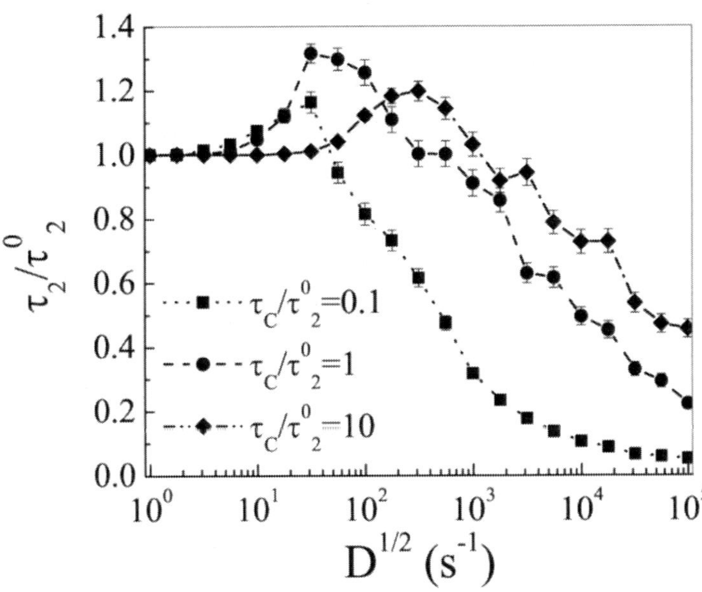

Figure 4. Ratio between the lifetime of the first excited level of the Er^{3+} ions τ_2 in the presence of noise and τ_2^0 (obtained in the absence of noise) as a function of the noise amplitude of the external colored noise $D^{1/2}$. Curves were obtained for different values of τ_C, namely, $\tau_C = 0.1\tau_2^0,\ \tau_2^0,\ 10\tau_2^0$.

For $\tau_C \sim \tau_2^0$, both values of the noise intensity, $D^{1/2}$ □ $10\ s^{-1}$ and $D^{1/2}$ □ $10^2\ s^{-1}$, determine a clear slowing down of the deexcitation process of the level 2 of Er atoms (see dashed-line curves in both panels of Figure 3). Differently, when $\tau_C >> \tau_2^0$, the deexcitation process becomes quasi-deterministic and the curves (dashed-dotted line in both panels of Figure 3) of the population N_2 approach those obtained in the absence of external noise ($D^{1/2}$ □ 0).

To understand better the effects of the external colored noise on the deexcitation process of the metastable level ${}^4I_{13/2}$, we evaluated (performing 10^4 numerical realizations) both the average values and error bars of the deexcitation times of the ${}^4I_{13/2}$ energy level.

In particular, for the ${}^4I_{13/2}$ level of the Er atoms, we calculate the ratio between the deexcitation time τ_2 in the presence of noise and that obtained in the absence of noise (τ_2^0), as a function of the amplitude of the external colored noise, $D^{1/2}$, and for different values of the ratio τ_C/τ_2^0, namely $\tau_C/\tau_2^0 = 0.1$, 1, 10. The results, shown in Figure 4, indicate that the addition of a source of correlated noise, characterized by $0.1\tau_2^0 < \tau_C < 10\tau_2^0$, can determine an enhancement of the lifetime τ_2 with respect to τ_2^0 up to 30% for $\tau_C = \tau_2^0$. We note that τ_2/τ_2^0 is a nonmonotonic function of $D^{1/2}$, which exhibits a maximum whose position depends on the value of the correlation time τ_C. In particular the position of the maximum is shifted towards right for increasing values of τ_C/τ_2^0. The nonmonotonic behaviour of τ_2/τ_2^0 as a function of $D^{1/2}$, which represents a further example of noise enhanced stability (NES) [17-21, 32-38], determines an improvement of the upconversion process. In order to quantify the effect of the external noise on the upconversion mechanism, we introduce the quantities η and η_0 defined as the efficiency of the upconverting process from the level ${}^4I_{13/2}$ in the presence and in the absence of noise, respectively. The effect of noise on the upconversion mechanism can be therefore evaluated defining the efficacy η/η_0. Pilla et al. and Jacinto et al. have showed that the quantum efficiency of the fluorescence of a metastable state is related to the speed of ETU processes. [9-10] In particular they have obtained the following relation between the quantum efficiency and the lifetime of a metastable state

$$\eta = \frac{w_{21}\tau_2}{1 + C_{up} N_2^0 \tau_2}, \tag{11}$$

valid also in the absence of noise with the foresight to substitute τ_2 with $\tau^0{}_2$.

Using Eq. (11) to express the ratio η/η_0, one gets

$$\frac{\eta}{\eta_0} = \frac{\tau_2}{\tau^0{}_2} \cdot \frac{1 + C_{up} N_2^0 \tau^0{}_2}{1 + C_{up} N_2^0 \tau_2}. \tag{12}$$

Since the binomial terms are practically equal to 1, the ratio η/η_0 coincides with τ_2/τ_2^0, so that the latter provides a good measure of the efficacy for the system investigated. As a consequence, the results shown in Figure 4 represents the behaviour of the efficacy η/η_0 as a function of the noise amplitude $D^{1/2}$, for different values of the noise correlation time.

Conclusion

In this chapter we have investigated the influence of an external source of colored noise on the deexcitation process of the energy level $^4I_{13/2}$ in Er-doped Si nanocrystals, a system which is strongly interested by the upconversion phenomenon. Our findings show that the inclusion of a source of multiplicative colored noise, which acts on the population of the level $^4I_{13/2}$ of the Er atoms, can slow down the deexcitation process. As a consequence, the lifetime may increase up to 30% of the value observed in the absence of external noise. This enhancement, which is nonmonotonic with the amplitude of the external fluctuations, presents a maximum within a wide interval of intensities and correlation times of the noise source. The slowing down of the deexcitation process can be ascribed to the effective deexcitation rates of the Er atoms, during the depletion dynamics of $^4I_{13/2}$, i.e. the first excited level of Er^{3+} ions. In conclusion, the enhancement of the deexcitation time of the level $^4I_{13/2}$ of Er^{3+} ions embedded in Si nanocrystals, is strongly connected with both the correlation time and intensity of the noise source. Therefore, external fluctuations play a key role in modulating the average lifetime of the metastable state $^4I_{13/2}$, and can contribute to increase the efficiency of upconversion phenomena, such as the cooperative upconversion between Er atoms. By adjusting the intensity of the external noise, it is possible to select the most favorable working condition in view of obtaining more efficient upconverters integrated in photovoltaic devices.

Acknowledgments

This work was supported by MIUR through Grant. No. PON02_00355_3391233, “Tecnologie per l’ENERGia e l’Efficienza energeTIC - ENERGETIC”.

References

[1] W. Shockley, H. J. Queisser, *J. Appl. Phys.* 32, 510 (1961).
[2] A. De Vos, H. Pauwels, *Appl. Phys. A: Mater. Sci. Process*. 25, 119 (1981).

[3] T. Trupke, M. A. Green, P. Würfel, *J. Appl. Phys.* 92, 4117 (2002).
[4] B. S. Richards, *Sol. Energ. Mat. Sol. C.* 90, 2329 (2006).
[5] A. Shalav, B. S. Richards, M. A. Green, *Sol. Energ. Mat. Sol. C.* 91, 829 (2007).
[6] S. V. Boriskina, G. Chen, *Opt. Commun.* 314, 71 (2014).
[7] W. G. Van Sark et al., *Nanoscale Res. Lett.* 8, 81 (2013).
[8] J. F. Suyver et al., *Opt. Mater.* 27, 1111 (2005).
[9] V. Pilla, T. Catunda, H. P. Jenssen, A. Cassanho, *Opt. Lett.* 28, 239 (2003).
[10] C. Jacinto et al., *Opt. Express* 13, 2040 (2005).
[11] B.T. Grenfell et al., *Nature* 394, 674 (1998).
[12] O. N. Bjørnstad, B. T. Grenfell, *Science* 293, 638 (2001).
[13] F. Principato, B. Spagnolo, G. Ferrante, A. Caddemi, *Fluct. Noise Lett.* 4, L425 (2004).
[14] N. Pizzolato, A. Fiasconaro, B. Spagnolo, *J. Stat. Mech.* P01011, 1 (2009).
[15] B. Spagnolo et al., *J. Phys.: Conf. Ser.* 174, 012037 (2009).
[16] N. Pizzolato, D. Persano Adorno, D. Valenti, B. Spagnolo, *Theor. Biosci.* 130, 203 (2011).
[17] S. Spezia, D. Persano Adorno, N. Pizzolato, B. Spagnolo, *J. Phys. Condens. Matter* 24, 052204 (2012).
[18] S. Spezia, D. Persano Adorno, N. Pizzolato, B. Spagnolo, *Acta Phys. Pol. B* 43, 1191 (2012).
[19] D. Persano Adorno, N. Pizzolato, B. Spagnolo, *J. Stat. Mech.* P01039, 1 (2008).
[20] D. Persano Adorno, N. Pizzolato, D. Valenti, B.Spagnolo, *Rep. Math. Phys.* 70, 171 (2012).
[21] S. Spezia, D. Persano Adorno, N. Pizzolato, B. Spagnolo, *Europhysics Lett.* 104, 47011 (2013).
[22] G. Bonanno, B. Spagnolo, *Fluct. Noise Lett.* 5, L325 (2005).
[23] G. Augello, D. Valenti, B. Spagnolo, *Int. J. Quantum Inform.* 6, 801 (2008).
[24] R. N. Mantegna, B. Spagnolo, *Phys. Rev. E* 49, R1792 (1994).
[25] L. Gammaitoni, P. Hänggi, P. Jung, F. Marchesoni, *Rev. Mod. Phys.* 70, 223 (1998).
[26] S. Spezia et al., *Eur. Phys. J. B* 65, 453 (2008).
[27] B. Spagnolo et al., *Eur. Phys. J. B* 69, 133 (2009).
[28] L. Gammaitoni, P. Hänggi, P. Jung, F. Marchesoni, *Eur. Phys. J. B* 69, 1 (2009).
[29] R. N. Mantegna, B. Spagnolo, *Phys. Rev. Lett.* 84, 3025 (2000).
[30] N. Pizzolato et al., *Phys. Biol.* 7, 034001 (2010).
[31] A. Fiasconaro, B. Spagnolo, *Phys. Rev. E* 83, 041122 (2011).
[32] R. N. Mantegna, B. Spagnolo, *Phys. Rev. Lett.* 76, 563 (1996).
[33] N. Agudov, B. Spagnolo, *Phys. Rev. E* 64, 035102 (2001).
[34] A. A. Dubkov, N. Agudov, B. Spagnolo, *Phys. Rev. E* 69, 061103 (2004).
[35] B. Spagnolo, N. Agudov, A. A. Dubkov, *Acta Phys. Pol. B* 35, 1419 (2004).
[36] A. Fiasconaro, B. Spagnolo, S. Boccaletti, *Phys. Rev. E* 72, 061110 (2005).
[37] E. S. Ross, *Br. J. Sports Med.* 41, 656 (2007).
[38] M. Trapanese, *J. Appl. Phys.* 105, 07D313 (2009).
[39] D. Pacifici et al., *Phys. Rev. B* 67, 245301 (2003).
[40] C. W. Gardiner, *Handbook of Stochastic Methods for Physics, Chemistry and the Natural Sciences*, 2nd edition, (Springer-Verlag, Berlin, 1985).
[41] K. S. Zhuravlev, A. Yu. Kobitski, *Semiconductors* 34, 1203 (2000).

In: Advances in Semiconductor Research
Editors: D. Persano Adorno and S. Pokutnyi
ISBN: 978-1-63321-755-3

Chapter 4

EXCITON STATES SPECTROSCOPY IN QUASI - ZERO - DIMENSIONAL SEMICONDUCTOR NANOSTRUCTURES: THEORY

Sergey I. Pokutnyi*
Chuiko Institute of Surface Chemistry,
National Academy of Sciences of Ukraine, Kyiv, Ukraine

Abstract

The review analyzes the results of theoretical investigations of exciton states (electron - hole pairs states) in quasi - zero - dimensional nanosystems consisting of spherical semiconductor nanocrystals (quantum dots) placed in transparent dielectric matrices. The theory of exciton states in a quantum dot under conditions of dominating polarization interaction of an electron and a hole with a spherical (quantum dot – dielectric matrix) interface are developed. It is shown, that the energy spectrum of a heavy hole in the valence band quantum dot is equivalent to the spectrum of hole carrying out oscillator vibrations in the adiabatic electron potential. We show that the absorption and emission edge of quantum dots is formed by two transitions of comparable intensity from different hole size – quantization levels and into a lower electron size – quantization level. A theoretical prospect of using hole transitions between equidistant series of quantum levels observed in nanocrystals for desining a nanolaser is proposed. The theory of an exciton formed from a spatially separated electron and a hole (the hole is in the quantum dot volume, and the electron is localized at the outer spherical quantum dot–dielectric matrix interface) is developed within the modified effective mass method. The effect of increasing the exciton binding energy significantly in quantum dots of zinc selenide, synthesized in a borosilicate glass matrix, relative to that in a zinc selenide single crystal is revealed. We developed the theory of biexcitons formed from spatially separated electrons and holes (the hole is in quantum dot volume, and the electron is localized at the outer surface of the quantum dot – dielectric matrix interface) in a nanosystem that consists of zinc selenide quantum dots synthesized in a glassy matrix.

Keywords: spectroscopy, exciton states, energy spectrum, exciton binding energy, absorption of light, optical nanolaser, biexcitons, semiconductor quantum dots

---

* E-mail address: Pokutnyi_Sergey@inbox.ru

Introduction

The optical properties of quasi – zero – dimensional structures consisting of spherical semiconductor nanocrystals (SNs) (quantum dots (QDs)) and dielectric nanocrystals of radius a ~ 1-10 nm, grown in transparent dielectric matrices, are currently being intensively studied [1-12]. The reason for these studies is that heterophase systems are promising materials for creating new nonlinear optoelectronic elements, in particular elements for controlling optical signals in optical lasers or optical computers [1-18].

Since the energy gap of a semiconductor is much smaller than in the dielectric host, the motion of carriers in the semiconductor microcrystal is restricted by its volume. Optical and nonlinear optical properties of heterophase systems are determined by the energy spectrum of spatially limited electron – hole pairs (EHP) and excitons) Refs. [1-22]. Quantum well effects have been found by optical spectroscopic studies in the energy spectrum of electrons and excitons [1-6] in such quasi – zero – dimensional structures.

The energy spectrum of an exciton in a small SN was obtained in Refs. [4-22] as a function of the SN radius a. The mentioned spectrum has been obtained using the adiabatic approximation, under the assumption that $m_e \ll m_h$, where m_e and m_h are the effective masses of an electron and a hole in a SN. In the exciton Hamiltonian, the Coulomb interaction between an electron and hole, as well as the polarization interaction of the electron and the hole with the surface of the SN, have been taken into account. The first order perturbation theory on the electron wave function of a spherical potential well of infinite depth has been used.

In Ref. [22] the exciton energy spectrum of a small SN, under the assumption that the electron could leave the SN volume for the ambient dielectric medium, has been obtained. The problem solved in Ref. [22] was that of the electron energy spectrum in small SN. In these studies the finite of the potential barrier (V_0) at the boundary between the SN and the dielectric matrix, and the associated penetration of the ambient host by the electron, have been taken into account. The electron wave functions were obtained in Refs. [20-22] for an electron travelling both in the SN and in the ambient dielectric host.

The motion of charge carriers in semiconductor nanocrystals, like in quantum dots, is limited in three directions. Therefore the energy spectrum of charge carriers is discrete in SNs with linear dimension of the order of 1-10 nm. In our opinion this property may be used for designing optical nanolasers or optical computers [1, 13]. In this context we point out that the size (i.e., radius) a of a quantum dot should be in the range of a few nanometers in order to provide that the difference, i.e., energy splitting$\Delta E_{e(h)}$, between electron and hole levels is of the order of a few $k_B T_0$ at the room temperature T_0(k_B is the Boltsman constant).

In experiments reported in Ref. [2], an interesting energetic structure was found which consisted of an equidistant series of quantum levels. The above structure is caused by quantization of the heavy hole energy spectrum in the adiabatic potential of the electron. Presents the theoretical prospect of using semiconductor nanocrystals for designing nanolasers [13].

The theory of exciton states in a QD under conditions of dominating polarization interaction of an electron and a hole with a spherical (quantum dot – dielectric matrix) interface is developed [10-27].

In [28], a new modified effective mass method was proposed to describe the exciton energy spectrum in semiconductor QDs with radii of $a \approx a_{ex}$ (a_{ex} is the exciton Bohr radius in the semiconductor material contained in the QD volume). It was shown that, within a model in which the QD is represented as an infinitely deep potential well, the effective mass approximation can be applied to the description of an exciton in QDs with radii a comparable to the exciton Bohr radius a_{ex}, assuming that the reduced effective exciton mass is a function of the radius a, $\mu = \mu(a)$ [28]. Contributions of a kinetic energy of an electron, energy of a Coulomb interaction between an electron and a hole, and energy of their polarization interaction with a QD surface to an energy spectrum of an exciton in a QD are analyzed [26-29].

In the adiabatic approximation and within the modified effective mass method [28], an expression for the binding energy of an exciton whose electron and hole move in the semiconductor QD volume was derived in [29]. In [29], the effect of significantly increasing the exciton-binding energy in cadmium selenide and sulfide QDs with radii a, comparable to the exciton Bohr radii *aex*, relative to the exciton- binding energy in cadmium selenide and sulfide single crystals (by factors of 7.4 and 4.5, respectively) was also detected.

A theory of the size quantization Stark effect in SNs has been developed for the case of dominating polarization interaction of an electron and a hole with the nanocrystal surface [17]. A shift of electron and hole quantum well levels in a SN in the interband absorption range in a uniform external electric field is determined by the quantum-confinement quadratic Stark effect. An electrooptical method is proposed, making it possible to estimate the characteristic quantum dot radius at which three-dimensional excitons can exist [17]. Electrooptical properties of quasi-zero-dimensional nanosystems containing SNs are governed by the quantum-confinement quadratic Stark effect in the range of interband absorption [17].

The theory of an exciton formed from spatially separated electron and a hole (the hole is in the quantum dot volume, and the electron is localized at the outer spherical quantum dot–dielectric matrix interface) is developed within the modified effective mass method. The effect of increasing the exciton binding energy significantly in quantum dots of zinc selenide, synthesized in a borosilicate glass matrix, relative to that in a zinc selenide single crystal is revealed.

It was shown that the short-wavelength shift of the peak of the low-temperature luminescence spectrum of samples containing zinc-selenide quantum dots, observed under the experimental conditions, is caused by quantum confinement of the ground-state energy of the exciton with a spatially separated electron and hole [23, 24].

The paper is organized as follows. In Section 2 we discuss the energy spectrum of exciton states. In section 3 we consider interband absorption and emission of light in SNs. The concept of using CdS nanocrystals as active lasing medium for constructing nanolasers is presented in Section 4. The theory of an exciton and biexciton formed from spatially separated electrons and holes is developed within the modified effective mass method in Section 5.

2. Spectrum of an Exciton in a Quasi –Zero –Dimensional Semiconductor Nanosystem

A simple model of a quasi – zero – dimensional structure in the form of a neutral spherical SN of radius a and permittivityε_2, embedded in a medium with permittivityε_1, was discussed in Refs. [7-12]. An electron (e) and a hole (h) with effective masses m_e and m_h were assumed to travel within the SN; we use $\boldsymbol{r_e}$ and $\boldsymbol{r_h}$ to denote the distances of an electron and a hole from the center of the SN, respectively (see Figure 1). We assume that the permittivities satisfy the relation $\varepsilon_2 \gg \varepsilon_1$, and that the conduction and valence bands are parabolic. In this model, in the framework of the effective mass approximation, the exciton (EHP) Hamiltonian takes the form [19-22]

$$H = -\frac{\hbar^2}{2m_e}\nabla_e^2 - \frac{\hbar^2}{2m_h}\nabla_h^2 + E_g + V_{eh}(\boldsymbol{r_e},\boldsymbol{r_h}) + V_{h\acute{h}}(r_h, a) + V_{e\acute{e}}(\boldsymbol{r_e}, a) + \\ + V_{e\acute{h}}(\boldsymbol{r_e},\boldsymbol{r_h}, a) + V_{h\acute{e}}(\boldsymbol{r_e},\boldsymbol{r_h} a) \qquad (1)$$

where the first two terms represent the kinetic energy of the electron and the hole, $V_{eh}(\boldsymbol{r_e},\boldsymbol{r_h})$ is the energy of the Coulomb interaction between the hole and the electron

$$V_{eh}(\boldsymbol{r_e},\boldsymbol{r_h}) = -\frac{e^2}{\varepsilon_2 a}\frac{a}{(r_e^2 - 2r_e r_h cos\theta + r_h^2)^{1/2}} \qquad (2)$$

where the angle θ between the vector $(\boldsymbol{r_e})$ and $(\boldsymbol{r_h})$ and $V_{h\acute{h}}(\boldsymbol{r_h}, a)$ are the energies of interaction of the electron and the hole with their own images, $V_{eh`}(,\boldsymbol{r_e},\boldsymbol{r_h}, a)$ and $V_{he`}(\boldsymbol{r_e},\boldsymbol{r_h}, a)$ are the energies of interactions with "strange" images, and E_g is the bandgap in the infinite semiconductor with permittivity ε_2. (see Figure 1).

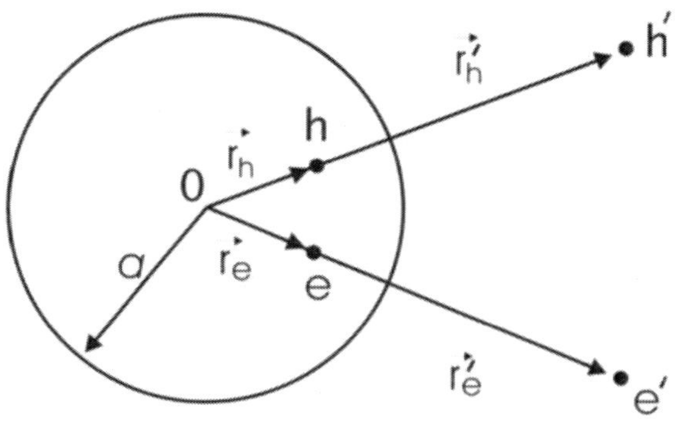

Figure 1. Schematic diagram of an exciton in a spherical semiconductor nanocrystal. The radius vectors $\boldsymbol{r_e}$ and $\boldsymbol{r_h}$ determine the distance of the electron e and the hole h from the center of a nanocrystal of radius a. The image charges $e' = (a/r_e)e$ and $h' = -(a/r_h)e$ are located at the distances $r_e' = (a^2/r_e)$ and $r_h' = (a^2/r_h)$ from the center of the nanocrystal 0 and represent point image charges of the electron and hole, respectively.

For arbitrary ε_1 and ε_2, the terms in Eq. (1) that describe the energy of the polarization interaction of the electron and hole with the SN surface can be written in an analytic form [19-22] which is particularly simple for the case $\varepsilon_2 \gg \varepsilon_1$:

$$V_{h\acute{h}}(\boldsymbol{r_h}, a) = \frac{e^2}{2\varepsilon_2 a}\left(\frac{a^2}{a^2 - r_h^2} + \frac{\varepsilon_2}{\varepsilon_1}\right), \tag{3}$$

$$V_{e\acute{e}}(\boldsymbol{r_e}, a) = \frac{e^2}{2\varepsilon_2 a}\left(\frac{a^2}{a^2 - r_e^2} + \frac{\varepsilon_2}{\varepsilon_1}\right), \tag{4}$$

$$V_{e\acute{h}} = V_{h\acute{e}} = -\frac{e^2}{2\varepsilon_2 a}\frac{a}{[(r_e r_h/a)^2 - 2r_e r_h \cos\theta + a^2]^{1/2}}. \tag{5}$$

The spectrum of an exciton (EHP) in a small semiconductor microcrystal was investigated in Refs. [19-22] in the case where its size was restricted by the condition

$$a_0 \ll a_h \ll a \leq a_e \cong a_{ex}, \tag{6}$$

where $a_h = \varepsilon_2\hbar^2/(m_h e^2)$, $a_e = \varepsilon_2\hbar^2/(m_e e^2)$ and $a_{ex} = \varepsilon_2\hbar^2/(\mu e^2)$ are the Bohr radii of the hole, electron and exciton in the semiconductor with permittivity ε_2, respectively; e is the electron charge, $\mu = m_e m_h/(m_e + m_h)$ is the reduced effective mass of the EHP, and a_0 is a characteristic size that is of the order of the interatomic distance. When the condition (6) is satisfied, the polarization interaction is a significant part in the potential energy of the Hamiltonian (1). The above inequalities also enable us to examine the motion of an electron and a hole in the effective mass approximation. The validity of Eq. (6) further enables us to consider the motion of a heavy hole ($m_h \gg m_e$) in the electron potential averaged over the motion of the electron (adiabatic approximation).

The adiabatic approximation can be used if we assume that the electron kinetic energy is the largest term and take the last four terms in Eq. (1) together with nonadiabatic operator as a perturbation. Then, taking into account only the first order of the perturbation theory, one can easily obtain the spectrum of an exciton (EHP), $E_{n_e,l_e,m_e}^{n_h,l_h,m_h}(a)$, in the state $(n_e, l_e, m_e;\ n_h, l_h, m_h)$ (here n_e, l_e, m_e and n_h, l_h, m_h are the main, orbital and magnetic quantum numbers of an electron and a hole) [19-22, 26-29]:

$$E_{n_e,l_e,m_e}^{n_h,l_h,m_h}(a) = \frac{\hbar^2}{2m_e a^2}u_{n_e,l_e}^2 + \bar{V}_{e\acute{e}}(a) + \lambda_{n_e,l_e,m_e}^{n_h,l_h,m_h}(a) + E_g, \tag{7}$$

where the first term is the kinetic energy of an electron in an infinite spherical well, and $\bar{V}_{ee^{\prime}}(a)$ is the average value of the electron interaction with the self – image with the infinitely deep spherical well functions

$$\Psi_{n_e}, l_e, m_e = \frac{\sqrt{2/r_e}}{a} Y_{l_e,m_e}(\theta, u)\frac{J_{l_e+1/2}(u_{n_e,l_e} r_e/a)}{J_{l_e+3/2}(u_{n_e,l_e})}, \tag{8}$$

where Y_{l_e,m_e} are normalized spherical functions, u_{n_e,l_e} are roots of the Bessel function $J_{l_e+1/2}(u_{n_e,l_e}) = 0$. The quantity $\lambda_{n_e,l_e,m_e}^{n_h,l_h,m_h}(a)$ is an eigenvalue of the heavy hole Hamiltonian

$$H_h = -\frac{\hbar^2}{2m_h}\nabla_h^2 + V_{h\acute{h}}(r_h, a) + \bar{V}_{n_e,l_e,m_e}(r_h, a), \quad (9)$$

where

$$\bar{V}_{n_e,l_e,m_e}(r_h, a) = \bar{V}_{eh}(r_h, a) + \bar{V}_{e\acute{h}}(r_h, a) + \bar{V}_{h\acute{e}}(r_h, a) \quad (10)$$

is the average value of the Coulomb interaction and of the electron and hole interaction with “alien” images on “free” electron states (8).

Quantitative results for the EHP spectrum (7) are obtained here only for the simple case $l_e = 0$. Using expressions (3) – (5), (8) and (10) one can obtain

$$\bar{V}_{eh}^{n_e,0,0}(x, s) = s^{-1}[\frac{\sin(2\pi n_e x)}{\pi n_e x} - 2Ci(2\pi n_e x) + 2Ci(2\pi n_e) + 2\ln x - 2], \quad (11)$$

$$\bar{V}_{e\acute{h}}^{n_e,0,0}(x, s) + \bar{V}_{h\acute{e}}^{n_e,0,0}(x, S) = -2s^{-1}, \quad (12)$$

$$\bar{V}_{e\acute{e}}^{n_e,0,0}(x, s) = s^{-1}\left(Z_{n_e,0} + \frac{\varepsilon_2}{\varepsilon_1}\right), \bar{V}_{h\acute{h}}(x, s) = s^{-1}(\frac{1}{1-x^2} + \frac{\varepsilon_2}{\varepsilon_1}) \quad (13)$$

where

$$Z_{n_e,0} = 2\int_0^1 dx_e \frac{sin^2(\pi n_e x_e)}{1 - x_e^2},$$

and Ci(y) is the integral cosine. Here and below the energy is measured in units $Ry_h = \hbar^2/2m_h a_h^2$ and nondimensional values of the length $x = (r_h/a)$ and $s = (a/a_h)$ are used.

Note that in the approximation considered the interaction of an electron with the images (its own and “alien”) (12), (13) and the interaction of a hole with an “alien” image (12) yield a constant addition $\sim s^{-1}$ to the hole energy.

Taking into account formulae (3) and (10) – (12), we write the potential energy in the heavy hole Hamiltonian (9) as follows

$$\bar{U}^{n_e,0,0}(x, s) = V_{h\acute{h}}(x, s) + \bar{V}^{n_e,0,0}(x, s) = \quad (14)$$

$$= s^{-1}[\frac{1}{1-x^2} + \frac{\sin(2\pi n_e x)}{\pi n_e x} - 2Ci(2\pi n_e x) + 2Ci(2\pi n_e) + 2\ln x + \frac{\varepsilon_2}{\varepsilon_1} - 4].$$

The minimum of the potential energy (14)

$$\bar{U}_{min}^{n_e,0,0}(s) = \bar{U}^{n_e,0,0}(x = 0, s) = \frac{P_{n_e,0}}{s},$$

where

$$P_{n_e,0} = 2Ci\,(2\pi n_e) - 2\ln(2\pi n_e) - 2\gamma + \frac{\varepsilon_2}{\varepsilon_1} - 1,$$

($\gamma = 0{,}577$ is the Euler constant) is reached at the point $x = 0$. A series expansion of the potential (14) with respect to the parameter $x^2 \ll 1$ with accuracy up to the first two terms gives the hole spectrum $\lambda^{t_h}_{n_e,0,0}(\bar{s})$ in oscillator form [19-22]:

$$\lambda^{t_h}_{n_e,0,0}(\bar{s}) = \frac{P_{n_e,0}}{\bar{s}} + \bar{\omega}(\bar{s},n_e)\left(t_h + \frac{3}{2}\right), \quad (15)$$

where

$$\bar{\omega}(\bar{s},n_e) = 2.232(1 + \frac{2}{3}\pi^2 n_e^2)^{1/2}\bar{s}^{-3/2}, \quad (16)$$

and $\bar{\omega}(\bar{s},n_e)$is the frequency of the hole oscillator vibrations, $t_h = 2_{n_r} + l_h = 0,1, ...$ is the hole main quantum number (n_r is the hole radial quantum number).

Taking into account formulae (13), (15) and (16) we obtain the EHP spectrum $[E^{t_h}_{n_e,l_e=0}(\bar{s})]$ in the state $(n_e, l_e = 0; t_h)$ for the SN of radius $(\bar{s} = \bar{a}/a_h)$ [19-22]:

$$E^{t_h}_{n_e,l_e=0}(\bar{s}) = E_g + \frac{\pi^2 n_e^2}{\bar{s}^2}K\beta + \bar{s}^{-1}\left(Z_{n_e,0} + P_{n_e,0} + \frac{\varepsilon_2}{\varepsilon_1}\right) + \bar{\omega}(\bar{s},n_e)(t_h + \frac{3}{2})\,, \quad (17)$$

where K=0.67 represents the spread of the SN radii [4] and $\beta = m_e/m_h$. Moreover, numerical analysis of x – ray data [2], which took the spread of the SN radii into account, shows that the mean radius evaluated over the Lifshits – Slezov distribution was $\bar{s} = 0{,}86s$ [2], where s is the SN radius obtained in the small variance approximation.

The main contribution to the EHP spectrum (17) is provided by second term (electron kinetic energy), which comes from the purely spatial restrictions imposed on the quantization region. The last two terms, which are associated with the Coulomb and polarization interactions between the electron and the hole, are regarded as corrections. The polarization interaction between hole and electron, on the one hand, and the surface of the SN, on the other, provides, like the size quantization of carriers, a contribution to the renormalization of the energy gap of the SN given by Eq. (17). The polarization interaction, which is of the same order of magnitude as the exciton binding energy in the SN, in this case is found to be much smaller than the size quantization energy of the charge carriers in the SN.

We also note that the exciton spectrum (17) is valid only for low – lying states of the exciton $(n_e, 0; t_h)$ in the SN, for which $(E_{n_e,0} - E_g) \ll V_0$, where V_0 is the depth of the potential well for an electron in the SN (e.g., in the cadmium sulphide SN, V_0=2.3- 2.5eV when the condition (6) is satisfied [30]).

In Refs. [2] interband absorption spectra of CdS SNs ($\varepsilon_2 = 9.3$) of sizes varying from 1 to 10^2 nm dispersed in a transparent dielectric matrix of silicate glass were investigated. The effective masses of electrons and holes in CdS were, respectively, $m_e/m_0 = 0.205$ and $m_h/m_0 = 5$ (i.e., $m_e \ll m_h$). The dimensional quantization phenomenon found in ref. [2] for the electron energy spectrum was qualitatively described by the formula

$$E_{n_e,l_e}(a) = E_g + \frac{\hbar^2}{2\mu a^2} u^2_{n_e,l_e} \tag{18}$$

In this case in the region of SN sizes a comparable with the exciton radius, $a \le a_{ex} \cong 3$ nm, a deviation of the experimental points from the theoretical dependence (18) was observed. The spectrum of electron – hole pairs in the interval of SN sizes $a \le a_{ex}$ calculated in Refs. [19-22] gave a qualitatively correct description of the dependence of the experimental spectrum [2] on the SN radius a.

In the experiment reported in Ref. [2], an energy structure consisting of an equidistant series of levels was found in the region if transitions to the lower level of dimensional quantization of the electron ($n_e = 1$). The above structure is caused by quantization of the hole energy spectrum in the adiabatic potential of the electron.

From the comparison of formula (16) (for $n_e = 1$) with the experimental dependence of the splitting magnitude on the SN size $\bar{a}$ obtained in Ref. [5], it follows that for SNs of radii $1.5 \le \bar{a} \le 3.0$ nm, the splitting $\bar{\omega}(\bar{s})$ (Eq. (16)) is in good agreement with the experimental data [2] and differs from the latter only slightly ($\le 4\%$).

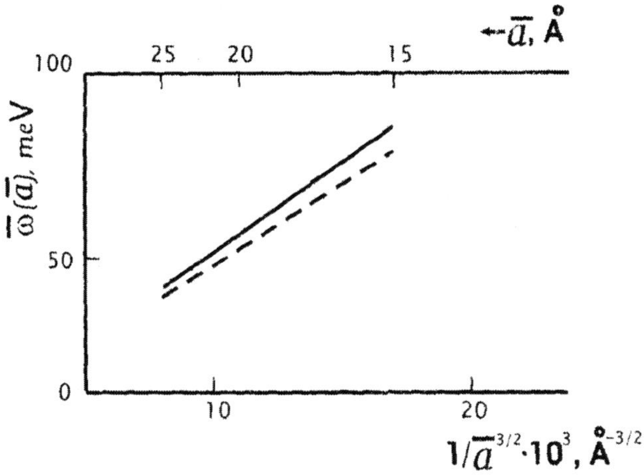

Figure 2. The dependence of the splitting $\bar{\omega}\left(\bar{a}\right)$ (16) of the radius of the nanocrystal $\bar{a}$. The solid line shows the experimental dependence $\bar{\omega}\left(\bar{a}\right)$ [2].

3. Interband Absorption and Emission of Light in a Nanocrystal

The interband absorption of light in SNs was studied theoretically in [14-16] using the dipole approximation in the framework of the model considered here, and under the assumption that the absorption length $\lambda \gg a$.

An expression for the quantity $K(\bar{s}, \omega)$ defined by the hole optical transition from the energy level $t_h = 2n_h$ to the lowest electron level ($n_e = 1, l_e = m_e = 0$) was derived in the following form [15]:

$$K(\bar{s},\omega) = A\sum_{n_h} L_{n_h}(\bar{S})\delta[\Delta(\omega) - \frac{\pi^2}{\bar{s}^2}\frac{m_h}{m_e} - \frac{1}{\bar{s}}\left(Z_{1,0} + P_{1,0} + \frac{\varepsilon_2}{\varepsilon_1}\right) - \bar{\omega}(\bar{s}, n_e = 1)(2n_h + 3/2)] \quad (19)$$

where $\Delta(\omega) = \hbar\omega - E_g, \omega$ - incident light frequency, and *A* is proportional to the square of the absolute value of the dipole moment matrix element calculated with Bloch functions. The quantity $K(\bar{s},\omega)$ (19) connects the energy absorbed by SN in a time unit with the time average of electric field square of incident wave. Moreover, the product of $K(\bar{s},\omega)$ and SN concentration in the dielectric matrix gives electric conductivity of the considered quasi – zero – dimensional system for the frequency ω, which is connected with light absorption coefficient in the usual way.

We determine the quantity $K(\bar{s},\omega)$ (19) corresponding to hole optical transition from the energy level $t_h = 2n_h$ to the lowest electron level $(n_e = 1, l_e = m_e = 0)$. In this case, the expression for the quantity $L_{n_h}(\bar{s})$, given by the square of the overlap integral of the electron and hole wave functions, takes the form [14-16]

$$L_{n_h}(\bar{s}) = \frac{2\pi^{5/2}}{\left(1+\frac{2}{3}\pi^2\right)^{3/4}}\frac{n_h+1}{2^{2n_h}n_h!}(\bar{s})^{-3/4} \quad (20)$$

In the interband optical absorption spectrum of SN, the radius of which fulfills condition (7), each line corresponding to given values of the radial n_e and orbital l_e quantum numbers turns into a series of close lying equidistant levels, corresponding to various values of the main hole quantum number t_h. This conclusion follows from (15), (16) and (19), and is a direct consequence of the Coulomb and polarization interactions of an electron and a hole in SN.

We can estimate value of the overlap integral square $(K(\bar{s},\omega)/A)$ using (19), (20) and the experimental data taken from Ref. [2]. For the hole transitions from the equidistant quantum levers: $(n_h = 0;\ l_h = m_h = 0), (n_h = 1; l_h = m_h = 0)$,

$(n_h = 2; l_h = m_h = 0), (n_h = 3; l_h = m_h = 0)$, to the lowest electron size –quantized level $(n_e = 1, l_e = m_e = 0)$, we have

$$\frac{K(\bar{s},\omega)}{A} = L_1(\bar{s}) + L_2(\bar{s}) + L_3(\bar{s}), \quad (21)$$

where $L_0 = 7.66\,\bar{s}^{-3/4}, L_1 = 0.5\,L_0, L_2 = 9.4\cdot 10^{-2}L_0, L_3 = 10^{-2}L_0$. From above expression, it follows that the main contribution in the light absorption coefficient is a cadmium sulphide SN with the radius *s* (6) comes from the hole spectral lines corresponding to quantum numbers $(n_h = 0;\ l_h = m_h = 0)$ and $(n_h = 1; l_h = m_h = 0)$, the transition oscillator strengths of which are dominant. The contribution of higher exited hole lines $(n_h \geq 2; l_h = m_h = 0)$ is negligible.

This way, in the framework of the considered model of the quasi – one – dimensional system it was shown that the absorption and emission edge of a cadmium sulphide SN is formed by two transitions of comparable intensities.

4. Nanolaser on Heavy Hole Transition in Semiconductor Nanocrystals

We assume that CdS microcrystals have a direct band structure, like bulk CdS. The extrema of the conduction and valence bands are placed in the centre of the Brillouin zone. Moreover, CdS is wide – gap semiconductor ($E_g = 2.58\ eV$) in which nonparabolicity effect is weak. At the same time, the dispersion laws near the bottom of the conduction band and the top of the valence band can be considered as parabolic [2].

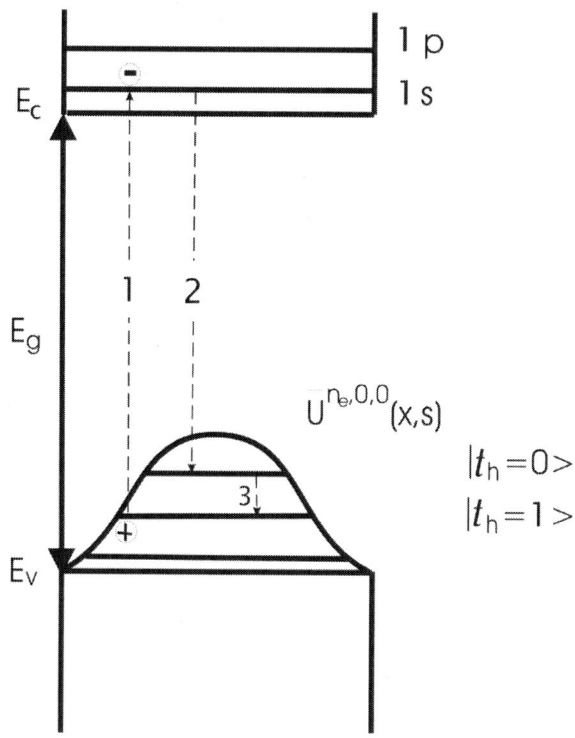

Figure 3. The scheme of the optical transition of an electron and a hole in a CdS nanocrystal. The electron energy spectrum $E_{n_e,l_e}(a)$ (18) and the hole potential energy $\bar{U}^{n_e,0,0}(x,s)$ (14) are shown; E_g – bandgap width. The electron states in the conduction band are labeled as (1 s) and (1 p). The hole states in the valence band are denoted as $|t_h = 0\rangle$ and $|t_h = 1\rangle$. Transitions are labelled as follows: "1" - pumping, "2"- creating inverse population, "3" - lasing

In the construction of the theory, it was assumed that the electron and hole bands are parabolic. The nonparabolicity parameter j of such bands for EHP energy $E_{1,0}^{t_h}(a)$ (17) in the group state, which is obtained from the experiments of Refs. [2] in SN of size $a \sim a_{ex}$ takes a small value

$$j(a) = \frac{E_{1,1}^{t_h} - E_g}{E_g} \leq 8\%.$$

This relation gives a basis for the assumption that the valence and conduction bands are, to a high degree of accuracy, parabolic.

In Figure 3, the energy diagram of the optical transitions in CdS SN of radius a fulfilling the condition (6) is shown Ref. [13]. The electron energy spectrum $E_{n_e,l_e}(a)$ (18) in the conduction band of a SN of the radius a comparable to the linear dimension of quasiparticles in semiconductors is determined only by the size quantization effect. The energy spectrum $\lambda_{n_e,0,0}^{t_h}$ of heavy hole in the valence band is equivalent to the spectrum of hole carrying out oscillator vibrations with frequency $\bar{\omega}(\bar{S},n_e)$ (Eq.(16)) in the adiabatic electron potential.

On the basis of presented results we can formulate an action scheme for a laser designed on CdS quantum dots [13] grown in boric – silicate glass matrices [1-6]:

1. Pumping: the hole transition from energy level $\lambda_{n_e,0,0}^{t_h}(a)$ to electron state $E_{n_e=1,l_e=0}(a)$ (1s - state) in conduction band with energy (transition “1” in Figure 3):

$$h\nu_1(\bar{s}) = E_g + E_{1,0}(\bar{s}) + \lambda_{1,0,0}^{1}(\bar{s}) \quad (22)$$

2. Creation of inverse population on the hole level $|t_h = 0\rangle$ by means of electron transition from the energy level $E_{1,0}(\bar{s})$, with spin flip [31], to the hole energy level $\lambda_{1,0,0}^{0}(\bar{s})$ (transition “2” in Figure 3). The energy of this transition

$$\hbar\nu_2(\bar{s}) = E_g + E_{1,0}(\bar{s}) + \lambda_{1,0,0}^{0}(\bar{s}) \quad (23)$$

3. Lasing step: transition of the hole in the valence band between equidistant levels $|t_h = 0\rangle \rightarrow |t_h = 1\rangle$ (transition “3” in Figure 3) separated by the energy

$$\Delta(\bar{s}) = \hbar\bar{\omega}(n_e = 1, \bar{s}). \quad (24)$$

In the experimental conditions from Refs. [1-6] at liquid nitrogen temperatures (T=77K), the distance ΔE_e between electron energy levels ($n_e = 1$, $l_e = 1$) (1p - state) and ($n_e = 1, l_e = 0$) (1s state), according to Eq. (18)

$$\Delta E_e = \frac{(u_{1,1}^2 + \pi^2)(m_h/m_e)Ry_h}{S^2} \cong 480meV,$$

where $u_{1,1}$=4.493 [32]. The distance $\Delta(\bar{s})$ between the hole energy levels (23) is $\Delta(\bar{s}) = \hbar\bar{\omega}(n_e = 1, \bar{s}) \cong 53meV$.

For SN of the radius a = 2nm, $\Delta(\bar{s})$ is several times greater than the thermal energy k_BT at T = 77 K. In a nanolaser on CdS nanocrystals, the pumping energy $h\nu_1 \cong 3.11eV$ (22) would correspond to visible light, and lasing at energy $\Delta(\bar{s})$ ≅53 meV would occur in the infrared range [13].

5. Excitons and Biexcitons Formed from Spatially Separated Electrons and Holes on Quasi - Zero - Dimensional Semiconductor Nanosystems

In [28], a new modified effective mass method was proposed to describe the exciton energy spectrum in semiconductor QDs with radii of $a \approx a_{ex}$ (a_{ex} is the exciton Bohr radius in the semiconductor material contained in the QD volume). It was shown that, within a model in which the QD is represented as an infinitely deep potential well, the effective mass approximation can be applied to the description of an exciton in QDs with radii a comparable to the exciton Bohr radius a_{ex}, assuming that the reduced effective exciton mass is a function of the radius a, $\mu = \mu(a)$.

In the adiabatic approximation and within the modified effective mass method [28], an expression for the binding energy of an exciton whose electron and hole move in the semiconductor QD volume was derived in [29]. In [29], the effect of significantly increasing the exciton-binding energy in cadmium selenide and sulfide QDs with radii a, comparable to the exciton Bohr radii aex, relative to the exciton- binding energy in cadmium selenide and sulfide single crystals (by factors of 7.4 and 4.5, respectively) was also detected.

In the experimental study [6], it was found that excess electrons produced during interband excitation of the cadmium sulfide QD have a finite probability of overcoming the potential barrier and penetrating into the borosilicate glass matrix into which the QD is immersed. In experimental studies [6] of glass samples with cadmium-sulfide and zinc-selenide QDs, it was found that the electron can be localized in the polarization well near the outer QD surface, while the hole moves within the QD volume.

In [6], the optical properties of borosilicate glass samples containing QD zinc selenide are experimentally studied. The average radii of such QDs are in the range $a \approx 2.0$–4.8 nm. In this case, the values of a are comparable to the exciton Bohr radius $a_{ex} \approx 3.7$ nm in a ZnSe single crystal. At low QD concentrations, when the optical properties of the samples are mainly controlled by those of individual QDs in the borosilicate glass matrix, a shift of the peak of the low temperature luminescence spectrum to the short wavelength region (with respect to the band gap E_g of the zinc selenide - single crystal) was observed. The authors of [6] assumed that this shift is caused by quantum confinement of the energy spectra of the electron and hole localized near the spherical surface of the QD.

The use of semiconductor nanosystems as the active region of nanolasers is prevented by the low binding energy of the QD exciton [6,13]. Therefore, studies directed toward the search for nanostructures in which a significant increase in the binding energy of QD excitons would be observed are of importance.

The theory of an exciton with a spatially separated electron and hole was developed within the modified effective mass method in which the reduced effective exciton mass is a function of the semiconductor QD radius a [23, 24]. The effect of significantly increasing the binding energy $|E_{ex}(a)|$ of the exciton ground state in a nanosystem containing zinc-selenide QDs with radii $4{,}0 \leq a \leq 29{,}8$ nm was detected; in comparison with the exciton binding energy in a zinc-selenide single crystal, the increase factor is 4.1–72.6. It was shown that the effect of significantly increasing the binding energy $|E_{ex}(a)|$ of the exciton ground state in the nanosystem under study is controlled by two factors: (i) a substantial increase in the electron–hole Coulomb interaction energy and an increase in the energy of the interaction of the electron and hole with "foreign" images (the "dielectric enhancement" effect); (ii) spatial confinement of the quantization region by the QD volume; in this case, as the QD radius a increases, starting from $a \geq 29.8$ nm, the exciton becomes two-dimensional with a ground-state energy that exceeds the exciton binding energy in a zinc-selenide single crystal by almost two orders of magnitude [23, 24].

The average zinc- selenide QD radius was determined by comparing the dependence of the exciton ground-state energy on the QD radius, obtained by the variational method within the modified effective mass method, with the experimental peak of the low-temperature luminescence spectrum [6]. It was shown that the short-wavelength shift of the peak of the low-temperature luminescence spectrum of the samples containing zinc-selenide QDs, which was observed under the experimental conditions, is caused by renormalization of the electron–hole Coulomb interaction energy and also the energy of the polarization interaction of the electron and hole with the spherical QD–dielectric matrix interface, related to spatial confinement of the quantization region by the QD volume. In this case, the hole moves in the QD volume, and the electron is localized at the outer spherical QD–dielectric matrix interface [23, 24].

To apply semiconductor nanosystems containing zinc- selenide QDs as the active region of lasers, it is required that the exciton binding energy $|E_{ex}(a)|$ (QD radius $4{,}0 \leq a \leq 29{,}8$ nm) in the nanosystem be on the order of several kT_0 at room temperature T_0 (k is the Boltzmann constant). Nanosystems consisting of zinc- selenide QDs grown in a borosilicate glass matrix can be used as the active region of semiconductor QD lasers. The parameter $|E_{ex}(a)/kT_0|$ take significant values in the range from 3.1 to 56 [23, 24].

We developed the theory of a biexcitons formed from spatially separated electrons and holes (the hole is in QD volume, and the electron is localized at the outer surface of the QD–matrix interface) in a nanosystem that consists of ZnSe QDs synthesized in a glassy matrix [25]. The dependences of the total energy and binding energy of the singlet ground biexciton state in such a system on the spacing between the QD surfaces and the QD radius are derived by the variational method. It is shown that biexciton formation is of the threshold character and possible in nanosystems, in which the spacing between the QD surfaces is larger than a certain critical spacing [25]. It is shown that the major contribution to the biexciton binding energy is made by the energy of the exchange interaction of electrons with holes and this contribution is much more substantial that the contribution of the energy of Coulomb interaction between the electrons and holes. It is established that the spectral shift of the low-temperature luminescence peak in such QDs is due to quantum confinement of the energy of the biexciton ground state [25].

Conclusion

The review provides the results of theoretical investigations of excitons states (electron - hole pairs states) in a quasi - zero - dimensional nanosystems consisting of spherical semiconductor nanocrystals (quantum dots) placed in transparent dielectric matrices. The theory of exciton states in a QDs under conditions of dominating polarization interaction of an electron and a hole with a spherical (quantum dot – dielectric matrix) interface are developed. An shown, that the energy spectrum of a heavy hole in the valence band QD is equivalent to the spectrum of the hole carrying out oscillator vibrations in the adiabatic electron potential.

In the framework of the dipole approximation interband absorption of light in quasi - zero - dimensional nanosystems is studied. An expression is obtained for the light absorption coefficient under conditions where the polarization interaction of the electron and a hole with the surface of the SNs plays role. Finally it is shown on the basis of the given model of quasi – zero – dimensional nanosystems that the absorption and emission edge of SNs is formed by two transitions of comparable intensity from different hole size – quantization levels and into a lower electron size – quantization level. A theoretical prospect of using hole transitions between equidistant series of quantum levels observed in nanocrystals for desining a nanolaser has been proposed.

The theory of excitons formed from a spatially separated electron and hole is developed within the modified effective mass method. The effect of significantly increasing the exciton binding energy in quantum dots of zinc selenide, synthesized in a borosilicate glass matrix, relative to that in a zinc selenide single crystal is revealed. It was shown that the short-wavelength shift of the peak of the low-temperature luminescence spectrum of the samples containing zinc-selenide QDs, which was observed under the experimental conditions, is caused by renormalization of the electron–hole Coulomb interaction energy and also the energy of the polarization interaction of the electron and hole with the spherical QD–dielectric matrix interface, related to spatial confinement of the quantization region by the QD volume. Nanosystems consisting of zinc- selenide QDs grown in a borosilicate glass matrix can be used as the active region of semiconductor QD lasers.

We developed the theory of biexcitons formed from spatially separated electrons and holes in a nanosystem that consists of ZnSe QDs synthesized in a glassy matrix. It is shown that the major contribution to the biexciton binding energy is made by the energy of the exchange interaction of electrons with holes and this contribution is much more substantial that the contribution of the energy of Coulomb interaction between the electrons and holes. It is established that the spectral shift of the peak in such QDs is due to quantum confinement of the energy of the biexciton ground state.

References

[1] Alferov, J.I., (2002). Progress development in semiconductor nanostructures. *Physics Uspekhi*. 172, 1068 - 1074.

[2] Ekimov, A.I., Onushchenko, A.A., Efros, Al.L., (1986). Zize quantization of excitons in nanocrystals. *Sov. Phys. JETP Lett*. 43, 292-294.

[3] Efros, A.L., Efros, Al.L., Interband absorption light in semiconductor sphere. (1982).*Sov. Phys. Semiconductors*. 16, 955- 962.

[4] Kulchin, Y.N., Scherbakov, A.V., Dzyuba, V.P., Voznesenskiy, S.S., (2008). Nonlinear optical properties of heterogeneous nanocomposites. *Quantum Electronics*. 38, 154 - 158.

[5] Dzyuba,V.P., Kulchin, Y.N., Krasnok, A.E., (2010). Nonlinear refractive index of heterogeneous nanocomposites in weak optical fields. *Technical Physics Letters*. 36, 1-9.

[6] Bondar, V.N., Brodin, M.S., (2010). Optical properties semiconbductor quantum dots. *Semiconductors.* 44, 884-890.

[7] Pokutnyi, S.I., *Theory of excitons in quasi – zero – dimensional semiconductors systems;* Academperiodika: Kiev, Ukraine, 2003; pp. 24-35, 82-94.

[8] Pokutnyi, S.I., *Spectroscopy of electron and exciton states in low – dimensional systems;* Academperiodika: Kiev, Ukraine, 2005; pp. 44-55, 142-152.

[9] Pokutnyi, S.I., *Optics of nanosystems*; Academperiodika: Kiev, Ukraine, 2007; pp. 54-64.

[10] Pokutnyi, S.I., Shpak, A.P., (2005). Influence of polarizing interfction on spectroscopy of excitons in quasi – zero – dimensional semiconductors systems. *Progr. Phys. Metal.* 6, 105-134.

[11] Pokutnyi, S.I., (2005). Optical of spectroscopy of exciton states in quasi – zero – dimensional semiconductors nanosystems (review). *Ukr. J. Phys. Rev.* 3, 46- 74.

[12] Pokutnyi, S.I., Shpak, A.P., Uvarov, V.N., (2007). Bulk local states in quasi – zero – dimensional nanosystems. *Progr. Phys. Metal.* 8, 1-20.

[13] Pokutnyi, S.I., (2005). Optical nanolaser heavy hole transitions in in quasi – zero – dimensional semiconductors nanosystems. *Phys. Lett. A.* 342, 347-352.

[14] Pokutnyi, S.I., (2006). Absorption of light in quantum dots. *Semiconductors.* 40, 217-224.

[15] Pokutnyi, S.I., (2011). Interband absorption of light in semiconductor quantum dots. *J. Nanoscience Lett.* 1, 191-198.

[16] Pokutnyi, S.I., Gorbyk, P.P., (2013). Absorption of light in positron and electron states in quasi – zero – dimensional nanosystems. *Optics*, 2, 47-50.

[17] Pokutnyi, S.I., (2004). Size quantization Stark effect in semiconductor quantum dots. *J. Appl. Phys.* 96, 1115-1122.

[18] Pokutnyi, S.I., (1997). Absorption of light in in quasi – zero – dimensional nanosystems. *Phys. Solid State.* 39, 604- 612.

[19] Efremov, N.A., Pokutny, S.I., (1990). Energy spectrum of exciton in spherical particle. *Sov. Phys. Solid. State.* 32, 955- 964.

[20] Pokutnyi, S.I., (1992). Size quantization of electron-hole pair in semiconductor quantum dots. *Phys. Lett. A.* 168, 433-438.

[21] Efremov, N.A., Pokutny, S.I., (1985). Macroscopic local charge states in ultradispersion media. *Sov. Phys. Solid. State.* 27, 27- 35.

[22] Pokutnyi, S.I., (1995). Size quantization of exciton in semiconductor quantum dots. *Phys. Lett. A.* 203, 384-401.

[23] Pokutnyi, S.I., (2013). On an exciton with a spatially separated electron and hole in quasi – zero – dimensional semiconductor nanosystems. *Semiconductors.* 47, 791-798.

[24] Pokutnyi, S.I., (2013). Binding energy of the exciton of a spatially separated electron and hole in quasi – zero – dimensional semiconductor nanosystems .*Technical Physics Letters.* 39, 233-235.

[25] Pokutnyi, S.I., (2013). Biexcitons formed from spatially separated electrons and holes in quasi – zero – dimensional semiconductor nanosystems. *Semiconductors.* 47, 1626-1635.

[26] Pokutny, S.I., Shpak, A.P., Uvarov, V.N., Pokutnyi, M.S., (2008). Theory of exciton states in quasi – zero – dimensional nanosystems. *Metal Phys. Advanc. Technol.* 30, 459-478.

[27] Pokutnyi, S.I., (2005). Exciton states in quasi – zero – dimensional semiconductor nanosystems. *Semiconductors.* 39, 1066-1074.

[28] Pokutnyi, S.I., (2007). Exciton states in semiconductor quantum dots in the framework of the modified effective mass method. *Semiconductors.* 41, 1323-1331.

[29] Pokutnyi, S.I., (2012). Exciton states in semiconductor quantum dots. *Semiconductors.* 46, 174-184.

[30] Grabovskis, V., Dzenis, Y., Ekimov, A., (1989). Nonlinear optics of semiconductor – doped glasses . *Sov. Phys. Solid State*, 31, 272- 275.

[31] Bogdankevich, O., Darznek, S., Eliseev, P., *Popuprovodnikovye lazery*; Nauka: Moskva, Russia, 1976; pp. 94-98.

[32] Flugge, S., *Practical Quantum Machanics I*, Springer – Verlag: Berlin, Germany, 1971; pp. 111-114.

In: Advances in Semiconductor Research
Editors: D. Persano Adorno and S. Pokutnyi

ISBN: 978-1-63321-755-3

Chapter 5

THEORY OF EXCITONS AND QUASIMOLECULES FORMED FROM SPATIALLY SEPARATED ELECTRONS AND HOLES IN QUASI - ZERO - DIMENSIONAL SEMICONDUCTOR NANOSYSTEMS

Sergey I. Pokutnyi*
Chuiko Institute of Surface Chemistry, National Academy of Sciences of Ukraine, Kyiv, Ukraine

Abstract

The theory of an exciton formed from spatially separated electron and hole (the hole is in the quantum dot volume, and the electron is localized at the outer spherical quantum dot–dielectric matrix interface) is developed within the modified effective mass method. The effect of significantly increasing the exciton binding energy in quantum dots of zinc selenide, synthesized in a borosilicate glass matrix, relative to that in a zinc selenide single crystal is revealed. It was shown that the short-wavelength shift of the peak of the low-temperature luminescence spectrum of samples containing zinc-selenide quantum dots, observed under the experimental conditions, is caused by quantum confinement of the ground-state energy of the exciton with a spatially separated electron and hole.

A review devoted to the theory of excitonic quasimolecules (biexcitons) (formed of spatially separated electrons and holes) in a nanosystem that consists of ZnSe quantum dots synthesized in a borosilicate glass matrix is developed within the context of the modified effective mass approximation. It is shown that biexciton (exciton quasimolecule) formation is of the threshold character and possible in nanosystem, in with the spacing between the quantum dots surfaces is larger than a certain critical spacing.

Keywords: excitons, modified effective mass method, exciton binding energy, quantum dots

* E-mail address: Pokutnyi_Sergey@inbox.ru (Corresponding Author)

Introduction

At present, quasi-zero-dimensional semiconductor nanosystems consisting of spherical semiconductor nanocrystals, i.e., quantum dots with radii of a =1–10 nm, containing cadmium sulfide and selenide, gallium arsenide, germanium, silicon, and zinc selenide in their volume, synthesized in a borosilicate glass matrix, attract particular attention due to their unique photoluminescence properties, i.e., the ability to efficiently emit light in the visible or near infrared ranges at room temperature [1–10]. The optical and electro-optical properties of such quasi-zero dimensional nanosystems are controlled to a large extent by the energy spectrum of the spatially confined electron–hole pair (exciton) [4–16].

In most theoretical models for calculating the energy spectra of quasiparticles in quantum dots (QDs), the effective mass approximation is used, which was considered to be applicable to QDs by analogy with bulk single crystals [11–13]. However, the problem of the applicability of the effective mass approximation to the description of semiconductor QDs is still unsolved [4–18].

In [14], a new modified effective mass method was proposed to describe the exciton energy spectrum in semiconductor QDs with radii of $a \approx a_{ex}$ (a_{ex} is the exciton Bohr radius in the semiconductor material contained in the QD volume). It was shown that, within a model in which the QD is represented as an infinitely deep potential well, the effective mass approximation can be applied to the description of an exciton in QDs with radii a comparable to the exciton Bohr radius a_{ex}, assuming that the reduced effective exciton mass is a function of the radius a, $\mu = \mu(a)$.

In the adiabatic approximation and within the modified effective mass method [14], an expression for the binding energy of an exciton whose electron and hole move in the semiconductor QD volume was derived in [15]. In [15], the effect of significantly increasing the exciton-binding energy in cadmium selenide and sulfide QDs with radii a, comparable to the exciton Bohr radii aex, relative to the exciton- binding energy in cadmium selenide and sulfide single crystals (by factors of 7.4 and 4.5, respectively) was also detected.

In the experimental study [7], it was found that excess electrons produced during interband excitation of the cadmium sulfide QD have a finite probability of overcoming the potential barrier and penetrating into the borosilicate glass matrix into which the QD is immersed. In experimental studies [10, 19] (as well as in [7]) of glass samples with cadmium-sulfide and zinc-selenide QDs, it was found that the electron can be localized in the polarization well near the outer QD surface, while the hole moves within the QD volume.

In [10, 19], the optical properties of borosilicate glass samples containing QD zinc selenide are experimentally studied. The average radii of such QDs are in the range $a \approx 2.0$–4.8 nm. In this case, the values of a are comparable to the exciton Bohr radius $a_{ex} \approx 3.7$ nm in a ZnSe single crystal. At low QD concentrations, when the optical properties of the samples are mainly controlled by those of individual QDs in the borosilicate glass matrix, a shift of the peak of the low temperature luminescence spectrum to the short wavelength region (with respect to the band gap E_g of the zinc selenide - single crystal) was observed. The authors of [10] assumed that this shift is caused by quantum confinement of the energy spectra of the electron and hole localized near the spherical surface of the QD. In this case, the following problem remained open: the quantum confinement of which electron and hole states (the hole moving in the QD volume and the electron localized at the outer spherical QD–dielectric matrix interface or the electron and hole localized in the QD volume) caused such a shift of the luminescence spectrum peak?

The use of semiconductor nanosystems as the active region of nanolasers is prevented by the low binding energy of the QD exciton [8, 9, 13]. Therefore, studies directed toward the search for nanostructures in which a significant increase in the binding energy of QD excitons would be observed are of importance.

Currently, the theory of exciton states in quasi- zero- dimensional semiconductor nanosystems has not been adequately studied, in particular, no theory exists for an exciton with a spatially separated electron and hole in quasi- zero- dimensional nanosystems. Therefore, in this study, we developed the theory of an exciton formed from spatially separated electron and hole (the hole is in the semiconductor QD volume, and the electron is localized at the outer spherical surface of the QD–dielectric matrix interface) [20 – 22]. It was shown that the short wavelength shift of the peak of the low temperature luminescence spectrum of samples containing zinc-selenide QDs, observed under the experimental conditions of [10], is caused by quantum confinement of the ground state energy of the exciton with a spatially separated electron and hole. The effect of significantly increasing the binding energy of an exciton (with a spatially separated electron and hole) in a nanosystem containing zinc-selenide QDs in comparison with the binding energy of an exciton in a zinc-selenide single crystal (by a factor of 4.1–72.6) was detected [20 – 22].

In [10,19], a shift of the spectral peak of the low- temperature luminescence wasobserved also for samples with a QD concentrations from $x = 0.003$ to 1%. It was noted [10, 19] that, at such a QD content in the samples, one must take into account the interaction between charge carriers localized above the QD surfaces. Therefore, in [23, 24] we develop the theory of a excitonic quasimolecules (biexcitons) (formed from spatially separated electrons and holes) in a nanosystem that consists of ZnSe QDs synthesized in a borosilicate glass matrix.

Exciton Ground-State Energy in the Nanosystem

Let us consider the simple model of a quasi - zero - dimensional system, i.e., a neutral spherical semiconductor QD of the radius a, which contains semiconductor material with the permittivity ε_2 in its volume, surrounded by a dielectric matrix with the permittivity ε_1. A hole h with the effective mass m_h moves in the QD volume, while an electron e with the effective mass $m_e^{(1)}$ lies in the matrix (r_e and r_h are the distances from the QD center to the electron and hole). Let us assume that the QD valence band is parabolic. Let us also assume that there is an infinitely high potential barrier at the spherical QD – dielectric matrix interface; therefore, the hole h cannot leave the QD volume and the electron e cannot penetrate into the QD volume in the model under study [20 – 22].

The characteristic dimensions of the problem are the quantities

$$a_h = \varepsilon_2 \hbar^2 / m_h e^2, \; a_{ex} = \varepsilon_2 \hbar^2 / \mu e^2, \quad a_e = \varepsilon_1 \hbar^2 / m_e^{(1)} e^2, \tag{1}$$

where a_h and a_{ex} are the hole and exciton Bohr radii in the semiconductor with the permittivity ε_2, e is the elementary charge, $\mu = m_e^{(2)} m_h / (m_e^{(2)} + m_h)$ is the reduced effective mass of the exciton, $m_e^{(2)}$ is the effective mass of an electron in the semiconductor with the permittivity ε_2, and a_e is the electron Bohr radius in the dielectric matrix with the permittivity ε1. The fact that all characteristic dimensions of the problem are significantly larger than the interatomic distances a_0,

$$a,\ a_e,\ a_h,\ a_{ex} >> a_0 \tag{1a}$$

allows us to consider the electron and hole motion in the quasi - zero - dimensional nanosystem in the effective mass approximation [11–13].

We analyzed the conditions of carrier localization in the vicinity of a spherical dielectric particle of the radius a with the permittivity ε2 in [25–27]. There the problem of the field induced by the carrier near a dielectric particle immersed in a dielectric medium with the permittivity ε_1 was solved in a final analytical form, and analytical expressions for the potential energy of the interaction of the carrier with the spherical interface of two media are presented.

Solving the Poisson equation with usual electrostatic boundary conditions

$$\varphi(r',j\,|\,\boldsymbol{r},i)\big|_{r'=a} = \varphi(\boldsymbol{r}',j)\big|_{r'=a},$$
$$\varepsilon_1 \frac{\partial\varphi(\boldsymbol{r}',j\,|\,\boldsymbol{r},i)}{\partial\boldsymbol{r}'}\Big|_{r'=a} = \varepsilon_2 \frac{\partial\varphi(\boldsymbol{r}',j)}{\partial\boldsymbol{r}'}\Big|_{r'=a}, \tag{2}$$

the potential $\varphi(\boldsymbol{r}',j|\boldsymbol{r},i)|$ at the observation point $\boldsymbol{r}'$ in a medium with the permittivity ε_j, induced by the charge e at the point $\boldsymbol{r}$ in a medium with the permittivity ε_i, can be presented as a sum of the potentials induced by the image point charge $e'(\boldsymbol{r}_{ij}\,|\boldsymbol{r})$ at the point $\boldsymbol{r}_{ij} =(a/r)^2\boldsymbol{r}\delta_{ij} + \boldsymbol{r}(1 - \delta ij)$ and the linear distribution with the density $\rho_{ij}(y, r)$ of the image charge along a straight line passing through the center of the dielectric particle with the radius a and the charge at the point $\boldsymbol{r}$ [25-27]:

$$\varphi(\boldsymbol{r}',j\,|\,\boldsymbol{r},i) = \frac{e}{\varepsilon_j\,|\boldsymbol{r}'-\boldsymbol{r}|} + \frac{e'(\boldsymbol{r}_{ij}\,|\,\boldsymbol{r})}{\varepsilon_j\,|\boldsymbol{r}'-\boldsymbol{r}_{ij}|} + \frac{1}{\varepsilon_j}\int_0^\infty \frac{dy\rho_{ij}(y,\boldsymbol{r})}{|\boldsymbol{r}'-y(\boldsymbol{r}/r)|}, \tag{3}$$

where

$$\boldsymbol{r}_{11} = (a^2/r^2)\boldsymbol{r},\ e'(\boldsymbol{r}_{11}\,|\,\boldsymbol{r}) = -\beta(a/r)e,$$
$$\rho_{11}(y,r) = \beta\alpha(a^2/ry)^{(1-\alpha)}(e/a)\theta((a^2/r)-y); \tag{3a}$$

$$\boldsymbol{r}_{22} = (a^2/r^2)\boldsymbol{r},\ e'(\boldsymbol{r}_{22}\,|\,\boldsymbol{r}) = \beta(a/r)e,$$
$$\rho_{22}(y,r) = \beta(1-\alpha)(a^2/ry)^{\alpha}(e/a)\theta(y-(a^2/r)); \tag{3b}$$

$$\boldsymbol{r}_{12} = \boldsymbol{r}\ ,\ e'(\boldsymbol{r}_{12}\,|\,\boldsymbol{r}) = \beta e\ ,\tag{3c}$$
$$\rho_{12}(y,r) = \beta(1-\alpha)(r/y)^{\alpha}(e/r)\theta(y-r)\ ;$$

$$\boldsymbol{r}_{21} = \boldsymbol{r}\ ,\ e'(r_{12}\,|\,r) = -\beta e\ ,\tag{3d}$$
$$\rho_{21}(y,r) = \beta\alpha(r/y)^{(1-\alpha)}(e/r)\theta(r-y)\ ,$$

where $\theta(x)$ is the Heaviside unit-step function,

$$\beta = \frac{(\varepsilon_2-\varepsilon_1)}{(\varepsilon_2+\varepsilon_1)},\ \alpha = \frac{\varepsilon_1}{(\varepsilon_2+\varepsilon_1)}\tag{4}$$

Using expressions (3)–(3d), the energy $U(\boldsymbol{r}_e,\boldsymbol{r}_h,a)$ of the polarization interaction of the electron and hole with the spherical QD–matrix interface at the relative permittivity $\varepsilon = (\varepsilon_2/\varepsilon_1) \gg 1$ can be presented as an algebraic sum of the energies of the interaction of the hole and electron with self- $V_{hh'}(r_h,a)$, $V_{ee'}(r_e, a)$ and "foreign" $V_{eh'}(\boldsymbol{r}_e, \boldsymbol{r}_h, a)$, $V_{he'}(\boldsymbol{r}_e, \boldsymbol{r}_h, a)$ images, respectively [15,16,26- 28],

$$U(\boldsymbol{r}_e,\boldsymbol{r}_h,a,\varepsilon) = V_{hh'}(r_h,a,\varepsilon) + V_{ee'}(r_e,a,\varepsilon) + V_{eh'}(\boldsymbol{r}_e,\boldsymbol{r}_h,a,\varepsilon) + V_{he'}(\boldsymbol{r}_e,\boldsymbol{r}_h,a,\varepsilon)\ ,\tag{5}$$

where

$$V_{hh'}(r_h,a,\varepsilon) = \frac{e^2\beta}{2\varepsilon_2 a}\left(\frac{a^2}{a^2-r_h^{\ 2}} + \varepsilon\right),\tag{6}$$

$$V_{ee'}(r_e,a,\varepsilon) = -\frac{e^2\beta}{2\varepsilon_1 a}\cdot\frac{a^4}{r_e^2(r_e^2-a^2)},\tag{7}$$

$$V_{he'}(\boldsymbol{r}_e,\boldsymbol{r}_h,a,\varepsilon) = \frac{e^2\beta}{2\varepsilon_2 a}\cdot\frac{a^2}{r_e\left|\boldsymbol{r}_h-(a/r_e)^2\boldsymbol{r}_e\right|},\tag{8}$$

$$V_{eh'}(\boldsymbol{r}_e,\boldsymbol{r}_h,a,\varepsilon) = -\frac{e^2\beta}{2\varepsilon_1 a}\cdot\frac{a^2}{r_h\left|\boldsymbol{r}_e-(a/r_h)^2\boldsymbol{r}_h\right|}.\tag{9}$$

In the studied simple model of a quasi zero dimensional nanostructure within the above approximations and in the effective mass approximation using the triangular coordinate system [14-16], $r_e = |\mathbf{r}_e|$, $r_h = |\mathbf{r}_h|$, $r = |\mathbf{r}_e - \mathbf{r}_h|$, with the origin at the center of the QD, the exciton Hamiltonian (with a spatially separated hole moving within the QD volume and an electron in the dielectric matrix) takes the form [20 – 22, 29 - 32]:

$$H(\mathbf{r}_e,\mathbf{r}_h,\mathbf{r},a)=-\frac{\hbar^2}{2m_e^{(1)}}\left(\frac{\partial^2}{\partial r_e^2}+\frac{2}{r_e}\cdot\frac{\partial}{\partial r_e}+\frac{r_e^2-r_h^2+r^2}{r_e r}\cdot\frac{\partial^2}{\partial r_e\partial r}\right)-$$
$$-\frac{\hbar^2}{2m_h}\left(\frac{\partial^2}{\partial r_h^2}+\frac{2}{r_h}\cdot\frac{\partial}{\partial r_h}+\frac{r_h^2-r_e^2+r^2}{r_h r}\cdot\frac{\partial^2}{\partial r_h\partial r}\right)-\frac{\hbar^2}{2\mu_0}\left(\frac{\partial^2}{\partial r^2}+\frac{2}{r}\cdot\frac{\partial}{\partial r}\right)+$$
$$+V_{eh}(\mathbf{r})+U(\mathbf{r}_e,\mathbf{r}_h,a,\varepsilon)+V_e(r_e)+V_h(r_h)+E_g, \tag{10}$$

where the first three terms are the operators of the electron, hole, and exciton kinetic energy, Eg is the band gap in the semiconductor with the permittivity ε_2, $\mu_0 = m_e^{(1)} m_h/(m_e^{(1)} + m_h)$ is the reduced effective mass of the exciton (with a spatially separated hole and electron). In the Hamiltonian (10), the polarization interaction energy $U(\mathbf{r}_e, \mathbf{r}_h, a, \varepsilon)$ (5) is defined by formulas (6)–(9), and the electron–hole Coulomb interaction energy $Veh(r)$ is described by the formula

$$V_{eh}(r) = -\frac{1}{2}\left(\frac{1}{\varepsilon_1}+\frac{1}{\varepsilon_2}\right)\frac{e^2}{r} \tag{11}$$

In the exciton Hamiltonian (10), the potentials

$$V_h(r_h) = \begin{cases} 0, & r_h \le a \\ \infty, & r_h > a \end{cases}, \tag{12}$$

$$V_e(r_e) = \infty, \quad r_e \le a$$

describe the quasiparticle motion using the models of an infinitely deep potential well.

As the QD radius a increases (so that $a \gg a_{ex}^0$), the spherical interface of the two media (QD–matrix) passes to the plane ⟨semiconductor material with the permittivity ε_2⟩–matrix interface. In this case, the exciton with the spatially separated electron and hole (the hole moves within the semiconductor material and the electron lies in the borosilicate glass matrix) becomes two-dimensional [20 – 22].

The main contribution to the potential energy of the Hamiltonian (10) describing exciton motion in a nanosystem containing a large-radius QD, $a \gg a_{ex}^0$, is made by the electron–hole Coulomb interaction energy $V_{eh}(r)$ (11). The energy of the hole and electron interaction with self- $V_{hh'}(r_h, a, \varepsilon)$ (6), $V_{ee'}(r_h, a)$ (7) and “foreign” $V_{eh'}(\mathbf{r}_e, \mathbf{r}_h, a)$ (9), $V_{he'}(\mathbf{r}_e, \mathbf{r}_h, a)$ (8) images make a significantly smaller contribution to the potential energy of the Hamiltonian (10).

In the first approximation, this contribution can be disregarded. In this case, only the electron–hole Coulomb interaction energy (11) remains in the potential energy of the Hamiltonian (10) [20 – 22]. The Schrodinger equation with such a Hamiltonian describes a two-dimensional exciton with a spatially separated electron and hole (the electron moves within the matrix, and the hole lies in the semiconductor material with the permittivity ε_2), whose energy spectrum takes the form [33, 34]:

$$E_n = -\frac{Ry_{ex}^0}{(n+1/2)^2}, \qquad (13)$$

$$Ry_{ex}^0 = \frac{(\varepsilon_1+\varepsilon_2)^2}{4\varepsilon_1^2\varepsilon_2^2}\left(\frac{\mu_0}{m_0}\right)Ry_0$$

where $n = 0, 1, 2\ldots$ is the principal quantum number of the exciton and $Ry_0 = 13.606$ eV is the Rydberg constant. The Bohr radius of such a two- dimensional exciton is described by the formula

$$a_{ex}^0 = \frac{2\varepsilon_1\varepsilon_2}{\varepsilon_1+\varepsilon_2}\frac{\hbar^2}{\mu_0 e^2}, \qquad (14)$$

and the binding energy of the ground state of such a two-dimensional exciton, according to (13), is written as

$$E_{ex}^0 = -4Ry_{ex}^0 \qquad (15)$$

The binding energy (15) of the exciton ground state is understood as the energy required for bound electron and hole state decay (in the state with $n = 0$).

To determine the ground-state energy of an exciton (with a spatially separated electron and hole) in a nanosystem containing QDs of the radius a, we use the variational method. When choosing the variational exciton wave function, we use an approach similar to that developed in [14]. Let us write the variational radial wave function of the exciton ground - state (1s electron state and 1s hole state) in the nanosystem under study in the form [20 – 22]:

$$\psi_0(r_e, r_h, r, a) = A\exp\left(-\frac{\mu(a)}{\mu_0}\frac{r}{a_{ex}^0}\right)\frac{(a^2-r_h^2)}{a^2} \times \frac{(r_e^2-a^2)}{a^2}\frac{r}{a}\frac{|r_e-(a/r_h)^2 r_h|r_h}{a^2}\frac{|r_h-(a/r_e)^2 r_e|r_e}{a^2} \qquad (16)$$

Here the coefficient A is determined from the condition of normalization of the exciton wave function (16),

$$\int_a^\infty r_e dr_e \int_0^a r_h dr_h \int_r^{r_e+r_h} \psi_0^2(r_e, r_h, r, a) r dr = 1,$$

and the effective reduced exciton mass $\mu(a)$ is the variational parameter.

As the QD radius a increases (so that $a \gg a_{ex}^0$), a two-dimensional exciton is formed in the nanosystem. This factor leads to the fact that the variational exciton wave function (16) contains the Wannier–Mott two-dimensional exciton wave eigenfunction [33, 34]. Furthermore, the polynomials from r_e and r_h enter the exciton variational function (16), which make it possible to eliminate singularities in the functional $E_0(a,\mu(a))$ in the final analytical form.

To determine the exciton ground-state energy $E0(a, \varepsilon)$ in the nanosystem under study by the variational method, we write the average value of the exciton Hamiltonian (10) in wave functions (16) in the form

$$E_0(a,\mu(a)) = \langle \psi_0(\mathbf{r}_e,\mathbf{r}_h,r,a)|H(\mathbf{r}_e,\mathbf{r}_h,r,a)|\psi_0(\mathbf{r}_e,\mathbf{r}_h,r,a)\rangle = \int_a^\infty dr_e \int_0^a dr_h \int_r^{r_e+r_h} dr\, r\, r_e\, r_h\, \psi_0(\mathbf{r}_e,\mathbf{r}_h,r,a)\, H(\mathbf{r}_e,\mathbf{r}_h,r,a)\, \psi_0(\mathbf{r}_e,\mathbf{r}_h,r,a) \quad (17)$$

The dependence of the energy $E_0(a)$ of the exciton ground state ($n_e = 1$, $l_e = m_e = 0$; $n_h = 1$, $l_h = m_h = 0$), (n_e, l_e, m_e and n_h, l_h, m_h are the principal, orbital, and magnetic quantum numbers of the electron and hole, respectively) on the QD radius a is calculated by minimizing the functional $E_0(a, \mu(a))$ (17),

$$\frac{\partial E_0(a,\mu(a))}{\partial \mu(a)} = F(\mu(a), a) \quad (18)$$

Without writing cumbersome expressions for the first derivative of the functional $\partial E_0(a, \mu(a))/\partial \mu(a) = F(\mu(a), a)$, we present the numerical solution to the equation $F(\mu(a), a) = 0$ (18) in tabulated form.

It follows from the table that the solution to this equation is the function $\mu(a)$ which monotonically weakly varies within the limits [20 – 22]:

$$0.304 \leq \mu(a)/m_0 \leq 0.359 \quad (19)$$

as the QD radius a varies within the range

$$2.0 \leq a \leq 29.8 \text{ nm} \quad (20)$$

(m_0 is the electron mass in vacuum). In this case, the reduced exciton effective mass $\mu(a)$ (19) in the nanosystem slightly differs from the effective mass of an exciton (with a spatially separated hole and electron) $\mu = 0.304m_0$ by the value $(\mu(a) - \mu_0)/\mu_0 \leq 0.18$ when the QD radii vary within the range (20).

Substituting the values of the variational parameter $\mu(a)$ (19) from the table simultaneously with the corresponding QD radii from the range (20) into the functional $E_0(a, \mu(a))$ (17), we obtain the exciton ground-state energy $E_0(a, \varepsilon)$ (17) as a function of the QD radius a. [20 – 22].

The results of variational calculation of the energy of the ground state of an exciton $E_0(a, \varepsilon)$ (17) in the nanosystem under study containing zinc-selenide QDs of the radius a (20) are shown in the figure [20 – 22]. Here, the values of function $\mu(a)$ (19) and the results of variational calculation of the exciton ground-state energy $E_0(a, \varepsilon)$ (17) are obtained for a nanosystem containing zinc-selenide QDs, synthesized in a borosilicate glass matrix, studied in the experimental works [10, 19].

In the experimental work [10], borosilicate glass samples doped with zinc selenide with concentrations from x = 0.003 to 1%, obtained by the sol–gel method were studied. According to X-ray diffraction measurements, the average radii a of ZnSe QDs formed in the samples are within $a \approx 2.0$–4.8 nm.. In this case, the values of $\tilde{a}$ are comparable to the exciton Bohr radius $a_{ex} \approx 3.7$ nm in a zinc-selenide single crystal. At low QD concentrations (x = 0.003 and 0.06%), their interaction can be disregarded. The optical properties of such nanosystems are mainly controlled by the energy spectra of electrons and holes localized near the spherical surface of individual QDs synthesized in the borosilicate glass matrix.

In [10, 19], a peak in the low-temperature luminescence spectrum at an energy of $E1 \approx$ 2.66 eV was observed at the temperature T = 4.5 K in samples with x = 0.06%; this energy is lower than the band gap of a zinc-selenide single crystal (E_g = 2.823 eV). The shift of the peak of the low-temperature luminescence spectrum with respect to the band gap of the ZnSe single crystal to the short-wavelength region is $\Delta E1 = (E1 - E_g) \approx -165$ meV. The authors of [10] assumed that the shift $\Delta E1$ is caused by quantum confinement of the energy spectra of electrons and holes localized near the spherical surface of individual QDs and is associated with a decrease in the average radii a of zinc- selenide QDs at low concentrations (x = 0.06%). In this case, the problem of the quantum confinement of which electron and hole states (the hole moving within the QD volume and the electron localized at the outer spherical QD–dielectric matrix interface or the electron and hole localized in the QD volume) caused such a shift of the luminescence-spectrum peak remained open.

Comparing the exciton ground-state energy $(E_0(a,\varepsilon) - E_g)$ (17) with the energy of the shift in the luminescence-spectrum peak $\Delta E1 \approx -165$ meV, we obtain the average zinc-selenide QD radius $a1 \approx 4.22$ nm (see the figure) [20 – 22]. The QD radius $a1$ may be slightly overestimated, since variational calculation of the exciton ground-state energy can give slightly overestimated energies [33, 34]. The determined average QD radius $a1$ lies within the range of the average radii of zinc-selenide QDs ($a \approx 2.0$–4.8 nm), studied under the experimental conditions of [10, 19].

It should be noted that the average Coulomb interaction energy $\overline{V_{eh}}(a, \varepsilon) = \langle \psi_0(r_e, r_h, r, a) | V_{eh}(r) | \psi_0(r_e, r_h, r, a) \rangle$, between the electron and hole mainly contribute to the ground-state energy (17) of the exciton in the nanosystem containing zinc-selenide QDs with radii $a1$ comparable to the exciton Bohr radius in a zinc- selenide single crystal ($a_{ex} \approx 3.7$ nm). In this case, the average energy of the interaction of the electron and hole with self- and "foreign" images,

$$\left(\overline{V}_{ee'}(a,\varepsilon) + \overline{V}_{hh'}(a,\varepsilon) + \overline{V}_{eh'}(a,\varepsilon) + \overline{V}_{he'}(a,\varepsilon) \right) =$$
$$= \langle \psi_0(r_e,r_h,r,a) | V_{ee'}(r_e,a,\varepsilon) + V_{hh'}(r_h,a,\varepsilon) + V_{eh'}(r_e,r_h,a,\varepsilon) + V_{he'}(r_e,r_h,a,\varepsilon) | \psi_0(r_e,r_h,r,a) \rangle$$

makes a significantly smaller contribution to the exciton ground-state energy (17), $0.04 \leq |[V_{ee'}(a, \varepsilon) + V_{hh'}(a, \varepsilon) + V_{eh'}(a, \varepsilon) + V_{he'}(a,\varepsilon)]/ (a, \varepsilon)| \leq 0.12$ [20 – 22].

Thus, the short-wavelength shift $\Delta E1$ of the low temperature luminescence spectrum peak is caused by renormalization of the electron–hole Coulomb interaction energy $V_{eh}(r)$ (11) and also renormalization of the energy $U(r_e, r_h, r, a, \varepsilon)$ (5) of the polarization interaction of the electron and hole with the spherical QD–dielectric matrix interface, which is associated with spatial confinement of the quantization region by the QD volume. In this case, the hole moves within the QD volume, and the electron is localized at the outer spherical QD–dielectric matrix interface.

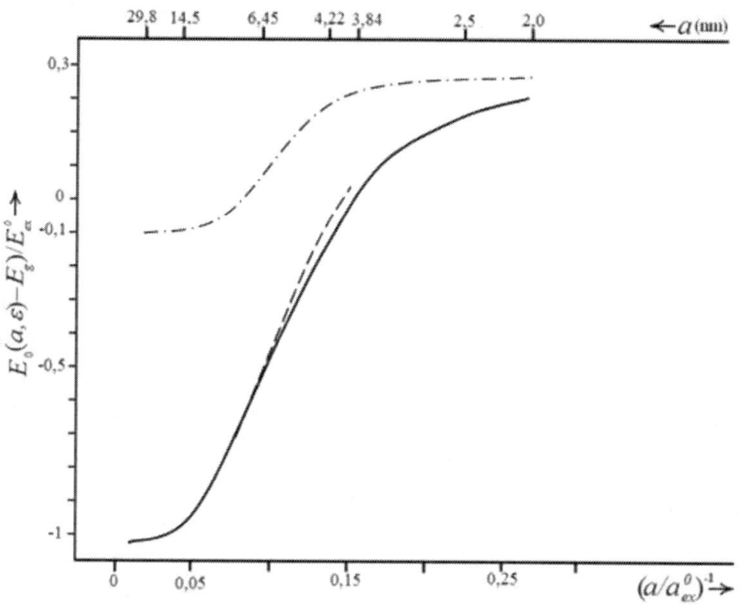

Figure 1. Dependences of the exciton ground state-energy (E_0 (a, ε) $-E_g$) (17) (solid curve) and the binding energy of the exciton ground state (E_{ex} (a, ε) – E_g) (21) (dashed curve) on the zinc-selenide QD radius a in the model of an exciton with a spatially separated electron and hole. The dash-dotted curve is the dependence of the exciton ground-state energy (E_0 (a, ε) – E_g) on the zinc-selenide QD radius a in the exciton model, in which the electron and hole move within the zinc-selenide QD volume [16]. E_g = 2.823 eV is the band gap in a zinc-selenide single crystal; E^0_{ex} = 1.5296 eV (15) and a^0_{ex}, = 0.573 nm (14) are, respectively, the binding energy of the ground state and the Bohr radius of a two-dimensional exciton with a spatially separated electron and hole.

Binding Energy of the Exciton Ground State in the Nanosystem

The binding energy of the ground state of an exciton (with a spatially separated electron and hole) $E_{ex}(a, \varepsilon)$ in a nanosystem containing zinc-selenide QDs of the radius a is the solution to the radial Schrodinger equation with a Hamiltonian containing, in contrast to Hamiltonian (10), only the terms $V_{he'}(r_e,r_h, a, \varepsilon)$ (8) and $V_{eh'}(r_e, r_h, a, \varepsilon)$ (9) in the polarization interaction energy $U(\mathbf{r}_e, \mathbf{r}_h, a, \varepsilon)$ (5), which describe the energies of the hole and electron interaction with "foreign" images, respectively [15, 27, 28]. Therefore, the exciton ground-state binding energy $Eex(a, \varepsilon)$ is defined by the expression [20 – 22]:

$$E_{ex}(a,\varepsilon)=E_0(a,\varepsilon)-\langle\psi_0(r_e,r_h,r,a)|(V_{hh'}(r_h,a,\varepsilon)+V_{ee'}(r_e,a,\varepsilon))|\psi_0(r_e,r_h,r,a)\rangle, \quad (21)$$

where the term $\langle\psi_0(r_e,r_h,r,a)|(V_{hh'}(r_h,a,\varepsilon)+V_{ee'}(r_e,a,\varepsilon))|\psi_0(r_e,r_h,r,a)\rangle$ describes the average energy of hole and electron interaction with self-images.

Since the average energies of the interaction of the hole with its image and the average energies of the interaction of the electron with its image make contri-butions with opposite signs to expression (21), they significantly compensate each other. Therefore, the binding energies of the exciton ground state $E_{ex}(a, \varepsilon)$ (21) slightly differ from the corresponding total energies of the exciton ground state $E_0(a, \varepsilon)$ (17). This difference

$$\Delta=|(E_{ex}(a,\varepsilon)-E_0(a,\varepsilon))/E_{ex}(a,\varepsilon)|,$$

varies within $\Delta \leq 4\%$ as QD radii a vary within the range $3.84 \leq a \leq 8.2$ nm (see the figure) [20 – 22].

The figure shows the dependences of the total energy $E_0(a, \varepsilon)$ (17) and the binding energy $E_{ex}(a, \varepsilon)$ (21) of the ground state of the exciton with a spatially separated electron and hole on the QD size for a nanosystem containing zinc-selenide QDs of the radius a. We can see that bound states of electron–hole pairs arise near the spherical surface of the QD starting from the QD critical radius $a \geq a_c^{(1)} \approx 3.84$ nm. In this case, the hole is localized near the QD inner surface, and the electron is localized at the outer spherical QD–dielectric matrix interface. Starting from the QD radius $a \geq a_c^{(1)}$, the electron–hole pair states are in the region of negative energies (counted from the top of the band gap E_g for a zinc-selenide single crystal), which corresponds to the electron–hole bound state [20 -22, 29 – 23]. In this case, the electron–hole Coulomb interaction energy $V_{eh}(r)$ (11) and the energy $U(r_e, r_h, r, a, \varepsilon)$ (5) of the polarization interaction of the electron and hole with the spherical QD–dielectric matrix interface dominate over the energy of quantum confinement of the electron and hole in the nanosystem under study.

The total energy $|E_0(a, \varepsilon)|$ (17) and the binding energy $|E_{ex}(a, \varepsilon)|$ (21) of the ground state of the exciton with a spatially separated electron and hole increase with QD radius a. In the range of radii

$$4.0 \leq a \leq 29.8 \text{ nm} \quad (22)$$

the binding energy $|E_{ex}(a, \varepsilon)|$ (21) of the exciton ground state significantly (by a factor of 4.1–76.2) exceeds the exciton binding energy in a zinc-selenide single crystal, $\overline{E_{ex}^0} \approx -21.07$ meV Starting from the QD radius $a \geq a_c^{(2)} \approx 29.8$ nm, the total energies (17) and binding energies (21) of the exciton ground state asymptotically tend to the value E^0_{ex} = –1.5296 eV which characterizes the binding energy of the ground state of a two-dimensional exciton with a spatially separated electron and hole (see the figure) [20 – 22, 29 – 32].

The obtained values of the total energy $E_0(a, \varepsilon)$ (17) of the exciton ground state in the nanosystem satisfy the inequality

$$E_0(a, \varepsilon) - E_g \ll \Delta V(a) \tag{23}$$

where $\Delta V(a)$ is the potential-well depth for the QD electron. For a large class of II–VI semiconductors in the region of QD sizes $a \geq a^0_{ex}$, $\Delta V(a)$ = 2.3–2.5 eV [7]. Satisfaction of condition (23), probably, makes it possible to disregard the effect of the complex structure of the QD valence band on the total energy (17) and the binding energy (21) of the exciton ground state in the nanosystem under study when deriving these quantities.

The effect of a significant increase in the binding energy $|E_{ex}(a, \varepsilon)|$ (21) of the exciton ground state in the nanosystem under study, according to formulas (5)–(9), (11), (13)–(15), (17), (21) is controlled by two factors [20- 22, 29 – 32]: (i) a significant increase in the energy of the electron–hole Coulomb interaction $|V_{eh}(r)|$ (11) and an increase in the energy of the interaction of the electron and hole with "foreign" images $|V_{eh'}(r_e, r_h, r, a, \varepsilon)|$ (9), $|V_{he'}(r_e, r_h, r, a, \varepsilon)|$ (8) (the "dielectric enhancement" effect [34]); (ii) spatial confinement of the quantization region by the QD volume; in this case, as the QD radius a increases, starting from $a \geq a_c^{(2)} \approx 52\, a^0_{ex} \approx$ 29.8 nm, the exciton becomes two-dimensional with a ground-state energy E^0_{ex} (15) that exceeds the exciton binding energy E_{ex} in the zincselenide single crystal by almost two orders of magnitude ($\left|E^0_{ex}/\overline{E^0_{ex}}\right| \approx 72.6$).

The "dielectric enhancement" effect is caused by the following factor. When the matrix permittivity ε_1 is significantly smaller than the QD permittivity ε_2, the most important role in the electron–hole interaction in the nanosystem under study is played by the field induced by these quasiparticles in the matrix. In this case, electron–hole interaction in the nanosystem appears to be significantly stronger than in an infinite semiconductor with the permittivity ε_2 [34].

In [16], in the nanosystem experimentally studied in [10], an exciton model in which the electron and hole move within the zinc-selenide QD volume is studied. Using the variational method, within the modified effective mass method, the dependence of the exciton ground-state energy $E_0(a, \varepsilon)$ on the QD radius a in the range (20) was obtained in [16] (see the figure). It was shown that, as the QD radius increases, starting from $a \geq a_c = 3.90\, \overline{a^0_{ex}}$ = 1.45 nm, a bulk exciton appears in the QD; its binding energy

$$\overline{E^0_{ex}} = -\frac{\hbar^2}{2\mu\left(\overline{a^0_{ex}}\right)^2} \tag{24}$$

is 21.07 meV ($\mu = 0.132 m_0$ and $\overline{a^0_{ex}}$ = 3.7 nm are the reduced effective mass and Bohr radius of the exciton in the zinc selenide forming the QD volume). The bulk exciton in the QD is understood as an exciton whose structure (reduced mass, Bohr radius, and binding energy) in the QD does not differ from the structure of an exciton in an infinite semiconductor material. As the QD radius a increases ($a \geq a_c$), the exciton ground-state energy $E_0(a)$ asymptotically follows the binding energy of the bulk exciton (24) (see the figure) [20 – 22, 29 – 32].

Thus, using the exciton model in which an electron and hole move in the QD volume, it is impossible to interpret the mechanism of the appearance of the nanosystem luminescence - spectrum peak with the shift $\Delta E1 \approx -165$ meV, obtained in [10, 19].

A comparison of the dependences of the exciton ground- state energy $E_0(a)$ in the nanosystem [10], obtained using two - exciton models (see the figure) (the electron and hole move within the zinc-selenide QD volume [16] (model **I**); the hole moves within the zinc-selenide QD volume, and the electron is localized in the boron silicate glass matrix near the QD spherical surface (model **II**) allows the following conclusion. In model **I**, as the QD radius a increases, starting from $a \geq a_c \approx 14.5$ nm, the exciton ground- state energy $E_0(a)$ asymptotically follows the binding energy of the bulk exciton $\overline{E^0_{ex}} \approx -21.07$ meV (24); in model II, as the QD radius increases, starting from $a \geq a_c^{(2)} \approx 29.8$ nm, the exciton ground-state energy (17) asymptotically follows $E^0_{ex} = -1.5296$ eV (15) (characterizing the binding energy of the ground state of a two- dimensional exciton with a spatially sep arated electron and hole), which is significantly lower than $\overline{E^0_{ex}} \approx -21.07$ meV [20- 22, 29 – 32].

Excitonic Quasimolecule Formed from Spatially Separated Electrons and Holes

We consider a model nanosystem [23, 24] that consists of two spherical semiconductor QDs, A and B , synthesized in a borosilicate glass matrix with the permittivity ε_1. Let the QD radii be a , the spacing between the spherical QD surfaces be D. Each QD is formed from a semiconductor material with the permittivity ε_2. For simplicity, without loss of generality, we assume that the holes h (A) and h (B) with the effective masses m_h are in the QD (A) and QD (B) centers and the electrons $e(1)$ and $e(2)$ with the effective masses $m_e^{(1)}$ are localized near the spherical QD(A) and QD (B) surfaces, respectively. The above assumption is reasonable, since the ratio between the effective masses of the electron and hole in the nanosystem is much smaller that unity: $\left(\left(m_e^{(1)}/m_h\right) \ll 1\right)$. Let us assume that there is an infinitely high potential barrier at the spherical QD – matrix interface. Therefore, in the nanosystem, holes do not leave the QD bulk, whereas electrons do not penetrate into the QDs.

In the context of the adiabatic approximation and effective mass approximation, using the variational method, we obtain the total energy $E_0(\tilde{D},\tilde{a})$ and the binding energy $E_B(\tilde{D},\tilde{a})$ of the biexciton singlet ground state (the spins of the electrons $e(1)$ and $e(2)$ are antiparallel) in such system as functions of the spacing between the QD surfaces be D and the QD radius a [23, 24]:

$$E_0(\tilde{D},\tilde{a}) = 2E_{ex}(\tilde{a}) + E_B(\tilde{D},\tilde{a}), \tag{25}$$

Here, the binding energy $E_{ex}(\tilde{a})$ (17) of the ground state of the exciton (formed from an electron and a hole spatially separated from the electron) localized above the QD(A) (or QD(B)) surface is determined by in [23, 24] (parameters $\tilde{a} = (a/a_{ex}^0)$ ($a_{ex}^0 = 3{,}7$ nm - the exciton Bohr radius in a single crystal ZnSe, $\tilde{D} = (D/a_{ex}^0)$). For the nanosystem under study, the values of the binding energies $E_{ex}(\tilde{a})$ are calculated in [23, 24] for the experimental conditions of [10, 19].

The results of variational calculation of the binding energy $E_B(\tilde{D},\tilde{a})$ of the biexciton singlet ground state in the nanosystem of ZnSe QDs with average radii of $\bar{a}_1 = 3{,}88$ nm, synthesized in a borosilicate glass matrix are shown in [23, 24]. Such a nanosystem was experimentally studied in [10, 19]. In [10, 19], the borosilicate glassy samples doped with ZnSe to the content x from $x = 0.003$ to 1% were produced by the sol-gel technique. At a QD content of $x = 0.06$ %, one must take into account the interaction of charge carriers localized above the QD surfaces.

The binding energy $E_B(\tilde{D},\tilde{a})$ of the biexciton singlet ground state in the nanosystem of ZnSe QDs with average radii of $\bar{a}_1 = 3{,}88$ nm has a minimum of $E_B^{(1)}(D_1,\bar{a}_1) \approx -4{,}2$ meV (at the spacing $D_1 \cong 3{,}2$ nm) [23, 24]. The value of $E_B^{(1)}$ corresponds to the temperature $T_c \approx 49$ K). In [23, 24], it follows that a biexciton (excitonic quasimolecule) is formed in the nanosystem, starting from a spacing between the QD surfaces of $D \geq D_c^{(1)} \cong 2{,}4$ nm. The formation of such a excitonic quasimolecule (biexciton) is of the threshold character and possible only in a nanosystem with QDs with average radii $\bar{a}_1$ such that the spacing between the QD surfaces D exceeds a certain critical spacing $D_c^{(1)}$. Moreover, the exciton quasimolecule (biexciton) can exist only at temperatures below a certain critical temperature: $T_c \approx 49$ K [23, 24].

As follows from the results of variational calculation [23, 24], the binding energy of an exciton (formed from an electron and a hole spatially separated from the electron) localized above the surface of the QD(A) (or a QD(B)) with an average radius of $\bar{a}_1 = 3{,}88$ nm is $E_{ex}(\bar{a}_1) \cong -54$ meV. In this case the energy of the biexciton singlet ground state $E_0(\tilde{D},\tilde{a})$ (25) takes the value $E_0(\tilde{D},\tilde{a}) = -112$meV.

From the results of variational calculation [23, 24], of the biexciton (exciton quasimolecule) binding energy $E_B(\tilde{D},\tilde{a})$, it follows that the major contribution to the binding energy (25) is made by the average energy of the exchange interaction of the electrons $e(1)$ and $e(2)$ with the holes h (A) and h (B). At the same time, the energy of Coulomb interaction makes a much smaller contribution of the biexciton binding energy $E_B(\tilde{D},\tilde{a})$ (25).

The major contribution to the exchange – interaction energy is made by the energy of the exchange interaction of the electron $e(1)$ with the holes h (B), as well as of the electron $e(2)$ with the holes h (B), as well as of the electron $e(2)$ with the holes h (A). The major contribution to the Coulomb – interaction energy is made by the energy of Coulomb interaction of the electron $e(1)$ with the holes h (B), as well as of the electron $e(2)$ with the holes h (A) [23, 24].

As the spacing D between the QD(A) and QD(B) surfaces is increased, starting from $D \geq D_c^{(2)} \cong 16{,}4$ nm, the average Coulomb – interaction energy substantially desreases. In addition, because of the decrease in the overlapping of the electron wavefunction, the average exchange interaction energy substantially decreases as well. As a consequence, the average Coulomb – interaction energy and the average energy of the exchange interaction of the electrons $e(1)$ and $e(2)$ with the holes h (A) and h (B) sharply desrease in comparison with the exciton binding energy $E_{ex}(\tilde{a})$ (17) [23, 24], resulting in decomposition of the biexciton in the nanosystem into two excitons (formed of spatially separated electrons and holes) localized above the QD(A) and QD(B) surfaces.

Conclusion

The theory of an exciton with a spatially separated electron and hole is developed within the modified effective mass method [14] in which the reduced effective exciton mass is a function of the semiconductor QD radius a. The average zinc- selenide QD radius was determined by comparing the dependence of the exciton ground-state energy (17) on the QD radius, obtained by the variational method within the modified effective mass method [14], with the experimental peak of the low-temperature luminescence spectrum [10, 19]. It was shown that the short-wavelength shift of the peak of the low-temperature luminescence spectrum of the samples containing zinc-selenide QDs, which was observed under the experimental conditions of [10, 19], is caused by renormalization of the electron–hole Coulomb interaction energy (11) and also the energy of the polarization interaction (5) of the electron and hole with the spherical QD–dielectric matrix interface, related to spatial confinement of the quantization region by the QD volume. In this case, the hole moves in the QD volume, and the electron is localized at the outer spherical QD–dielectric matrix interface [20 -.22, 29 – 32].

To apply semiconductor nanosystems containing zinc- selenide QDs as the active region of lasers, it is required that the exciton binding energy $|E_{ex}(a, \varepsilon)|$ (21) in the nanosystem be on the order of several kT_0 at room temperature T_0 (k is the Boltzmann constant) [13]. Nanosystems consisting of zinc- selenide QDs grown in a borosilicate glass matrix can be used as the active region of semiconductor QD lasers. In the range of zinc-selenide QD radii a (22), the parameter $|E_{ex}(a, \varepsilon)/kT_0|$ take significant values in the range from 3.1 to 56 [20 – 22, 29 – 32].

The effect of significantly increasing the binding energy (21) of the exciton ground state in a nanosystem containing zinc-selenide QDs with radii a (22) was detected; in comparison with the exciton binding energy in a zinc-selenide single crystal, the increase factor is 4.1–72.6. [20 – 22, 29 – 32]. It was shown that the effect of significantly increasing the binding energy (21) of the exciton ground state in the nanosystem under study is controlled by two factors [20 – 22, 29 – 32]: (i) a substantial increase in the electron–hole Coulomb interaction energy (11) and an increase in the energy of the interaction of the electron and hole with "foreign" images (8), (9) (the "dielectric enhancement" effect [34]); (ii) spatial confinement of the quantization region by the QD volume; in this case, as the QD radius a increases, starting from $a \geq a_c^{(2)} \approx 29.8$ nm, the exciton becomes two-dimensional with a ground-state energy (15) that exceeds the exciton binding energy in a zinc-selenide single crystal by almost two orders of magnitude.

A review devoted to the theory of excitonic quasimolecule (biexciton) (formed of spatially separated electrons and holes) in a nanosystem that consists of ZnSe QDs synthesized in a borosilicate glass matrix is developed within the context of the modified effective mass approximation. Using the variational method, we obtain the total energy and the binding energy of the biexciton singlet ground state in such system as functions of the spacing between the QD surfaces and the QD radius. It is established that, in a nanosystem composed of ZnSe QDs with the average radii $\bar{a}_1$, the formation of a biexciton (exciton quasimolecule) is of the threshold character and possible in a nanosystem, in with the spacing D between the QD surfaces is defined by the condition $D_c^{(1)} \leq D \leq D_c^{(2)}$ [23, 24]. Moreover, the exciton quasimolecule (biexciton) can exist only at temperatures below a certain critical temperature: $T_c \approx 49$ K [23, 24]. It is established that the spectral shift of the low- temperature luminescence peak [10, 19] in such a nanosystem is due to quantum confinement of the energy of the biexciton singlet ground state.

References

[1] Ekimov, A., Onushchenko, A., (1981). Zize quantization of electrons in microcrystals. *Journal Experimental Theoretical Physics Letters*. 34, 345-348.

[2] Ekimov, A., Onushchenko, A., (1984). Zize quantization of electrons in semiconductor microcrystals. *Journal Experimental Theoretical Physics Letters*. 40, 1136-1139.

[3] Ekimov, A., Efros, A., (1985). Excitons in semiconductor microcrystals. *Solid State Communication*. 56, 921-924.

[4] Ekimov, A., Onushchenko, A., Efros, A., (1986). Zize quantization of electron-hole pairs in nanocrystals. *Journal Experimental Theoretical Physics Letters*. 43, 376-379.

[5] Chepik, D., Efros, A., Ekimov, A., (1990). Spectroscopy of excitons in semiconductor nanocrystals. *Journal of Luminescence*. 47, 113-118.

[6] Ekimov, A., Hache, F., Schanne-Klein, M., (2003). Optical properties semiconductor quantum dots. *Journal Optical Society American*. B 20, 100 -108.

[7] Grabovskis, V., Dzenis, Y., Ekimov, A., (1989). Fotoluminescences of excitons in semiconductor nanocrystals. *Physics Solid State*. 31, 149 -152.

[8] Alferov, J.I., (2002). Progress development in semiconductor nanostructures. *Physics Uspekhi*. 172, 1068 - 1074.

[9] Alferov, J.I., (1998). Optical properties in semiconductor nanostructures. *Semiconductors*. 32, 1-8.

[10] Bondar, V.N., Brodin, M.S., (2010). Optical properties semiconoductor quantum dots. *Semiconductors*. 44, 884-890.

[11] Efros, A.L., Efros, Al.L., Interband absorption light in semiconductor sphere. (1982).

[12] *Sov. Phys. Semiconductors*. 16, 955- 962.

[13] Pokutnyi, S.I., (2004). Size quantization Stark effect in semiconductor quantum dots. *J. Appl. Phys*. 96, 1115-1122.

[14] Pokutnyi, S.I., (2005). Optical nanolaser heavy hole transitions in quasi – zero – dimensional semiconductors nanosystems. *Physics Lett. A*. 342, 347-352.

[15] Pokutnyi, S.I., (2007). Exciton states in semiconductor quantum dots in the framework of the modified effective mass method. *Semiconductors*. 41, 1323-1331.

[16] Pokutnyi, S.I., (2010). Exciton states in semiconductor quantum dots. *Semiconductors*. 44, 488-493.

[17] Pokutnyi, S.I., (2012). Exciton states in semiconductor nanosystems. *Semiconductors*. 46, 174-184.

[18] Soloviev, V., Eeichofer, A., (2001). Approximation of effective mass in nanosystems. *Physica Status Solidi B*. 224, 285-291.

[19] Yeh, C., Zhang, S., Zunger, A., (2004). The effective mass method in nanosystems. *Physical Review B*. 62, 14408-14416.

[20] Bondar, N., Brodyn, M., (2010). Spectroscopy of semiconbductor quantum dots. *Physics E*. 42, 1549-1555.

[21] Pokutnyi, S.I., (2013). Binding energy of the exciton of a spatially separated electron and hole in quasi – zero – dimensional semiconductor nanosystems. *Technical Physics Letters*. 39, 233-235.

[22] Pokutnyi, S.I., (2013). On an exciton with a spatially separated electron and hole in quasi – zero – dimensional semiconductor nanosystems. *Semiconductors*. 47, 791 -798.

[23] Pokutnyi, S.I., (2014). Theory of excitons formed from spatially separated electrons and holes in quasi – zero – dimensional semiconductor nanosystems. *SOP Transactions Theoretical Physics*. 1, 24 -35.

[24] Pokutnyi, S.I., (2013). Biexcitons formed from spatially separated electrons and holes in quasi – zero – dimensional semiconductor nanosystems. *Semiconductors*. 47, 1626 – 1635.

[25] Pokutnyi, S.I., (2014). Quasi – zero – dimensional nanostructures: Excitonic quasimolecules. *J. Appl. Chem*. 2, 1- 4.

[26] Efremov, N.A., Pokutny, S.I., (1985). Macroscopic local charge states in ultradispersion media. *Sov. Phys. Solid. State*. 27, 27- 35.

[27] Efremov, N.A., Pokutny, S.I., (1990). Energy spectrum of exciton in spherical particle. *Sov. Phys. Solid. State*. 32, 955- 964.

[28] Pokutnyi, S.I., (1992). Size quantization of electron-hole pair in semiconductor quantum dots. *Physics Lett. A*. 168, 433-438.

[29] Pokutnyi, S.I., (2011). Theory of exciton states in semiconductor nanosystems. *Physics Express*. 1, 158-164.

[30] Pokutnyi, S.I., Gorbyk, P.P., (2013). Superatoms in quasi – zero – dimensional nanostructures (review). *Progr. Phys. Metal*. 14, 144 - 168.

[31] Pokutnyi, S.I., Gorbyk, P.P., (2013). Superatoms in quasi – zero – dimensional nanosystems. *J. Appl. Chem*. 1, 44 - 47.

[32] Pokutnyi, S.I., (2012). Exciton states in quasi – zero – dimensional semiconductor nanosystems: Theory. *Phys. Express*. 2, 20 - 26.

[33] Pokutnyi, S.I., (2013). Exciton states in quasi – zero – dimensional nanostructures. *J. Appl. Chem.* 1, 5 - 18.

[34] Lozovik, Y., Nishanov, V., (1976). Exciton Wannier-Mott near the interface. *Physics Solid State*. 18, 1905-1911.

[35] Keldysh, L., (1979). The interaction of the electrons in a thin films. *Journal Experimental Theoretical Physics Letters*. 29, 621-624.

SECTION 2.
SPINTRONICS

In: Advances in Semiconductor Research
Editors: D. Persano Adorno and S. Pokutnyi
ISBN: 978-1-63321-755-3

Chapter 6

Spin Dynamics in Semiconductors in the Streaming Regime

*L. E. Golub** and E. L. Ivchenko*
Ioffe Physical-Technical Institute of the Russian Academy of Sciences, St. Petersburg, Russia

Abstract

We present results of a cross-disciplinary theoretical research at the interface of spin physics and hot-electron transport. A moderately strong electric field is assumed to provide the streaming regime where each free charge carrier, an electron or a hole, accelerates quasiballistically in the "passive" region until reaching the optical-phonon energy, then emits an optical phonon and starts the next period of acceleration. The inclusion of spin degree of freedom into the streaming-regime kinetics gives rise to rich and interesting spin-related phenomena. Firstly, in the streaming regime the spin relaxation is substantially modified, and the current-induced spin orientation is remarkably increased. Under short-pulsed photoexcitation at the bottom of conduction band the photoelectrons execute a periodic damped motion in the energy space with the period equal to the free flight time of an electron in the passive region. If the short optical pulse is circularly polarized so that the photocarriers are spin oriented, then the spin energy distribution is oscillating in time as well, which can be detected in the pump-probe time-resolved experiments. We show that the spin-orbit splitting of the conduction band becomes a source for additional spin oscillations, periodic or aperiodic depending on the value of electric field.

PACS: 72.25.Hg, 72.25.Pn, 72.25.Rb, 73.63.Hs

Keywords: Spin-orbit splitting, semiconductor heterostructures, streaming

1. Introduction

Spin physics is a rapidly growing area of research in condensed matter science aimed at the creation, manipulation and detection of spins in various systems. Important and interesting fundamental results, also promising for possible future applications, have been obtained

*E-mail address: golub@coherent.ioffe.ru

in semiconductors and semiconductor nanostructures [1]. Highly sensitive methods of Kerr and Faraday rotation spectroscopy give a possibility to probe spin orientation within subpicosecond resolution [2]. A challenging problem in the spin physics is how to affect the spin by instantaneous non-magnetic methods, in particular, by static or optical electric fields. In semiconductors, optical selection rules allow one to create spin-oriented carriers (optical orientation) which may then be accelerated by a static external electric field. Another possibility to monitor the spin by electrical means arises due to the spin-orbit interaction which couples the electron orbital and spin degrees of freedom. In semiconductor heterostructures, the spin-orbit interaction is described by the Rashba and Dresselhaus contributions to the electron effective Hamiltonian linear in the electron quasimomentum $\boldsymbol{p}$, for review see Ref. [3]. Generally, they can be presented as follows

$$H_{\rm so}(\boldsymbol{p}) = \beta_{ij}\sigma_i p_j = \frac{\hbar}{2}\boldsymbol{\sigma}\cdot\boldsymbol{\Omega}_{\boldsymbol{p}}\,, \qquad (1)$$

where σ_i $(i = x, y)$ are the Pauli matrices, $\boldsymbol{\beta}$ is a second-rank pseudotensor, and the effective momentum-dependent Larmor frequency is defined by $\Omega_{\boldsymbol{p},i} = 2\beta_{ij}p_j/\hbar$. Hereafter, for definiteness, we consider zinc-blende-lattice based quantum wells grown along the direction [001] and use the coordinate frame $x \parallel [1\bar{1}0], y \parallel [110]$ and $z \parallel [001]$.

We study here the electron spin dynamics in semiconductor heterostructures in the presence of an external in-plane electric field. The attention is focused on the so-called streaming regime [4–9] characterized by the periodic optical-phonon emission by electrons subjected to acceleration in the strong electric field. A number of interesting time-resolved spin effects are shown to be caused by the electric field. We also discuss a possibility to detect them in the pump-probe experiments, both in the absence and presence of an external magnetic field.

2. Streaming in 2D Semiconductors

In an external electric field $\boldsymbol{F}$, the carrier distribution is anisotropic in the momentum space. In particular, it is characterized by a nonzero electron drift momentum $p_{\rm dr}$ in the opposite to field direction. For comparatively weak fields $F \ll 1$ kV/cm, the anisotropy is weak: the ratio $p_{\rm dr}/p$ is smaller than 0.1, where p is the root-mean-square of the electron quasimomentum in the absence of the field. An application of the increasing field leads to a transformation from the linear dependence $p_{\rm dr} \propto F$ to the saturated behaviour. If the interparticle collisions play a minor role in the kinetics, the electron momentum-space distribution in strong electric fields becomes extremely anisotropic, *streaming*-like. Each electron accelerates quasiballistically in the "passive" region until reaching the threshold energy equal the optical-phonon energy $\hbar\omega_0$ and the corresponding threshold quasimomentum $p_0 = \sqrt{2m\hbar\omega_0}$, where m is the electron effective mass. Then it looses its energy by emitting an optical phonon and starts the next period of acceleration.

Thus, in the streaming regime realized in an appropriate range of the dc electric fields, the electron distribution in the momentum space has a steaming-like, or needle-like, form spread between the Γ-point $\boldsymbol{p} = 0$ and the point $\boldsymbol{p}_0 = p_0\hat{e}$, where $\hat{e}$ is the unit vector in the direction opposite to the electric field $\boldsymbol{F}$. The formation of such an anisotropic electron distribution requires the following hierarchy of times

$$\tau \ll t_{\rm tr} \ll \tau_p, \tau_{\rm ee}\,. \tag{2}$$

Here τ is the time of optical phonon emission by an electron in the "active" energy region exceeding $\hbar\omega_0$, $t_{\rm tr} = p_0/|eF|$ is the travel time through the passive region of the momentum space, τ_p is the momentum relaxation time due to electron scattering by acoustic phonons or static imperfections, and $\tau_{\rm ee}$ is the electron-electron collision time. The electron-electron scattering tends to convert the anisotropic distribution to the shifted Maxwellian distribution with an effective electron temperature and a drift velocity $p_{\rm dr}/m$. The corresponding time can be estimated by $\tau_{\rm ee}^{-1} \sim (e^2/\varkappa\hbar)^2 N/\omega_0$, where N is the electron 2D density and $\varkappa$ is the dielectric constant.

The electron distribution function $f_{\boldsymbol p}$ is given by [10]

$$f_{\boldsymbol p} = N\frac{2\pi^2\hbar^2}{p_0}\tilde{\delta}(p_y)\begin{cases} \tilde{\theta}\,(p_x), & \text{if} \quad p < p_0, \\ \exp\left(-\alpha\frac{p_x - p_{0x}}{p_0}\right), & \text{if} \quad p > p_0, \end{cases} \tag{3}$$

where $\alpha = p_0/|eF|\tau$, $p_{0x} = \sqrt{p_0^2 - p_y^2}$, and

$$\tilde{\delta}(p_y) = \sqrt{\frac{\alpha}{2\pi}}\frac{1}{p_0}\exp\left(-\frac{\alpha}{2}\frac{p_y^2}{p_0^2}\right), \qquad \tilde{\theta}\,(p_x) = \frac{1}{2}\left[1 + \text{erf}\left(\sqrt{\frac{\alpha}{2}}\frac{p_x}{p_0}\right)\right].$$

One can see from Eq. (3) that the penetration δp_x into the active region and the width $2\delta p_y$ of the distribution in the transverse direction p_y are given by $\delta p_x = p_0/\alpha$ and $\delta p_y = p_0/\sqrt{\alpha}$. Thus, the needle-like distribution is formed provided $\alpha \gg 1$ which establishes the upper limit for the electric field strength. In this case the function $\tilde{\theta}\,(p_x)$ can be approximated by the Heaviside function $\theta(p_x)$ and p_{0x} by $p_0 - (p_y^2/2p_0)$. It is also worth to mention that the value α gives the ratio between the numbers of particles in the passive and active regions as well as the ratio $t_{\rm tr}/\tau$ of the times spend by electrons in these regions.

If the electrons are spin polarized with the average spin $\boldsymbol s$ per particle, their spin distribution function in the $\boldsymbol p$-space is given by the main contribution

$$\boldsymbol S_{\boldsymbol p} = 2 f_{\boldsymbol p}\boldsymbol s \tag{4}$$

and an additional term of the order of $\Omega_{\boldsymbol p} t_{\rm tr}$.

2.1. Effect of Elastic Scattering on Streaming Distribution

In order to analyze consequences of elastic scattering by static imperfections, impurities and interface roughness, and/or quasi-elastic scattering by acoustic phonons, we start with the approximation of "ideal streaming" replacing the smoothed delta-function $\tilde{\delta}(p_y)$ in Eq. (3) by the Dirac delta-function and the smoothed theta-function $\tilde{\theta}$ by the Heaviside step function. Then the streaming distribution in the passive region reduces to

$$f_{\boldsymbol p}^0 = N\frac{2\pi^2\hbar^2}{p_0}\delta(p_y)\theta(p_x)\theta(p_0 - p_x). \tag{5}$$

In the following we include into consideration the elastic scattering by short-range imperfections and describe the scattering process $\boldsymbol{p}' \to \boldsymbol{p}$ by the rate $W_{\boldsymbol{pp}'} = (\pi\tau_p)^{-1}\delta(p^2 - p'^2)$ normalized on the inverse momentum relaxation time τ_p as follows

$$\tau_p^{-1} = \int d\boldsymbol{p}' W_{\boldsymbol{pp}'} .$$

Since the streaming regime is achieved provided τ_p is much longer than the travel time $t_{\rm tr}$ we treat the elastic scattering keeping the first order corrections in $t_{\rm tr}/\tau_p$ to the electron distribution function (5). Then the scatter of electrons over all azymuthal angles θ gives rise to a correction $f_{\boldsymbol{p}}^{(1)}$ which, in the passive region, satisfies the equation

$$eF \frac{\partial f_{\boldsymbol{p}}^{(1)}}{\partial p_x} = \int d\boldsymbol{p}' W_{\boldsymbol{pp}'} f_{\boldsymbol{p}'}^0 \qquad (6)$$

with the boundary condition $f_{\boldsymbol{p}}^{(1)} = 0$ at $p_x = -p_{0x} \equiv -\sqrt{p_0^2 - p_y^2}$. Here the right-hand side term represents the generation of electrons to the whole area $p = \sqrt{p_x^2 + p_y^2} \leq p_0$ due to the elastic scattering from the zeroth-approximation needle (3) while the term on the left-hand side describes the further acceleration of the scattered electrons in the electric field.

Substituting Eq. (5) into Eq. (6) we obtain

$$eF \frac{\partial f_{\boldsymbol{p}}^{(1)}}{\partial p_x} = \frac{\pi\hbar^2 N}{\tau_p p_0 p} .$$

The solution of this first-order differential equation leads to the formula

$$f_{\boldsymbol{p}}^{(1)} = \frac{t_{\rm tr}}{\tau_p} \frac{\pi\hbar^2 N}{p_0^2} \int\limits_{-p_{0x}}^{p_x} \frac{dp_x'}{\sqrt{p_x'^2 + p_y^2}} = \frac{t_{\rm tr}}{\tau_p} \frac{\pi\hbar^2 N}{p_0^2} \ln\left(\frac{p + p_x}{p_0 - p_{0x}} \right). \qquad (7)$$

The steady-state density of electrons scattered off the needle in the passive area is given by

$$\delta N = N \frac{t_{\rm tr}}{\tau_p} \int\limits_0^{2\pi} \frac{d\theta}{2\pi} \int\limits_0^1 ds\, s \ln\left[\frac{s(1+\cos\theta)}{1 - \sqrt{1 - s^2 \sin^2\theta}} \right] = N \frac{t_{\rm tr}}{\tau_p} \frac{2G+1}{\pi} \approx 0.9 N \frac{t_{\rm tr}}{\tau_p} , \qquad (8)$$

where $G \approx 0.9159\ldots$ is the Catalan constant. These electrons reach the sphere $p_x^2 + p_y^2 = p_0^2$, emit optical phonons and return to the point $p = 0$. Thus, in the first approximation, the total density N should be replaced by $N - \delta N$. The perturbation procedure can be continued by the replacements $f_{\boldsymbol{p}}^0 \to f_{\boldsymbol{p}}^0 + f_{\boldsymbol{p}}^{(1)}$ and $f_{\boldsymbol{p}}^{(1)} \to f_{\boldsymbol{p}}^{(2)}$ in the right- and left-hand sides of Eq. (6), respectively.

3. Spin Relaxation

The constriction of the distribution function $f_{\boldsymbol{p}}^0$ along the p_y axis and the electron homogeneous occupation of the p_x interval between 0 and p_0 substantially modifies the

Dyakonov-Perel spin relaxation of the s_y component controlled by the linear-$\boldsymbol{p}$ term $\beta_{xy}\sigma_x p_y$ in Eq. (1). Here, as above, we assume the electric field to be applied along the x axis. The spin-relaxation time τ_{sy} of this component is found from the equation, see [10]:

$$\frac{s_y}{\tau_{sy}} = \frac{1}{N}\sum_{\boldsymbol{p}} \Omega_{\boldsymbol{p},x}\delta S_{\boldsymbol{p},z}\,, \tag{9}$$

where $\Omega_{\boldsymbol{p},x} = 2\beta_{xy}p_y/\hbar$, $\boldsymbol{s}$ is the average electron spin per particle, $\delta \boldsymbol{S}_{\boldsymbol{p}}$ is the first-order correction (in $\boldsymbol{\Omega}_{\boldsymbol{p}}$) to the spin distribution function satisfying the spin kinetic equation, see Eq. (22) in Ref. [10],

$$eF\frac{\partial}{\partial p_x}\delta \boldsymbol{S}_{\boldsymbol{p}} + 2f^0_{\boldsymbol{p}}\,\boldsymbol{s}\times\boldsymbol{\Omega}_{\boldsymbol{p}} = 0\,. \tag{10}$$

Solving this equation with the streaming distribution (3) we obtain [10]

$$\frac{1}{\tau_{sy}} = 2\tau\left(\frac{\beta_{xy}p_0}{\hbar}\right)^2. \tag{11}$$

We remind that, in the state of equilibrium, the relaxation time τ_{sy} is given by the similar equation with τp_0^2 replaced by $\tau_p 2m\bar{\varepsilon}$, where $\bar{\varepsilon}$ is the equilibrium value of the electron average energy.

Now we consider effect of elastic scattering on the spin relaxation and show that even the first-order correction in the ratio $t_{\rm tr}/\tau_p \ll 1$ can modify τ_{sy}. For this purpose we replace $f^0_{\boldsymbol{p}}$ in Eq. (10) by $f^{(1)}_{\boldsymbol{p}}$ and find the scattering-induced correction $\delta \boldsymbol{S}^{(1)}_{\boldsymbol{p}}$. The latter is given therefore by the integral

$$\delta S^{(1)}_{\boldsymbol{p},z} = s_y\frac{2\Omega_{\boldsymbol{p},x}}{eF}\int_{-p_{0x}}^{p_x} dp'_x\, f^{(1)}_{p'_x,p_y}\,, \tag{12}$$

which can be solved analytically to give

$$\delta S^{(1)}_{\boldsymbol{p},z} = s_y\frac{t_{\rm tr}}{\tau_p}\frac{N\pi\hbar^2}{p_0^2}\frac{2\Omega_{\boldsymbol{p},x}}{eF}\left[p_0 - p + p_x\ln\left(\frac{p+p_x}{p_0-p_{0x}}\right)\right]. \tag{13}$$

Substituting this correction to Eq. (9) and summing over $\boldsymbol{p}$ we arrive at the contribution of the elastically scattered electrons to spin relaxation of the electron gas

$$\frac{1}{\tau^{\rm elast.}_{sy}} = \frac{7}{15}\frac{t^2_{\rm tr}}{\tau_p}\left(\frac{\beta_{xy}p_0}{\hbar}\right)^2. \tag{14}$$

In contrast to the electrons in the zeroth-approximation needle distribution (3) compressed along the p_y axis, the scattered electrons are spread over the whole passive area. The increased precession frequency $\Omega_{\boldsymbol{p},x} = 2\beta_{xy}p_y/\hbar$ can compensate smallness of the ratio $t_{\rm tr}/\tau_p$. For comparison of the contributions (11) and (14) it is instructive to sum them and present the total spin relaxation time in the following convenient form

$$\frac{1}{\tau_{sy}} = 2\tau\left(\frac{\beta_{xy}p_0}{\hbar}\right)^2\left(1+\frac{7}{30}\frac{t_{\rm tr}}{\tau_p}\alpha\right). \tag{15}$$

For $\alpha = 33$ and $\tau_p/t_{\rm tr} = 3$, we find the contribution (14) caused by elastic scattering to exceed that given by Eq. (11) by more than twice.

4. Spin Orientation by Electric Field

In systems with the effective Hamiltonian containing the $\boldsymbol{p}$-linear terms (1), the application of an electric field results in orientation of electron spins. Such electrical spin orientation is allowed by the symmetry only in gyrotropic systems where some components of the second-rank pseudotensor $\boldsymbol{\beta}$ are nonzero. In weak fields, this effect has been studied in many works, see the review [11], references therein, and the recent paper [12]. The spin polarization created by a weak electric field is given by

$$s_y = -c\frac{\beta_{yx}}{v}\frac{p_{\rm dr}}{p}, \tag{16}$$

where v and p are the zero-field root-mean-square electron velocity and momentum, and $p_{\rm dr} = eF\tau_p$. It is remarkable that the coefficient c depends on the ratio of spin and energy relaxation times. At slow energy relaxation, the spin is created independently at each energy ε resulting in a nonequilibrium spin distribution $S_y(\varepsilon) \propto f_0''(\varepsilon)$, where $f_0(\varepsilon)$ is the Fermi-Dirac distribution. In contrast, the fast energy relaxation establishes the equilibrium distribution within each spin-split subband, and the nonequilibrium spin distribution function $S_y(\varepsilon) \propto f_0'(\varepsilon)$. The integration over the energy yields different results for the spin polarization s_y. For a short-range momentum scattering potential, in two limiting cases of fast and slow energy relaxation the coefficient c in Eq. (16) differs by a factor of two [11]: $c_{\rm fast} = 1/2$ and $c_{\rm slow} = 1/4$. The results for these two extreme cases are shown by straight lines 1 in Fig. 1. One can see that the spin polarization in the fields $F \leq 10$ V/cm does not exceed 0.1%.

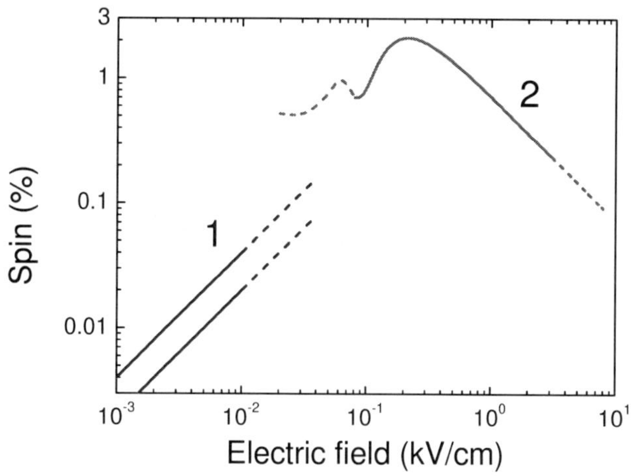

Figure 1. Dependence of the current-induced spin on the electric field. Lines 1 are calculated for weak fields in the limits of fast and slow energy relaxation, curve 2 presents the result of calculation in the streaming regime. Dashed ends of the curve 2 represent the results obtained on the edge of applicability of the streaming model. The parameters used are: $\hbar\beta_{yx} = 7$ meV Å, $\hbar\omega_0 = 36$ meV, $m = 0.067m_0$, $\tau = 0.1$ ps, $\tau_p = 10$ ps, $N = 10^{11}$ cm^{-2}, and zero temperature is assumed.

In strong fields where the streaming regime is realized, the situation changes drastically. Treating the term $\beta_{xy}p_y$ in Eq. (1) as the spin-orbit splitting of the spectrum we get the shifted needle-like distributions in the spin subbands with the projection $\pm 1/2$ onto the x axis. Then taking into account the terms $\Omega_y = 2\beta_{yx}p_x/\hbar$ and $\Omega_x = 2\beta_{xy}p_y/\hbar$ as the components of the precession frequency and neglecting elastic scattering processes in the passive area, we obtain the spin polarization in the form [10]

$$s_y = -\frac{1}{4\omega_0\tau}\frac{\mathrm{Si}(\xi)}{\xi}\mathrm{sign}(\beta_{yx}), \tag{17}$$

where $\xi = p_0\sqrt{2|\beta_{yx}|/\pi\hbar eF}$ and Si stands for the Fresnel sine integral. For small values of ξ the ratio $\mathrm{Si}(\xi)/\xi$ is approximated by $\pi\xi^2/6$ and Eq. (17) turns into $s_y = -m\beta_{yx}/6eF\tau$.

It follows from Eq. (17) that the electrically oriented spin is a non-monotonous function of the field strength in the streaming regime. The dependence (17) is shown by curve 2 in Fig. 1. The values of the spin orientation have an order of a few percent which is more than an order of magnitude larger than those at low fields. We note that the decrease in the spin relaxation time caused by elastic scattering and described by Eqs. (14), (15) is compensated by a corresponding increase in the generation rate.

5. Spin Beats in the Streaming Regime

Consider a undoped 2D structure excited by a short interband pulse creating photoelectrons at the bottom of the lowest conduction subband. Assuming a homogeneous pulse, the non-equilibrium distribution at the initial moment $t = 0$ is given by

$$f_{\boldsymbol{p}}(t=0) = 2(\pi\hbar)^2 N\delta(p_x)\delta(p_y), \tag{18}$$

where N is the photoelectron density. In the ideal streaming-regime model, the kinetic evolution of the photoelectrons is also described by a product of two delta-functions

$$f_{\boldsymbol{p}}(t) = 2(\pi\hbar)^2 N\,\delta\left(p_x - |eF|t_{\mathrm{tr}}\left\{\frac{t}{t_{\mathrm{tr}}}\right\}\right)\delta(p_y), \tag{19}$$

with the variable of one function being a function of time with the period t_{tr}. Here the symbol $\{\nu\}$ means the fractional part of the number ν. If the optical pulse is circularly polarized then the generated carriers are spin oriented, and the spin energy distribution is oscillating in time as well. Neglecting the splitting of spin states and the spin relaxation we can supplement Eq. (19) by the similar equation for the time-dependent spin distribution function, cf. Eq. (4)

$$\boldsymbol{S}_{\boldsymbol{p}}(t) = \boldsymbol{s}_0(2\pi\hbar)^2 N\delta\left(p_x - |eF|t_{\mathrm{tr}}\left\{\frac{t}{t_{\mathrm{tr}}}\right\}\right)\delta(p_y), \tag{20}$$

where $\boldsymbol{s}_0$ is the initial electron spin per particle. The both functions (19) and (20) are periodic until the elastic scattering is ignored. The simplest way to take into account the elastic scattering of the accelerating bunch of electrons is just multiply the above time-dependent functions by the exponential function $\exp(-t/\tau_p)$ turning them into damped oscillatory functions.

The temporary oscillations of $\boldsymbol{S_p}(t)$ can be detected in the pump-probe time-resolved experiments based on the spin Kerr or Faraday effects. For the probe light frequency slightly below the quantum-well fundamental edge $E_g/\hbar$, the Kerr rotation angle of polarization plane of the linearly polarized probe pulse is given by

$$\Theta_K(t) = \int d\varepsilon\, K(\varepsilon) S_{\boldsymbol{p},z}(\varepsilon, t)\,, \tag{21}$$

where ε is the electron energy referred to the bottom of the lowest 2D conduction subband, the energy-dependent coefficient $K(\varepsilon)$ is proportional to $[(E_g - \hbar\omega)(\mu_{eh}/m_e) + \varepsilon]^{-1}$, m_e is the in-plane electron effective mass, and μ_{eh} is the reduced mass of an electron and a heavy hole. The substitution of the spin distribution (20) multiplied by $\exp(-t/\tau_p)$ into Eq. (21) gives

$$\Theta_K(t) \propto \frac{s_{0,z} \exp(-t/\tau_p)}{(E_g - \hbar\omega)\mu_{eh}/m_e + \hbar\omega_0\{t/t_{\rm tr}\}^2} + \Delta\Theta_K^{\rm elast.}(t)\,, \tag{22}$$

where $\Delta\Theta_K^{\rm elast.}(t)$ is a correction arising due to the elastically scattered carriers and proportional to the ratio $t_{\rm tr}/\tau_p$. While deriving Eq. (22) we took into account that the electron energy in the bunch varies in time as $\varepsilon(t) = p_x^2(t)/2m = (p_0^2/2m)\{t/t_{\rm tr}\}^2 \equiv \hbar\omega_0\{t/t_{\rm tr}\}^2$.

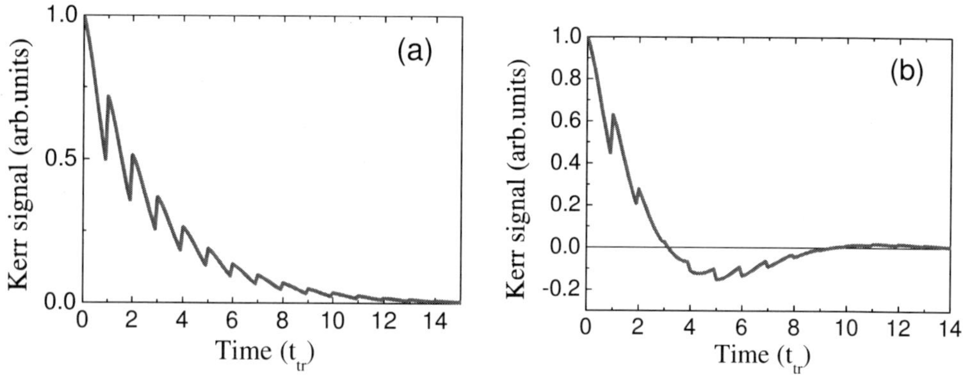

Figure 2. The damped oscillatory spin-Kerr signal at $(E_g - \hbar\omega)\mu_{eh}/m_e = 1.7\hbar\omega_0$, $\tau_p = 3t_{\rm tr}$ calculated neglecting the spin-orbit splitting (a) at zero magnetic field, and (b) in the external magnetic field with $\Omega_B t_{\rm tr} = 0.5$ which corresponds to $B = 1.5$ T at $|g| = 1$.

In the external in-plane magnetic field $\boldsymbol{B} \parallel y$ the electron spin component $\boldsymbol{s} \perp y$ rotates around the y axis with the angular frequency $\Omega_{\boldsymbol{B}} = g\mu_B B_y/\hbar$, where g is the electron g-factor and μ_B is the Bohr magneton. The Kerr rotation angle is obtained from Eq. (22) by multiplying the right-hand side by $\cos(\Omega_{\boldsymbol{B}} t)$. Indeed, in the absence of the spin-orbit splitting, the spin precession plays a role of the clock independent of the periodic acceleration and emission of optical phonons by the photoelectrons. The time dependence of the Kerr rotation signal at zero and nonzero magnetic field is plotted in Fig. 2, (a) and (b) respectively. Only the first main contribution in Eq. (22) is taken into account in the calculation. The curve in Fig. 2(a) has a period $t_{\rm tr}$ whereas the curve in Fig. 2(b) has two oscillations with two different periods, the longer one equal to the Larmor precession period $2\pi/\Omega_{\boldsymbol{B}}$ and the shorter one equal to the travel time $t_{\rm tr}$.

6. Spin Beats due to the Spin-Orbit Splitting

We consider again optical creation of the spin $\boldsymbol{s} \parallel z$ at the bottom of the conduction band at the moment $t = 0$. The time-dependent spin distribution is then given by

$$\boldsymbol{S}_{\boldsymbol{p}}(t) = N\boldsymbol{s}(t)(2\pi\hbar)^2\delta\left(p_x - |eF|t_{\rm tr}\left\{\frac{t}{t_{\rm tr}}\right\}\right)\delta(p_y)\,. \tag{23}$$

Taking into account the spin-orbit splitting (1) we obtain that the spin dynamics in the absence of external magnetic fields is described by [10]

$$s_z(t) = s_{0,z}\cos\Phi(t)\,, \qquad s_x(t) = s_{0,z}\sin\Phi(t)\,,$$

where $s_{0,z}$ is the initial spin, and the phase Φ changes in time according to

$$\Phi(t) = \Omega_{\rm dr}t_{\rm tr}\left(\left[\frac{t}{t_{\rm tr}}\right] + \left\{\frac{t}{t_{\rm tr}}\right\}^2\right), \tag{24}$$

with the symbol $[x]$ denoting the integer part of x. Here $\Omega_{\rm dr} = 2\sum_{\boldsymbol{p}}\Omega_{\boldsymbol{p},y}f_{\boldsymbol{p}}/N$ is the spin precession frequency in the electric field. In weak fields it is linear in F [13] while in the streaming regime one has [10]

$$\Omega_{\rm dr} = \frac{\beta_{yx}p_0}{\hbar}.$$

If $\Omega_{\rm dr}t_{\rm tr}$ is a rational part of 2π then the functions $s_z(t), s_x(t)$ are periodic. In particular, for $\Omega_{\rm dr}t_{\rm tr} = 2\pi n$ with an integer n, the period equals to $t_{\rm tr}/n$. For irrational values of $\Omega_{\rm dr}t_{\rm tr}/2\pi$, the variation $s_{z,x}(t)$ is aperiodic. The time dependence $s_z(t)$ for these two cases is shown in Fig. 3.

According to Eq. (24) the phase Φ is a quadratic function of $\{t/t_{\rm tr}\}$, this explains why in Fig. 3 the variation of oscillatory function $s_z(t)$ in positive and negative areas is not antisymmetric. For the average spin

$$\bar{\boldsymbol{s}}(T) = \frac{1}{T}\int_0^T \boldsymbol{s}(t)dt$$

we find a sharp electric-field dependence which can be called "mode-locking". Introducing $M = [T/t_{\rm tr}]$ and $\varphi = \Omega_{\rm dr}t_{\rm tr}$, we obtain

$$\frac{\bar{s}_z(T)}{s_{0,z}} = \frac{1}{M}\sum_{k=0}^{M}\int_0^1 dt\cos\left(k\varphi + \varphi t^2\right) = \frac{P_1{\rm Ci}\left(\sqrt{2\varphi/\pi}\right) + P_2{\rm Si}\left(\sqrt{2\varphi/\pi}\right)}{2M\sin\varphi\sqrt{2\varphi/\pi}}, \tag{25}$$

where

$$P_1 = \sin M\varphi + \sin\left(M+1\right)\varphi + \sin\varphi\,, \qquad P_2 = \cos M\varphi + \cos\left(M+1\right)\varphi - \cos\varphi - 1\,.$$

The expression for $\bar{s}_x(T)$ has a similar form. One can see that the average spin $\bar{s}_z$ is nonzero if the following condition is fulfilled:

$$\varphi = \Omega_{\rm dr}t_{\rm tr} = 2\pi n\,, \qquad \text{i.e.} \qquad F_n = \frac{\beta_{yx}p_0^2}{2\pi\hbar e n}. \tag{26}$$

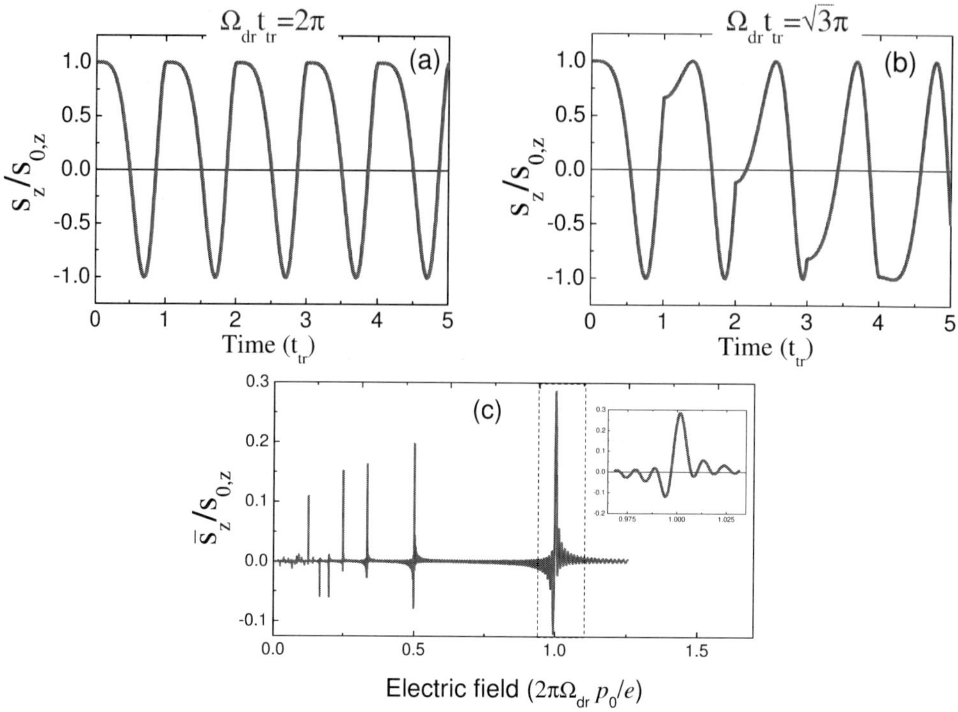

Figure 3. (upper panels) Spin beats caused by the spin-orbit splitting for $\Omega_{\rm dr}t_{\rm tr}$ equal to 2π (a) and $\sqrt{3}\pi$ (b); (lower panel, c) the time-averaged spin $\bar{s}_z(T)$ at $T = 100t_{\rm tr}$ calculated in the streaming regime neglecting elastic scattering. Inset highlights the narrow region of electric fields marked by dashed lines.

In this case the spin makes n full turns around the y axis during the time $t_{\rm tr}$, and the average spin polarization has a high value. The oscillatory behaviour of $s_z(t)$ in Fig. 3(a) corresponds to $n = 1$. Within the period for t lying between $mt_{\rm tr}$ and $(m+1)t_{\rm tr}$ $(m = 0, 1, 2\ldots)$ the spin z-component is decreasing from its initial value $s_{0,z}$, reaches the minimum value $-s_{0,z}$ and rapidly increases up to $s_{0,z}$. Nonlinear variation of $s_z(t)$ with $\{t/t_{\rm tr}\}$ results in a nonvanishing value of the average $\bar{s}_z(T)$. Moreover, for $T \gg t_{\rm tr}$, the average $\bar{s}_z(T)$ is still remarkable in the close proximity of φ to $2\pi n$ where $|\varphi - 2\pi n| \ll 1/M$. In the rest area of φ values the $\bar{s}_z$ have the order of magnitude of $1/M \ll 1$, see Eq. (25). The electric field dependence of $\bar{s}_z(T)$ for $T = 100t_{\rm tr}$ is shown in the lower panel of Fig. 3, the inset in Figure represents the selected area around $\Omega_{\rm dr}t_{\rm tr} = 2\pi$ in the enlarged scale.

Allowance for other mechanisms of electron scattering and spin relaxation leads to a decay of the oscillation amplitude. Figure 4 demonstrates the effect of elastic scattering on the dynamics of s_z and s_x components which is taken into account by the factor $\exp(-t/\tau_p)$. The Kerr rotation angle is given by Eq. (22) where $s_{0,z}$ is replaced by $s_z(t)$.

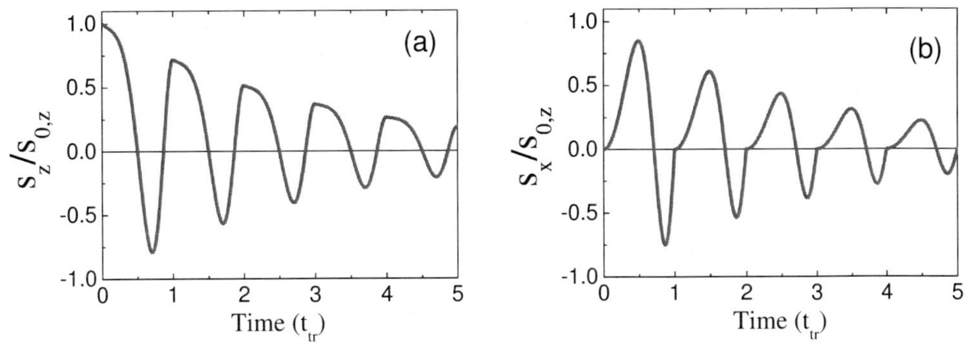

Figure 4. Spin beats of s_z (a) and s_x (b) with account for elastic scattering processes at $\Omega_{dr}t_{tr} = 2\pi$, $\tau_p/t_{tr} = 3$.

Conclusion

In this chapter we have demonstrated the rich variety of spin-dependent phenomena in the streaming regime that can be realized in moderate electric fields in semiconductor heterostructures. The proposed experiments, both steady-state and non-stationary, seem promising for studies of spin properties of two-dimensional electron (or hole) gas. Observation of photocarrier spin dynamics in the streaming regime by the pump-probe technique would allow the direct access to the time-dependent acceleration of charge carriers in the electric field.

Acknowledgments

Financial support of Russian Science Foundation (project 14-12-01067) and RF President Grant NSh-1085.2014.2 is gratefully acknowledged.

References

[1] *Spin Physics in Semiconductors*; Dyakonov, M. I.; Ed.; Springer, Berlin, 2008.

[2] Glazov, M.M. *Phys. Solid State* 2012, vol. 54, 1-27.

[3] Ganichev, S.D.; Golub, L.E. *Phys. Stat. Sol. (b)* 2014, DOI 10.1002/pssb.201350261.

[4] Andronov, A. A. in *Spectroscopy of Nonequilibrium Electrons and Phonons*; C.V. Shank and B.P. Zakharchenya; Ed.; Elsevier Science Publishers B.V., 1992; p. 169.

[5] Mickevičius, R.; Mitin V.; *Phys. Rev. B* 1995, vol. 51, 1609-1612.

[6] Vorob'ev, L. E.; Danilov, S. N.; Tulupenko, V. N.; Firsov D. A. *JETP Lett.* 2001, vol. 73, 219-222.

[7] Kim, K W; Korotyeyev, V V; Kochelap, V A; Klimov, A A; Woolard, D L *J. Appl. Phys.* 2004, vol. 96, 6488-6491.

[8] Syngayivska, G.I.; Korotyeyev, V.V. *Semicond. Phys. Quant. Electron. & Optoelectron.* 2007, vol. 10, 54-59.

[9] Syngayivska, G I ; Korotyeyev, V V; Kochelap, V A *Semicond. Sci. Technol.* 2013, vol. 28, 035007.

[10] Golub, L.E.; Ivchenko, E.L. *New J. Phys.* 2013, vol. 15, 125003 (15pp).

[11] Golub, L.E.; Ivchenko, E.L. *Phys. Rev. B* 2011, vol. 84, 115303 (1-8).

[12] Shen, Ka; Vignale, G.; Raimondi, R.; *Phys. Rev. Lett.* 2014, vol. 112, 096601 (1-5).

[13] Kalevich, V.K.; Korenev, V.L. *JETP Lett.* 1990, vol. 52, 230-235.

In: Advances in Semiconductor Research
Editors: D. Persano Adorno and S. Pokutnyi

ISBN: 978-1-63321-755-3

Chapter 7

SPIN DYNAMICS IN ONE-DIMENSIONAL SEMICONDUCTORS: UNUSUAL RELAXATION AND RESONANCES

***E. Ya. Sherman*[1], *D. V. Khomitsky*[2] and *V. K. Dugaev*[3,4]**
[1]Department of Physical Chemistry,
Universidad del País Vasco, Bilbao, Spain
and IKERBASQUE Basque Foundation
for Science, Bilbao, Spain
[2]Department of Physics, University of Nizhny Novgorod,
Nizhny Novgorod, Russian Federation
[3]Department of Physics, Rzeszów University of Technology,
Rzeszów, Poland
[4]Departamento de Física and CFIF, Instituto Superior Técnico,
Universidade de Lisboa, Lisbon, Portugal

Abstract

We present recent advances in spin dynamics, transport, and relaxation in one-dimensional semiconductors. Main attention will be paid to nanowire-based double quantum dots, spin-dependent transport and spin resonance in quantum wires, and to the qualitative role of disorder in these effects. We will show how the electron tunneling in double quantum dots influences the spin resonance and how the spin resonance modifies quantum charge transport in nanowires. Three different forms of disorder will be considered: conventional spin-independent disorder, and disorder in the spin-orbit coupling and in the hyperfine interaction between spin of electron and host lattice nuclei. We will demonstrate that randomness in one dimension leads to a spin freezing and to memory effects in spin dynamics.

1. Introduction

Semiconductors, known for variety of their optical, electrical, and thermal properties, became a clear manifestation of crucial importance of relativistic effects in solid state physics. Although the characteristic energies of carriers such as electrons and holes in solids are much less than the rest energy of an electron, the relativistic effects are clearly seen there. One of the most intriguing manifestations of these effects is spin-orbit coupling, introduced and first studied in seminal papers of Dresselhaus [1] and Rashba [2], where it was demonstrated that relativistic corrections are seen in the effective mass Hamiltonians as products of momentum and spin operators. This knowledge allowed researchers to explain many puzzling phenomena and predict new ones such as the ability to manipulate the electron spin by an external electric rather than by a magnetic field.

The progress in semiconductor technology at the end of 70th resulted in producing of a novel class of semiconductor systems: two-dimensional electron gas confined in heterostructures with a strong spatial asymmetry. The analysis of the symmetry of these structures showed that a linear in momentum spin-orbit coupling term, usually referred to as the Rashba coupling, should appear there [3]. In about a decade, this effect raised semiconductor spintronics. Relatively soon after that the Datta-Das spin field-effect transistor [4] was proposed. Although such a transistor can hardly be realized for various reasons, this work stimulated a rapid growth of the field. A seminal achievement, which made the semiconductors spintronics a truly experimental science was done by Nitta et al., who proved that in an inverted $In_{0.53}Ga_{0.47}As/In_{0.52}Al_{0.48}As$ quantum well the Rashba coupling can be controlled by applying a gate voltage and determined the relationship between the applied field and spin-orbit coupling [5]. In general, this control is strongly material and structure-dependent, as was demonstrated in recent experiments on n-type semiconductors [6, 7, 8, 9, 10, 11]. A similar effect of electric field has also been found in a p-type InAs, as reported by Matsuyama et al. [12].

The ability to produce new types of structures such as nanowires and quantum dots and apply them in spintronics stimulated further progress. Now it is impossible to describe the entire field of the semiconductor spintronics, and excellent publications can be recommended [13, 14, 15]. We will concentrate here on a discussion of two topics, narrowing even them considerably to their manifestation in one-dimensional systems.

First topic is spin manipulation by electric field and spin-dependent transport in a resonant time-dependent electric field. An essential requirement for spintronics devices is the ability to manipulate the electron spin. In the presence of spin-orbit coupling this can be done by the electric dipole spin resonance [16]. In Section 2 we will consider this effect for one-dimensional quantum dots. In addition, we will study the effect of spin resonance induced by a periodic electric field on the electron transport in one-dimensional nanowires.

Thus, the spin-orbit coupling is the key element in the spin manipulation. However, spin-orbit coupling is similar to the two-faced Janus (see Fig.1). On one hand, it helps the spin manipulation. On the other hand, it leads to a spin relaxation and dephasing, which will be the second topic of this Chapter, presented in Section 3. The most celebrated mechanism here is the Dyakonov-Perel' one, that is the diffusion in the spin subspace, similar to the motional narrowing of a spectral line. We will discuss novel mechanisms, related to the spin-orbit coupling, of the spin relaxation in one-dimensional systems. Another mechanism

Figure 1. Sad face of Janus looking in the past is related to the negative role of spin-orbit coupling such as the spin dephasing and relaxation. Joyful phase of Janus looking in the future is related to the positive role of spin-orbit coupling in spin manipulation and control.

of spin dephasing is related to the hyperfine interaction between electron and nuclear spins [17]. We will consider manifestations of this effect in nanowires as well.

2. Clean Systems: Driven Quantum Dots and Ballistic Transport

2.1. Spin Dynamics Driven by Electric Field

Driven quantum two-level systems attract attention due to the richness of phenomena occurring in different regimes of coupling to external field [18] and possible applications in quantum information devices. The two-level spins of electrons in semiconductor quantum dots (QDs) are widely expected as realizations of quantum information bits - "qubits" [19]. In the presence of spin-orbit (SO) coupling even simple systems such as single-electron QDs show a rich dynamics in coupled charge and spin channels. The spin dynamics in relatively weak electric fields in a single QD is well understood in terms of the electric dipole spin resonance (EDSR). In this process the electric field at the frequency of the Zeeman resonance for electron spin in magnetic field drives the orbital motion, and as a result, causes a spin dynamics. This coupling is in the core of the proposal by Rashba and Efros [16, 20] for a new technique to manipulate spin states by an external electric field. The EDSR spin-flip rate is orders of magnitude greater than the rate of the transitions due to the magnetic component of electromagnetic field. The efficiency of EDSR for GaAs QDs has been proven experimentally in Ref.[21] where the electric-field induced spin Rabi oscillations have been observed.

Another approach has been employed in Ref.[22], where the electric field was used for driving electrons in a nonuniform magnetic field, thus causing a spin dynamics. The coherent spin response of GaAs QDs shows that the problem with the spin dephasing induced by the SO coupling [23, 24] can be solved. Additional advantages for a reliable spin manipulation can be achieved by using hybrid structures such as quantum dot + superconducting stripline cavity [25] or quantum dot + multiferroic insulator in a superconducting resonator [26]. Recently, several approaches based on advanced quantum mechanics such as shortcuts to adiabaticity [27] and geometrical phases [28] have been developed to speed up the spin flips by electric field and to overcome the decoherence.

A generic system to study the driven behavior is a single parabolic one-dimensional QD where the electron states are the states of the harmonic oscillator. The coordinate and momentum can be expressed with the quantum mechanical ladder operators satisfying known commutation relations. As a result, well-known selection rules for coordinate and momentum matrix elements can be employed and the analysis can be sufficiently simplified. However, completely isolated QDs cannot be used for quantum information applications since an interdot interaction is necessary to produce and manipulate many-body states and a large scale of the system is required to make it work. Moreover, the experiments necessarily use at least a double QD [22, 29, 30] to detect the driven spin state relative to the spin of the reference electron, and the role of the induced interdot tunneling on the spin Rabi oscillations has been noticed [29]. This "photon-induced" tunneling can put a limitation on the abilities to manipulate the spins.

The knowledge of the coupled quantum dynamics of orbital and spin degrees of freedom when both are nontrivial, is still scarce. Here we make a step forward and address full driven spin/charge quantum dynamics in a one-dimensional double QD [31, 32, 33, 34] by considering a pumped motion of electron. These systems lavished attention due to their deceptive simplicity, where despite a well-established Hamiltonian, a rich variety of phenomena can be observed. In the classical regime, where only over-the-barrier motion can occur, the dynamics was studied in Ref.[35] and revealed interesting coupled irregular behavior of the spin and charge motion. In a more experimentally interesting quantum regime, the tunneling between single QDs plays the crucial role in the low-energy states. In the presence of SO coupling the electron states and the interdot tunneling become spin-dependent. These nonlinear systems driven by a time-dependent field can show a strongly irregular chaos-like behavior [36, 37, 38], and in the presence of spin-orbit coupling, a spin-orbital phase synchronization and full classical chaos have been predicted in certain regimes [39]. However, the existence of the quantum chaos, in contrast to the classical one, is disputed due to a finite set of energy eigenstates of the system. The interplay of irregularities for spin and charge degrees of freedom is important for the entire system dynamics: the driven motion being not rigorously chaotic, is, however, strongly irregular.

The important questions in the design of quantum gates are how fast can the gate operate, what limits the manipulation rate, and how efficient is the spin manipulation in terms of the achievable spin configurations. Driving by magnetic field has already been analyzed and understood (see, e.g. [40]). Spin manipulation by electric field still needs a thorough analysis. One can expect that a stronger driving field allows for a faster spin manipulation, as predicted in a simple Rabi picture of the driven oscillations. This picture is applicable for a single quantum dot with a parabolic confinement, where the electron displacement from the equilibrium is linear in the applied electric field [41]. However, double quantum dots where tunneling plays the crucial role for the orbital dynamics, and the corresponding energy scales are different from a single quantum dot, are more promising for the observation of new physics and applications in quantum information technologies [42]. The tunneling makes the description of the SOC puzzling since the electron momentum is imaginary at under-the-barrier motion, and the tunneling rate can be strongly spin-dependent [43, 44]. In addition, the double dots provide a possibility to study free and driven interacting qubits [29, 30, 45]. In this section we concentrate on one-dimensional systems attractive for spintronics [13, 14, 46, 47] and built-in there quantum dots [48, 49], and consider spin manipulation in a single-electron double quantum dot [33, 50, 51, 52, 53, 54] by a periodic electric field.

We mention for completeness another practically interesting possibility. One can produce a time-dependent field acting on the electron spin just by modifying the gate potential, which forms the quantum dot. Thus, the electron motion will be driven and the electron spin will precess either due to the spin-orbit coupling or due to a coordinate-dependent hyperfine field [55] inside the dot. Kato et al. [56] manipulated electron spins in a parabolic AlGaAs quantum well by a GHz bias applied to a single gate producing electric field $\mathcal{E}(t)$ perpendicular to the well. Another possibility is related to a time-dependent spin-orbit coupling, which can resonate with external magnetic field and drive spin-flip transitions. We will consider this effect in this section.

2.2. One-Dimensional Quantum Dot in a Periodic Electric Field: Combining Tunneling with Spin Motion

We begin with an example of driven coupled charge and spin dynamics in the double QD formed in a 1D quantum wire extended along the $x-$axis [57]. In the absence of driving, for a static magnetic field B parallel to the $z-$ axis, the Hamiltonian is the sum of two terms $H = H_0 + H_{\text{so}}$ with:

$$\begin{aligned} H_0 &= \frac{\hbar^2 k^2}{2m^*} + U(x) + \frac{\Delta}{2}\widehat{\sigma}_z, \\ H_{\text{so}} &= (\alpha_D \widehat{\sigma}_x + \alpha_R \widehat{\sigma}_y)\, k = \alpha\, (\widehat{\boldsymbol{\sigma}} \cdot \boldsymbol{n})\, k, \\ \alpha &\equiv \sqrt{\alpha_R^2 + \alpha_D^2}, \qquad \boldsymbol{n} \equiv \left(\frac{\alpha_D}{\alpha}, \frac{\alpha_R}{\alpha}, 0\right). \end{aligned} \tag{1}$$

Here k is the electron wavevector, $\widehat{\sigma}_i$ are the Pauli matrices, and Zeeman term $\Delta = g\mu_B B$, with g being the Landé factor, and μ_B being the Bohr magneton. In Eq.(1) we neglected the orbital effect of the magnetic field, as it is conventionally done in the analysis of one-dimensional systems.

The Hamiltonian H_{so} describes precession of a moving spin and yields spin precession length $L_{\text{so}} = \hbar^2/m^*\alpha$, that is the distance electron has to travel to flip its spin due to the spin-orbit coupling. The SO interaction is the sum of bulk-originated Dresselhaus (α_D) and structure-related Rashba (α_R) terms. The corresponding velocity operator

$$v \equiv \dot{x} = \frac{i}{\hbar}[H_0 + H_{\text{so}}, x] = \frac{\hbar k}{m^*} + \frac{\alpha}{\hbar}\,(\widehat{\boldsymbol{\sigma}} \cdot \boldsymbol{n}) \tag{2}$$

is spin-dependent and clearly demonstrates that the Gallilean invariance of the system is broken since the velocity is not simply proportional to the momentum.

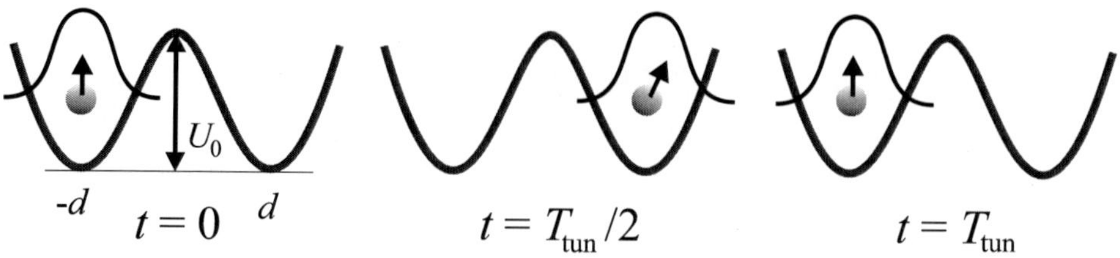

Figure 2. Qualitative picture of the spin precession induced by the interminima tunneling. Spin rotation is due to the electron displacement $\sim 2d$ in the tunneling process in the presence of the spin-orbit coupling. The spin state is restored upon electron return to the initial position. $T_{\rm tun}$ is the tunneling time.

The total Hamiltonian can be presented in a slightly different form, similar to motion in a gauge field as:

$$H = \frac{\hbar^2}{2m^*}(k - A)^2 + U(x) - \frac{m^*\alpha^2}{2\hbar^2}, \tag{3}$$

where $A = -m^*\alpha\left(\widehat{\boldsymbol{\sigma}} \cdot \boldsymbol{n}\right)/\hbar^2$. The SO interaction can be removed from $\hat{H}$ in Eq.(3) through a gauge transformation [58, 59] in the form of a SU(2) spin rotation:

$$\hat{S} = \exp\left(-i\left(\widehat{\boldsymbol{\sigma}} \cdot \boldsymbol{n}\right)x/L_{\rm so}\right). \tag{4}$$

After this transformation the Hamiltonian has, with the accuracy to a constant term, the form: $\hat{\tilde{H}} = \hbar^2k^2/2m^* + U(x)$, which we will use to describe spin relaxation in a disordered wire.

We describe a double QD by a quartic potential similar to that suggested in Ref.[60]

$$U(x) = U_0\left[-2\left(\frac{x}{d}\right)^2 + \left(\frac{x}{d}\right)^4\right], \tag{5}$$

where the minima at $-d$ and d are separated by a barrier of a height U_0 as shown in Fig.2. In the absence of spin-orbit coupling, the tunneling between the minima leads to the energy gap ΔE_g between the lowest spatially odd and even states. In the presence of the odd in the electron momentum SOC, the spatial parity of the eigenstates becomes only approximate, eventually resulting in the ability of the spin manipulation by electric field.

Now we include the time-dependent potential driving electron between the potential minima:

$$\widetilde{V} = -e\mathcal{E}(t)x, \tag{6}$$

where e is the electron charge, and $\mathcal{E}(t)$ is the electric field. In the spinor wavefunction

$$|\Psi\rangle \equiv \Psi(x,t) = \begin{bmatrix} \Psi_\uparrow(x,t) \\ \Psi_\downarrow(x,t) \end{bmatrix}, \tag{7}$$

each component can be expanded in a set of basis functions calculated in the absence of the spin-orbit coupling. The resulting spinor $|\Psi\rangle$ defined in Eq.(7) is, in general, related to the initial state as

$$|\Psi\rangle = \mathcal{T}\exp\left[-\frac{i}{\hbar}\int_0^t \hat{H}(t')dt'\right]|\Psi(0)\rangle, \tag{8}$$

where $\mathcal{T}$ stands for the time ordering. To calculate this evolution, we use the following approach. First, we diagonalize exactly the Hamiltonian $H_0 + H_{\rm so}$ in the basis of eigenspinors of H_0, $\psi_n(x)\,|\sigma\rangle$, taking it in a truncated form suitable for numerical calculations. After the diagonalization we obtain corresponding sets of eigenvalues $E_{\mathbf{m}}$ and of eigenstates $|\psi_{\mathbf{m}}\rangle$, where bold $\mathbf{m}$ incorporates the corresponding spin index. Then, we build in this basis the matrix of the Hamiltonian $\widetilde{V}$, that is, essentially, of the coordinate, $x_{\mathbf{m}_2\mathbf{m}_1} \equiv \langle\psi_{\mathbf{m}_2}|\,\widehat{x}\,|\psi_{\mathbf{m}_1}\rangle$ and study the full dynamics with the wavefunctions:

$$|\Psi\rangle = \sum_{\mathbf{m}} \xi_{\mathbf{m}}(t) e^{-iE_{\mathbf{m}}t/\hbar}\,|\psi_{\mathbf{m}}\rangle\,. \tag{9}$$

The evolution of a vector of coefficients $\xi_{\mathbf{m}}(t)$ is described by the system of first-order equations:

$$\frac{d}{dt}\boldsymbol{\xi}(t) = \boldsymbol{V}(t)\boldsymbol{\xi}(t), \qquad V_{\mathbf{m}_2\mathbf{m}_1} = i\frac{e}{\hbar}\mathcal{E}(t)x_{\mathbf{m}_2\mathbf{m}_1}e^{i\left(E_{\mathbf{m}_2}-E_{\mathbf{m}_1}\right)t/\hbar} \tag{10}$$

$$\boldsymbol{\xi}(t) = \mathcal{T}\exp\left[\int_0^t \boldsymbol{V}(t')dt'\right]\boldsymbol{\xi}(0). \tag{11}$$

Depending on the external perturbation, one can either use the direct calculation or resort to an improved Floquet method [34, 61].

The spin-dependence of the matrix element of coordinate responsible for the spin dynamics can be seen from expression $i(E_{\mathbf{m}_2} - E_{\mathbf{m}_1})x_{\mathbf{m}_2\mathbf{m}_1} = \hbar\,\langle\psi_{\mathbf{m}_2}|\,\widehat{v}\,|\psi_{\mathbf{m}_1}\rangle$ and the spin-dependent terms in the velocity in Eq.(2). There are two different types of $x_{\mathbf{m}_2\mathbf{m}_1}$: (i) matrix elements of the order of d due to the different parity of wavefunctions in the absence of SOC, and (ii) those proportional to the SOC strength. In the limit of weak SOC, $|\Delta - \Delta E_g| \gg \alpha|K|$, where K is the matrix element of k between the lowest-energy odd and even orbital states, one can use perturbation theory in the basis of four lowest-energy states $\psi_{\mathbf{m}}$. As a result, the SOC-determined matrix element of coordinate in the lowest spin-split doublet can be evaluated as

$$x_{\rm so} \sim 2dK\alpha\frac{\Delta}{\Delta^2 - (\Delta E_g)^2}. \tag{12}$$

We will consider below, using Eq.(8), the densities of probability and spin defined as

$$\rho(x,t) = \Psi^{\dagger}(x,t)\Psi(x,t), \quad \sigma_i(x,t) = \Psi^{\dagger}(x,t)\widehat{\sigma}_i\Psi(x,t). \tag{13}$$

With these distributions we find the gross quantities for, e.g., the right QD:

$$w_R(t) \equiv \int_0^{\infty}\rho(x,t)dx, \quad \sigma_R^i(t) \equiv \int_0^{\infty}\sigma_i(x,t)dx, \tag{14}$$

where $w_R(t)$ is the probability to find electron in the vicinity of the right minimum, and $\sigma_R^i(t)$ is the corresponding expectation value of the spin component. The full mean values for the spin components and the coordinate are defined correspondingly as

$$\langle\sigma_i(t)\rangle \equiv \int_{-\infty}^{\infty}\sigma_i(x,t)dx \equiv \mathrm{tr}\left(\widehat{\boldsymbol{\rho}}\widehat{\sigma}_i\right), \qquad \langle x(t)\rangle = \int_{-\infty}^{\infty}x\rho(x,t)dx, \tag{15}$$

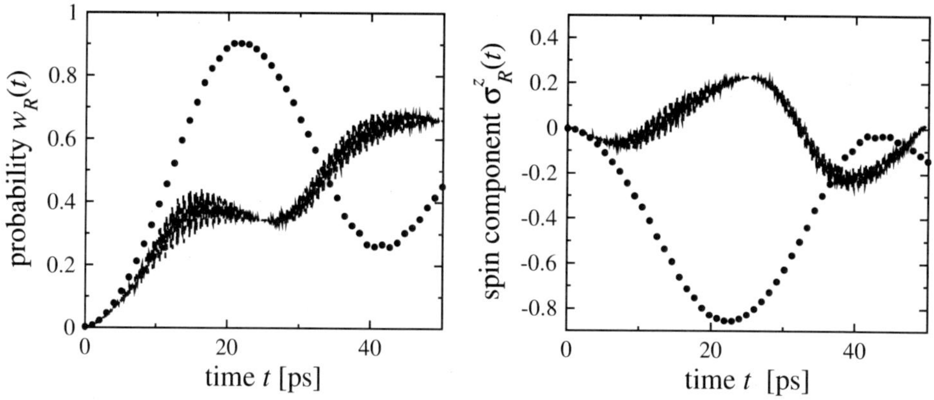

Figure 3. Evolution of the probabilities $w_R(t)$ (left panel) and $\sigma^z_R(t)$ (right panel) under the electric field pulse with $f = 1/8$ (filled circles) and $f = 1/2$ (rapidly oscillating) for the initial spin-down state. The pulse duration is $T = 25$ ps. The increase in the pulse field makes the motion less regular, and more high-energy components contribute into the dynamics, producing partial spin flip. We consider a structure with $d = 25\sqrt{2}$ nm and $U_0 = 10$ meV, see in Fig.2. The two lowest spin-degenerate energy levels are $E_1 = 3.938$ meV, $E_2 = 4.030$ meV counted from the potential minimum. The splitting $\Delta E_g = 0.092$ meV. The parameters of the SO coupling are $\alpha_R = 10^{-6}$ meVcm and $\alpha_D = 0.3\alpha_R$. The magnetic field is $B = 2$ T, producing the Zeeman spin splitting $\Delta = -0.054$ meV, comparable with ΔE_g, in GaAs with $g = -0.45$.

where $\sigma_i(x,t) = \Psi^\dagger(x,t)\widehat{\sigma}_i\Psi(x,t)$ is the spin density in Eq.(13) and $\widehat{\boldsymbol{\rho}}$ is the reduced spin density matrix, that is the full density matrix traced over the electron coordinate. It is important to mention that in the presence of spin-orbit coupling the reduced spin density matrix describes a mixed rather than a pure state. As a result, $1/2 \leq \mathrm{tr}\left(\widehat{\boldsymbol{\rho}}^2\right) \leq 1$, and the electron spin components satisfying condition $\sum_i \langle\sigma_i(t)\rangle^2 = 2\mathrm{tr}\left(\widehat{\boldsymbol{\rho}}^2\right) - 1$ form a Bloch ball of the radius equal to one rather than a Bloch sphere. The mixed character of the state has a strong influence on the entire spin dynamics, as we will see in what follows.

The effect of tunneling on electron spin is described by the spin precession angle defined for $\alpha_R = 0$ and a relatively weak Dresselhaus coupling as $\phi \approx \sigma^y_R(T_{\rm tun}/2)$. This value in Eq.(14) should be compared with the precession angle for the classical motion, where displacement of electron by the distance L leads to the precession angle $\phi_{\rm cl}(l) = 2L/L_{\rm so}$ for $L = 2d$, and is considerably different. Due to the enhanced role of the tunneling the precession angle becomes considerably smaller.

To consider the external perturbation we begin with a half-period electric field pulse $\mathcal{E}(t) = \mathcal{E}_0 \sin(\pi t/T)$ for $0 < t < T$, where $T = T_{\rm tun}/2$. In the case of our interest, where $\Delta \sim \Delta E_g$, the spectral width of the pulse, $\sim \hbar/T_{\rm tun}$, covers the energy scales of spin- and tunneling- splitting of the ground state, thus, driving the spin and the orbital dynamics simultaneously. The field strength is characterized by the dimensionless parameter f such that

$$|e|\,\mathcal{E}_0 \equiv f\,\frac{U_0}{2d}. \tag{16}$$

Three driving regimes are possible: (i) a relatively weak field, $f \ll 1$, where the shape of the quartic potential remains almost intact, and the tunneling is still crucially important, (ii) intermediate field, $f \leq 1$, and (iii) the strong field, $f > 1$, when the perturbation $-e\mathcal{E}(t)x$ considerably modifies the potential and the low-energy part of the spectrum. Here we will concentrate only on the first two regimes, where the electric field already causes qualitative changes in the spin dynamics.

An example of the observables for the pulses with different $\mathcal{E}_0$ is shown in Fig.3 for the initial states where a spin-down electron with $\langle\sigma_z(0)\rangle = -1$ is initially localized in the vicinity of the left, $-d$, minimum. For the initial spin-up state, the plots (not shown) look similar but demonstrate a more regular behavior. The difference between these two regimes of initial conditions is not trivial and can be understood as follows. The coupled spin-charge dynamics is determined by the difference in the energy of the nearest spin up and spin down states with opposite spatial parity, which can be coupled by the external electric field. The larger is the distance between the levels, the less efficient is the SOC for the coupled dynamics. For the low-energy spin-up state the distance is $|E_4 - E_1| = \Delta E_g + \Delta$, while for the initial spin-down state it is $|E_3 - E_2| = \Delta E_g - \Delta$. For $B = 2$ T, $(\Delta E_g - \Delta) / (\Delta E_g + \Delta) \approx 0.25$, making the spin flip for the latter case faster and the spin dynamics richer.

2.3. Periodic Electric Field: Strong Field Leads to Slow Rabi Spin Dynamics

In this subsection we consider spin dynamics under a perturbation [62] by resonant electric field at $t > 0$:

$$\mathcal{E}(t) = \mathcal{E}_0 \sin(\widetilde{\omega}_Z t). \tag{17}$$

Here $\widetilde{\omega}_Z$ is the exact frequency of the spin-flip transition. Although it includes the effects of SOC, it is usually very close to $|\Delta|$. The dimensionless amplitude (16) of the electric field $f = 1$ corresponds to the field of approximately 1.5×10^3 V/cm, similar to the values in the experiment of Ref.[21]. We will see how the spin precession during the tunneling and the mixed character of the spin state become crucially important for the spin dynamics. Even for a basic quantum system such as a single electron spin, the efficiency and the time scale of the manipulation strongly depend on the electron orbital motion and, as result, an unexpected dependence on the external electric field can be seen. An interesting example of such a dependence was recently presented for a promising for quantum information system such as nitrogen-vacancy pair in diamond [63]. It is the nonlinearity in the spin and charge dynamics which is expected to lead to unusual consequences on the driven spin behavior [64].

Under the resonance condition, behavior of the driven by electric field $\mathcal{E}(t)$ (17) double quantum dot can be approximately characterized as a superposition of two types of transitions. The first type is the resonant "spin-flip" transitions with the matrix element of coordinate determined by the SOC in Eq.(12). The other type is the off-resonant "spin-conserving" transitions with much larger matrix element [62] of the order of $e\mathcal{E}d$. Both types of transitions are crucially important for the understanding of the spin dynamics. With the estimate $K \approx m^*\Delta E_g d/\hbar^2$, we obtain $d \approx 10x_{\text{so}}$, as can be seen in Eq.(12). As a result, the off-resonant transitions are not weak compared to the required spin-flip ones and the entire dynamics involves more than one possible transition. We mention here that a

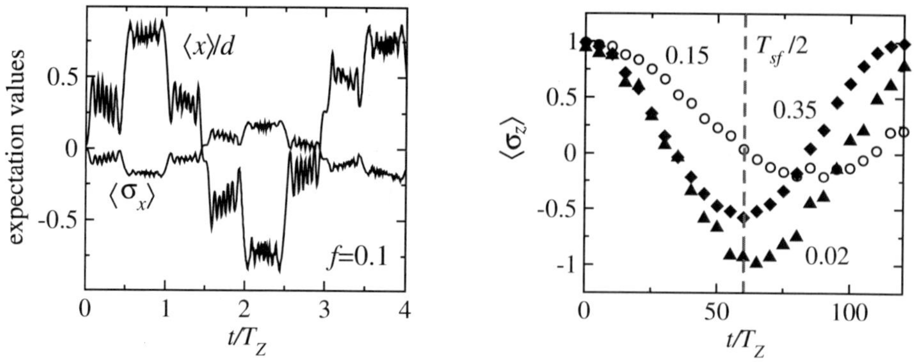

Figure 4. Left panel: Short-term expectation values of coordinate $\langle x\rangle/d$ and spin $\langle\sigma_x\rangle$, as marked near the plots, induced by external field with $f = 0.1$, B=1.73 T, T_Z=88 ps. Right panel: Stroboscopic time dependence of $\langle\sigma\rangle$ for different external driving fields f marked near the plots for $B = 6.93$ T with $\Delta = 2\Delta E_g$, vertical dashed line marks the operational definition of the spin-flip period. The structure is the same as in Fig.3.

similar state-leakage effect takes place in a completely different system such as superconducting qubits [65].

We begin with the short-time dynamics of coordinate $\langle x\rangle/d$ and spin $\langle\sigma_x\rangle$ for a moderate driving $f = 0.1$ (Fig.4 (left panel)) obtained by numerical integration of Eq.(8), for the initial state of the electron in the ground state of double QD. The other expectation value $\langle\sigma_z\rangle$ changes much slower, approximately as $\langle\sigma_x\rangle^2$ and will be treated later at a long timescale, which is the primary topic of our interest. It can be seen in the left panel of Fig.4 that fast oscillations accompany local variations of observables. Considerable changes such as Rabi oscillations of spin can be achieved only after many periods of the driving field.

To consider the long-term behavior we apply the Floquet theory [18, 34, 41, 61, 66] in the stroboscopic form [67]. In this approach, as a first step, the single-period propagator matrix $\mathbf{U}(T_Z)$ defined as

$$\mathbf{U}(T_Z) \equiv \mathcal{T}\exp\left[-\frac{i}{\hbar}\int_0^{T_Z} H(t')dt'\right] \tag{18}$$

is obtained by a high-precision numerical integration of the system (10) at one period of the driving $T_Z = 2\pi/\widetilde{\omega}_Z$. As a result, this propagator and stroboscopic evolution operator for integer N can be exactly presented as:

$$\mathbf{U}(T_Z) \equiv \sum_Q \boldsymbol{A}^Q\left(\boldsymbol{A}^Q\right)^\dagger e^{-iE_QT_Z/\hbar},$$
$$\mathcal{T}\exp\left[-\frac{i}{\hbar}\int_0^{NT_Z} H(t)dt\right] = \sum_Q \boldsymbol{A}^Q\left(\boldsymbol{A}^Q\right)^\dagger e^{-iE_QNT_Z/\hbar}, \tag{19}$$

where index Q runs over the full set of states described by quasienergies E_Q and eigenvectors $\boldsymbol{A}^Q$. Equation (19) provides a very accurate description since the N-period propagator is constructed explicitly if $\mathbf{U}(T_Z)$ is known.

Slowly evolving phenomena such as chaos development [67] and spin Rabi oscillations occur on the time scale $N \gg 1$, making the stroboscopic approach fully justified. In fact, even the experiment [21] uses stroboscopic approach with the intervals of the order of 100 ns to measure the slow dynamics of the driven electron spin.

The long-term stroboscopic spin dynamics for the z spin component is presented in Fig.4 (right panel) for different electric fields. The same Figure shows operational definition of the "spin-flip" time T_{sf}: The spin flip occurs when spin component shows a broad minimum accompanied by fast small oscillations, not shown there. There are two novel qualitative features of spin dynamics which can be extracted from Fig.4. They are different from the model of Ref.[21], where electron is always located near the minimum of the potential formed by the parabolic confinement, shifted periodically by a weak electric field. The differences are:

1) The spin-flip time dependence on the driving field f is nonmonotonic, as can be seen in Fig.5, in contrast to the $1/f$ behavior expected for the conventional Rabi oscillations.

2) In contrast to simple Rabi oscillations, the flips become incomplete, with $\langle \sigma_z \rangle = -1$ never reached.

These novel effects come from the enhanced interminima tunneling: the spin precession in the driven motion establishes corresponding spin dynamics and prevents the electric field from flipping the spin efficiently. This tunneling-induced spin precession results in producing a mixed rather than a pure spin state and in qualitative changes in the spin dynamics. In contrast to the linear dependence for a conventional two-level resonance Rabi formula, one can see a strongly different non-monotonous dependence in a multi-level structure, especially at high fields, where the Rabi frequency becomes much less than expected from the two-level model. A similar effect of an unexpectedly small Rabi frequency was recently observed [55] in driven by electric field double GaAs quantum dots, where the spin resonance occurs due to a coordinate-dependent hyperfine interaction. Although the origin of the resonance in [55] is different from the spin-orbit coupling we consider here, the spin-orbit interaction in GaAs establishes spin dynamics due to the interminima motion, thus, preventing fast spin flips. In addition, we mention that the regime in Fig.5 (left panel) shows more irregularities since at this choice of magnetic field all four lowest levels are equidistant and involved in the resonance while in Fig.5 (right panel) a more regular dependence is observed, reflecting a simpler structure of the resonances.

2.4. Measurement Theory, Strong Coupling, and Zeno Effect

As we have demonstrated, sometimes even a strong driving is not efficient for spin flips. Now we understand that the observed slowing down of spin dynamics can be seen on a more general ground, not restricted to the exact form of the equations of motion, as the Zeno effect (formulated in simple words as "a constantly watched pot never boils") of the frozen evolution of a measured quantity [68, 69, 70]. Indeed, the spin-orbit coupling operator $-i\widehat{\sigma}_i\partial/\partial x$ makes the orbital dynamics spin-dependent, and, as a result, performs the measurement of the σ_i component [71] in the sense of von Neumann procedure. This can be seen in the evolution of a two-component wave function:

$$e^{-\alpha t\widehat{\sigma}_z\partial/\partial x}\phi(x)\left(\zeta_1\left|1\right\rangle+\zeta_{-1}\left|-1\right\rangle\right)=\phi(x-\alpha t)\zeta_1\left|1\right\rangle+\phi(x+\alpha t)\zeta_-\left|-1\right\rangle, \qquad (20)$$

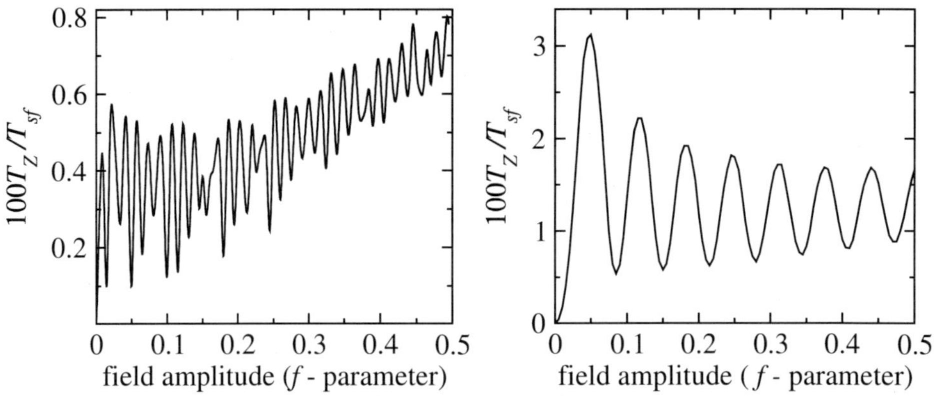

Figure 5. Nonlinear dependence of the Rabi spin-flip frequency on the electric field amplitude in multilevel system for two different magnetic fields. Left panel: $B = 1.73$ T, $\Delta = \Delta E_g/2$, the characteristic spin-flip rate of the order of 50 MHz. Right panel: $B = 6.93$ T, $\Delta = 2\Delta E_g$, the characteristic spin-flip rate is of the order of 400 MHz.

where we took $i = z$ as an example, and ζ_1 and ζ_{-1} correspond to ± 1 eigenvalues of $\widehat{\sigma}_z$, respectively. The SOC thus destroys the initial coherent superposition of spin-up and spin-down states and performs the von Neumann-like spin measurement by mapping the spin state on the electron position. In other words, to measure the spin state, we look at the particle coordinate, and associate the position with the spin [71]. The spin-orbit coupling forms a coherent superposition of spin components and at the same time, making a permanent measurement, destroys it leading to a slow spin dynamics. This quantum von Neumann measurement of the spin is, however, different from the experimental procedure applied, e.g. in Ref.[21], where the spin of electron in a quantum dot is measured by the probability of passing of another, spin-polarized, electron, determined by the Pauli blockade, dependent on the spin orientation.

Bearing in mind the Zeno effect in the measurement theory one can expect that with the increase in the spin-orbit coupling the spin manipulation by an external electric field can be less efficient. Indeed, in Ref. [72] a non-monotonous dependence of the Rabi frequency on the SOC strength has been obtained for a quantum wire-based quantum dot with a parabolic confinement. It was found for a driving by weak electric field, where the spin response is still linear, that with the increase in the coupling strength the Rabi frequency reaches a maximum and then decreases.

2.5. Ballistic Transport through a Long Wire: Gate-Driven Spin Resonance

As we have seen in previous subsections, the spatial dynamics of electron localized in a double quantum dot can strongly modify the Rabi frequency of the electric dipole spin resonance [30, 62]. Now one can ask: if we have an extended system such as an open quantum wire, where electrons are not localized, what are the signatures, if any, of the EDSR in the electron transport through such a system? Here we study ballistic electron transport in a single-channel quantum wire subject to dc and ac bias produced by a finger gate and to a Zeeman magnetic field and demonstrate the role of the EDSR in the conductance of such

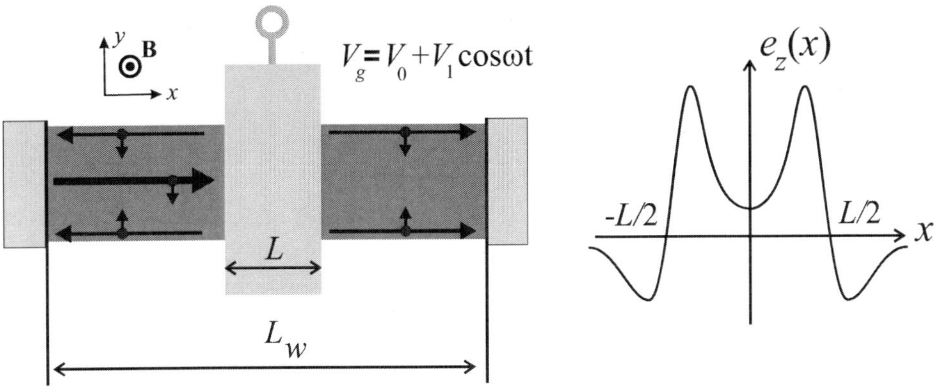

Figure 6. Left panel: View of a conducting channel subject to a dc potential of the quantum point contact gates (not shown here), and dc and ac potential of a finger gate of the width L. The inhomogeneous electric field created by the finger gate produces the scattering region. The distance between pairs of quantum point contacts is L_w. The spin-related scattering process is described by long horizontal arrows. Short vertical arrows show the directions of spins in the scattering/transmission process. Right panel: Schematic plot of coordinate-dependent z-component of electric field. Fast changes in $e_z(x)$ happen at $x \approx \pm L/2$ on the spatial scale of the distance between the finger gate and the conducting channel. The details of the Figure are not to scale.

a system [73]. We found that the spin flips induced by the periodic field lead to avoided crossings in the dependence of the conductance on the electron energy and the field frequency, or equivalently, on the Zeeman splitting. An ac biased finger gate contributes to the time-periodic Rashba spin-orbit coupling which may give rise to many interesting effects such as generation of spin current [74, 75] and spin-polarized wavepackets [76].

We take the geometry presented in Fig.6, where the electric field of the finger gate gives rise to a local Rashba coupling. Under the assumption of a stepwise Rashba coupling the electron ballistic transport in a quasi one-dimensional wire has undergone a thorough investigation [77, 78, 79, 80, 81, 82, 83, 84, 85]. Here we consider a realistic spatial dependence of the electric field of the biased gate. Some aspects of electron transport through the wires with spin-orbit interaction and subject to a time-periodic potential were studied in Refs.[86, 87].

The hybrid metal-semiconductor structure of our interest in the absence of the gate bias is symmetric with respect to the $z \to -z$ reflection. Then the conduction channel has only the Dresselhaus SOC [1, 88]. Next, similar to Ref.[89], we consider a metallic gate at height h from the conducting channel whose length L_w along the transport x-axis exceeds the gate width L, such that their ratio $L_w/L \geq 1$. The full coordinate dependence of the electrostatic potential of a biased metallic finger gate was found by Davies et al. [90]. For the structure in Fig.6, the gate potential has the form

$$\begin{aligned} V(x,z,t) &= (V_0 + V_1 \cos\omega t)\phi(x,z), \\ \phi(x,z) &= \frac{1}{\pi}\left[\arctan\frac{L/2+x}{z} + \arctan\frac{L/2-x}{z}\right], \end{aligned} \tag{21}$$

where the gate is chosen as the origin of z, and, therefore we take $z = -h$ in our calculation. The gate and lateral confining potentials form a Rashba coupling, which is nonuniform over x [91, 92]:

$$H_{\rm so} = -i\xi_R \left[(\mathbf{E} \cdot (\widehat{\boldsymbol{\sigma}} \times \boldsymbol{\nabla})) - \frac{1}{2} (\widehat{\boldsymbol{\sigma}} \cdot (\boldsymbol{\nabla} \times \mathbf{E})) \right], \tag{22}$$

where $\mathbf{E}(x,y,z,t) = -\boldsymbol{\nabla}(V(x,z,t)+V_{\rm LC}(y))$, and ξ_R is a material-dependent constant. Here $V_{\rm LC}(y)$ is the lateral confining potential [93], and the last contribution in Eq.(22) ensures the hermiticity of the Rashba coupling. The y-axis confinement is typically tens to hundreds of nanometers. This small length establishes an energy gap suppressing excitations of the transverse modes. For this reason we focus solely on the x-axis transmission and consider the ground state $\psi_0(y,z)$ well-localized on the scale much less than the wire length of the order of hundreds of nanometers. In addition, we assume that the magnetic length $\sqrt{\hbar c/|e|B}$ corresponding to the applied field B is much larger than the channel width and neglect the influence of the magnetic field on the orbital motion.

The effective 1D Hamiltonian now has the form [73]:

$$H_{1D} = \frac{\hbar^2 k^2}{2m^*} + \widetilde{U}_0(x,t) + \frac{1}{2}\{\alpha(x,t), k\}\widehat{\sigma}_y + \frac{\Delta}{2}\widehat{\sigma}_z, \tag{23}$$

where $\widetilde{U}_0(x,t) = e(V_0 + V_1\cos\omega t)\phi(x,-h)$. The shape of coordinate dependence of $\alpha(x,t) \sim e_z(x)$ is given by the coordinate dependence of the $z-$component of the electric field $e_z(x)$ presented in dimensionless units at $L \equiv 1$ as

$$e_z(x) = \frac{1}{\pi}\left[\frac{1+2x}{h^2+(1+2x)^2} + \frac{1-2x}{h^2+(1-2x)^2}\right], \tag{24}$$

as follows from the dc and ac gate potentials (21). The spin-orbit coupling term for coordinate-dependent $\alpha(x,t)$ is taken as anticommutator $\{\alpha(x,t),k\}\widehat{\sigma}_y$ to ensure the Hermitian Hamiltonian. Although the anticommutator of the momentum and coordinate-dependent spin-orbit coupling [31, 94] in Eq.(23) can be obtained directly with Eq.(22), it is often adopted by the application of the Dirac symmetrization rule for a product of noncommuting operators A and B as $AB \to \{A,B\}/2$. In addition, one can see that the Rashba coupling caused by the lateral confinement is excluded because the expectation value of the lateral components of the electric field vanishes.

Before proceeding with exact analysis of driven spin-dependent transmission, we address qualitatively an important question regarding whether a spin polarization defined as

$$P = \frac{G_{\uparrow\uparrow} + G_{\uparrow\downarrow} - G_{\downarrow\uparrow} - G_{\downarrow\downarrow}}{G_{\uparrow\uparrow} + G_{\uparrow\downarrow} + G_{\downarrow\uparrow} + G_{\downarrow\downarrow}}, \tag{25}$$

where $G_{\sigma\sigma'}$ is the corresponding component of the conductance, can appear here in the presence of the time-reversal symmetry at $\Delta = 0$. For stationary single-channel transmission the Rashba coupling cannot produce a spin polarization [95, 96]. However, the ac time-periodic Rashba coupling (24) opens additional spin dependent channels of electron transmission at the Floquet quasienergies $\varepsilon + \hbar m\omega$, $m = 0, \pm 1, \ldots$ Numerical calculations in Refs. [75, 94] show a nonzero spin polarization in the case of the step-wise time-periodic Rashba coupling. We argue that there is no spin polarization for a smooth, not a step-wise, coordinate behavior of the time-periodic Rashba coupling, at least, for zero Dresselhaus SOC. Indeed, for that case the only spin component in the Rashba coupling is $\sigma_y = \sigma$, which is an integral of motion. The analysis of the Hamiltonian (23) shows that the electron transmission with spin $\sigma = 1(\uparrow)$ is not mixed with the transmission with $\sigma = -1(\downarrow)$. It follows from Hamiltonian (23) that the time-resolved transmissions of opposite spin states are only shifted by half of the time period π/ω. Therefore, after time averaging, the spin-conserving conductances $G_{\sigma\sigma}$ do not depend on σ while there is no spin-flip conductance $G_{\sigma,-\sigma}$. This simple consideration shows that the spin polarization (25) vanishes. We show below that numerical computations agree with this consideration for a smooth coordinate dependence of the time-periodic Rashba coupling.

The procedure of calculating the transmission through the time-periodic potential is described in the literature [18, 97, 98, 99, 100, 101]. There are two time-dependent contributions in our system. The first one is the periodic oscillations in the gate potential, as considered in many publications for a spatially stepwise time-periodic potential [97, 98, 99, 101]. The second one is the oscillating Rashba coupling, considered for the case of stepwise spatial dependence in Refs.[75, 94].

To reduce the calculation of the conductance through the space- and time-dependent potential [18, 102, 103, 104] to the set of finite-order differential equations we use the tight-binding model on a chain. For the effective lattice constant l and intersite hopping t_{is} the kinetic energy is $E(k) = 2t_{\text{is}}(1 - \cos(kl))$, where k is the electron wavevector. The distance between two nearest sites should be chosen as $l = \hbar/\sqrt{2m^* t_{\text{is}}}$ to satisfy the value of effective mass in the compound of interest. If there is no SOC in the leads, the wave functions can be written as [101, 105, 106, 107],

$$\begin{aligned}
\text{incoming:}\quad \psi_{n\sigma}(t) &= \sum_{m\sigma'} \frac{e^{-i(\varepsilon/\hbar + m\omega)t}}{\sqrt{2\pi\rho(k_m)}}\left[\delta_{m,0}\delta_{\sigma\sigma'} e^{ik_0 ln} + r_{m\sigma\sigma'} e^{-ik_m ln}\right],\\
\text{outgoing:}\quad \psi_{n\sigma}(t) &= \sum_{m\sigma'} \frac{e^{-i(\varepsilon/\hbar + m\omega)t}}{\sqrt{2\pi\rho(k_m)}} t_{m\sigma\sigma'} e^{ik_m ln},
\end{aligned} \tag{26}$$

where

$$\varepsilon + \hbar m\omega = -2t_{\text{is}}\cos(k_m l), \qquad \rho(k_m) = \partial\varepsilon/\partial k_m, \tag{27}$$

and n enumerates the lattice sites.

The electron enters with the energy ε and spin σ and reflects and transmits with the energy $\varepsilon + \hbar m\omega$ and spin σ' with corresponding amplitudes $r_{m,\sigma\sigma'}$ and $t_{m,\sigma\sigma'}$, respectively. The spin-related conductance is

$$G_{\sigma\sigma'} = \frac{e^2}{2\pi\hbar}\sum_m \frac{\sin \text{Re}(k_m l)}{\sin k_0 l}|t_{m\sigma\sigma'}|^2. \tag{28}$$

This expression reduces to the standard formula in the continuum approximation [101]. Taking the real part of k_m in the nominator of Eq.(28) assures that the Floquet states with quasienergies $\varepsilon + \hbar m\omega$ beyond the propagation band having an imaginary wave vector k_m cannot participate in the conductance.

In the absence of magnetic field and ac driving, the resonance peaks in the transmission follow the eigenenergies of one-dimensional electron localized in a box with the Rashba

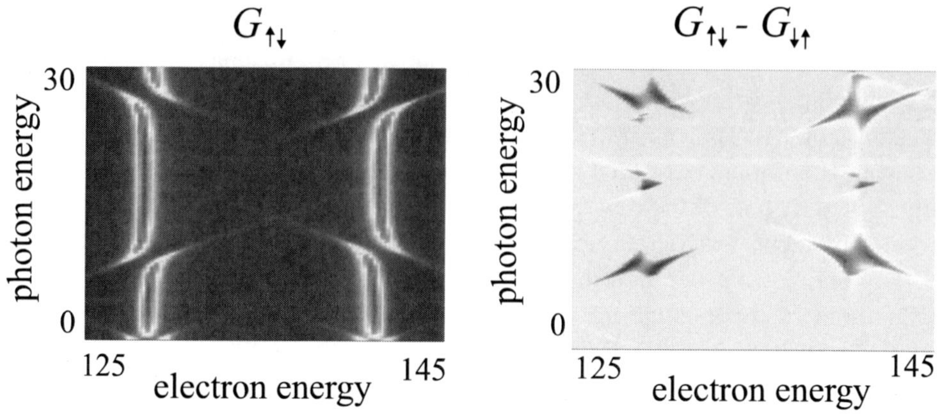

Figure 7. Left panel: Conductance $G_{\uparrow\downarrow}$ vs incident energy and frequency of the ac potential. The finger gate length $L = 100$ nm. Here we use L as the unit of length and $\hbar^2/2m^*L^2$ as the unit of energy. In these units we take parameters $eV_0 = 1$, $eV_1 = 0.25$, and $\widetilde{\alpha} = 1$. To compare with material parameters: $\widetilde{\alpha} \approx 0.7$ and $\widetilde{\alpha} \approx 3.0$ for InAs and InSb, respectively. The ratio $L_w/L = 4$ for L_w is shown in Fig.6, and the Zeeman splitting is $5\hbar^2/m^*L^2$. The picture clearly shows avoided crossings close to the resonances. The scale of the picture corresponds to maximum $G_{\uparrow\downarrow} \approx 0.25e^2/2\pi\hbar$. Right panel: the difference $G_{\uparrow\downarrow} - G_{\downarrow\uparrow}$, enhanced near the resonances. The maximum value of $|G_{\uparrow\downarrow} - G_{\downarrow\uparrow}|$ is of the order of $0.01e^2/2\pi\hbar$. To simulate the quantum point contacts we use the hopping matrix element $t = 0.5$ between the leads and the wire. For experimental realization the fields can be chosen in the range $B \sim 10$ T, $V_0 \sim 10^4$ V/cm, and $\omega/2\pi \sim 100$ GHz [73].

spin-orbit coupling. In full agreement with the rigorous results of Refs. [95, 96], numerically calculated spin polarization defined in Eq.(25) vanishes as expected from the time-reversal symmetry for 1D single-channel transmission.

Now we come to the full physical picture including driving and magnetic field. To be specific, we consider typical Rashba parameters, effective masses, and g-factors for the InAs- and InSb-based heterostructures [92, 108]. For these heavy compounds the Dresselhaus coupling is usually weak compared to the Rashba contribution, and we neglect it. Figure 7 shows the electron conductance of the quantum wire subject to a finger gate that is dc and ac biased for the incident energies substantially exceeding the effective potential barrier height. Switching off the ac potential, $V_1 = 0$, one obtains the resonance transmission for $\varepsilon \approx \varepsilon_n$ where ε_n are the eigenenergies of the closed wire under applied dc potential. The resonance positions, however, are slightly shifted because of the openness of the 1D wire. Then the application of the AC potential gives rise to the quasi energies $\varepsilon_n \pm \sqrt{2V_1^2 + \hbar^2\omega^2}$ [104]. The coincidence of these quasienergies with the basic eigenenergies $\varepsilon_{n'}$ results in avoided crossing.

Switching on the periodic Rashba coupling in this system gives rise to the new selection rules of the avoided crossing of the Floquet states. The time-periodic term $V_1 \{\widetilde{\alpha}(x), k\} \cos(\omega t)/2$ in the Rashba Hamiltonian (23) is odd with respect to the x-inversion. Therefore, it mixes the neighboring eigenstates with the opposite parity of the closed wire while the time-periodic potential $V_1\phi(x) \cos\omega t$ term mixes eigenstates with the same parity. As a result, we obtain the avoided crossings of nearest neighbor resonances shown in Fig.7 (left panel). With the growth of the potential amplitude the avoided crossings occur irrespective of the selection rules that give rise to a more complicated frequency-dependent behavior of the conductance.

Now we comment on the possibility of generating spin polarization in a single-channel transmission by the ac field. As it was argued above, the time-periodic Rashba SOC cannot lead to spin polarization for the transmission through a 1D wire at $B = 0$ even with frequency-shifted Floquet channels contributing into the transport. This general statement agrees with our calculations but disagrees with the numerical results of Ref. [94] where a nonzero spin polarization was found for zero Dresselhaus SOC. The difference is due to the coordinate dependence of the gate field. Numerical calculations show a decrease in the spin polarization with the decrease in the simulation lattice constant l. For a sufficiently small l the polarization vanishes. Thus, the stepwise approximation of a nonuniform Rashba coupling in Ref. [94] conceals a danger for numerical computations based on the finite difference schemes.

3. The Effects of Disorder

3.1. What to Expect from Disorder in One Dimension for Spin Dynamics ?

Finite spin polarization, once produced in a system with spin-orbit coupling, will not exist there forever. All kinds of randomness such as lattice defects, phonons and electron-electron collisions lead to spin relaxation since they put a moving electron in a randomly time-dependent environment. As a result, electron momentum becomes randomized, and, via spin-orbit coupling spin appears in a random in time field, and the initially prepared spin of a macroscopic system vanishes. This spin relaxation is undesirable for any applications in spintronics except for the cases when one needs to erase the spin-based memory. As a result, the quest for systems with ultralong spin relaxation times became the main problem for many spintronics researchers.

The most celebrated mechanism of spin relaxation in semiconductors is the Dyakonov-Perel‘ one, presenting essentially the diffusion in the spin subspace. It works excellently in systems where the momentum relaxation time is much less than the spin precession time, and surprisingly, the spin relaxation can slow down in the presence of external noise [109, 110]. Nowadays, clean systems where spin dephasing is due to a reversible momentum-dependent spin precession in the spin-orbit coupling field and has a large initial non-exponential time dependence, become available.

A completely opposite example is given by quantum dots, where electrons are fully localized, and, therefore, the Dyakonov-Perel’ mechanism disappears. The characteristic timescale of spin relaxation of electrons localized in quantum dots is expected to be several orders of magnitude longer than that of itinerant electrons. As a result, the spin relaxation through SO interaction requires phonon-induced coupling of different orbital states of the localized electron and a nonzero external magnetic field [44].

The quest for long spin relaxation time called for new techniques to study slow spin dynamics in systems close to the equilibrium. Near-equilibrium spin fluctuations [111] have correlations on the spin relaxation timescale. They can be Fourier-transformed to the frequency domain and characterized there by the spectral power, strongly related to the timescale of intrinsic spin fluctuations. This idea is in the background of the recently developed highly accurate low-frequency spin noise spectroscopy [112] aimed at the measurement of intrinsic equilibrium spin dynamics. Recent advances in the theoretical background of this method are given in Refs. [113, 114, 115, 116, 117]. Very recently, new techniques such as two-beam spectroscopy have been proposed [117] and it was shown that spin Hall effect can be seen in the spin noise experiments [118].

The spin noise spectroscopy experiments are based on the optical monitoring of spin fluctuations [119] in Faraday, Kerr or ellipticity signals measured with a weak linearly polarized probe beam incident on a sample. For the probe tuned with the frequency close to the absorption threshold [112, 115], the Kerr rotation angle $\kappa \propto \sigma_z$ [120]. Hence, its autocorrelation function is directly related to the spin noise: $\langle \kappa(t)\kappa(t')\rangle \propto \langle \sigma_z(t)\sigma_z(t')\rangle$, where $\sigma_z(t)$ is the density of the $z-$component of the total electron spin. As a result, this optical technique measures long-range correlations of equilibrium spin fluctuations in the laser spot. As an example we mention that the spin noise spectroscopy allows experimentalists to study slow spin dynamics in (110)-grown GaAs quantum wells [121] and in quantum dots [122].

Even if the spin-orbit coupling is not present in the system, another spin relaxation mechanism is almost always there. It is related to the hyperfine coupling of electron spin to the lattice nuclei [17, 123, 124, 125, 126, 127, 128]. Every nucleus having a nonzero total magnetic moment behaves like a tiny magnet with a point-like interaction with the surrounding electrons. The time evolution of initial spin states produced in the ensembles of quantum dots depends crucially on their sizes and becomes slower with the increase in the size of the electron cloud. This parameter is out of control since it follows statistical distributions related to the sample growth procedure. The spin dephasing times observed in the quantum dots [124] agree very well with those expected from theory [17]. In general, the mechanisms of spin dephasing in quantum dots due to hyperfine interaction and in clean electron gases are not very much different. Indeed, in a quantum dot the spin precession rate and the precession axis do not change in time on the timescale of spin dephasing. In the clean electron gas the rate and direction are determined by a constant in time electron momentum. The result averaged over the ensemble, whether of quantum dots or of electrons with different momenta, yields a reversible total spin dephasing [129, 130].

Quantum wires, being in between the two-dimensional electrons and quantum dots, demonstrate a width-dependent peculiar and slow spin dynamics [131, 132, 133], where, e.g., InAs, InSb as well as GaAs/AlGaAs systems are the prospective realizations for the spin transport applications. The effects of SO coupling on the transport were clearly demonstrated there [46, 77] and the nanowire based qubits were introduced [48, 134]. Extended nanowires are the natural channels for traveling, "flying" qubits [135, 136]. Recently Huang and Hu [141] proposed a formulation of the spin relaxation problem well suited for the quantum information purposes by studying the decoherence of electron spin enclosed in a quantum dot, which is transported in a random electric potential.

Although limits of free and strongly localized electrons are well understood and some aspects of the interplay of disorder-induced localization and spin relaxation have been addressed [137, 138, 139, 140], many open questions remain. One of the most peculiar features of one-dimensional systems is the strong role of disorder since an electron cannot

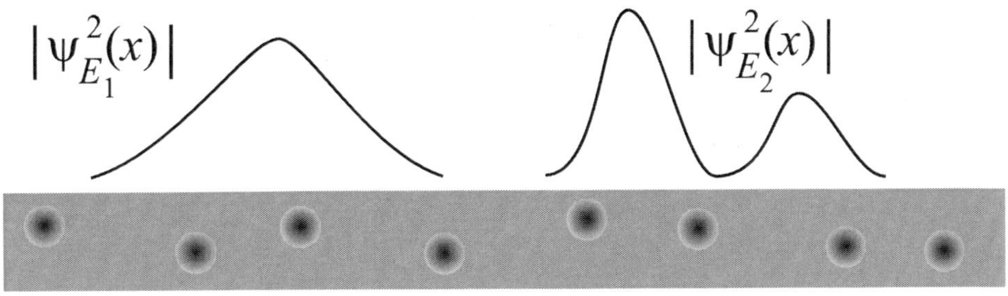

Figure 8. Semiconductor nanowire with random impurities shown as filled circles and a schematic plot of the electron density for low-energy localized states. Although we consider a strictly one-dimensional electron motion, the potential produced by impurities can be randomly distributed over the cross-section of the wire and averaged over quantized transverse motion to study the one-dimensional distributions. The Figure shows schematically the densities $|\psi^2_{E_1}(x)|$ and $|\psi^2_{E_2}(x)|$ for two states with $E_2 > E_1$ in a low-energy part of the spectrum.

avoid collisions with impurities. As a result, in an infinitely long system, all electron states become localized on the spatial scale determined by the disorder strength. This localization leads to important consequences for spin dynamics, bringing some aspects of it closer to ensembles of quantum dots than to gases of itinerant electrons. First, localization makes electron momentum a bad quantum number, and therefore, removes the Dyakonov-Perel' spin relaxation but does not prohibit the entire spin dynamics. As a nontrivial example of this interplay we mention that even the weak localization of two-dimensional electrons leads to a slow power-like $\sim t^{-1}$ [129] rather than exponential spin relaxation. Second, it strongly enhances the role of the hyperfine interaction for the spin dephasing. We will describe these two novel effects. In addition, we will discuss the role of disorder in the spin-orbit coupling.

3.2. Spin Freezing by Localization

A one-dimensional system of our interest provides on the one hand, the basic example of localization physics in a random potential [142, 143] and, on the other hand, is an example of a system, where spin-orbit coupling can be gauged away by a SU(2) transformation. Here we consider a system described by the Hamiltonian:

$$H = \frac{\hbar^2 k^2}{2m^*} + U_r(x) + \alpha\left(\widehat{\boldsymbol{\sigma}} \cdot \boldsymbol{n}\right) k, \tag{29}$$

where the potential $U_r(x)$ is a random function of coordinate, which can be characterized by correlation function $\langle U_r(x) U_r(x') \rangle = \langle U_r^2 \rangle F_{UU}(x - x')$, with $F_{UU}(x - x')$ being the corresponding range function. The randomness of $U_r(x)$ results in the electron localization (see upper panel of Fig.8) and spin relaxation.

Since in the absence of magnetic field, $(\widehat{\boldsymbol{\sigma}} \cdot \boldsymbol{n})$ commutes with the Hamiltonian and the corresponding observable is conserved, the spin density component along the $\boldsymbol{n}$-axis is time independent. It is useful here to gauge out the spin-orbit coupling by SU(2) spin rotation $\hat{S} = \exp\left(-i\left(\widehat{\boldsymbol{\sigma}} \cdot \boldsymbol{n}\right) x / L_{\text{so}}\right)$ in Eq.(4). A nontrivial dynamics of the transformed spin occurs for the perpendicular to $\boldsymbol{n}$ spin components $\langle \widetilde{s}_\gamma (x,t) \rangle$, where γ refers to the corresponding axis, and can be expressed in the most general way in terms of the spin diffusion [130] as

$$\langle \widetilde{s}_\gamma (x,t) \rangle = \int \sum_\beta D_{\gamma\beta}(x-x',t) \langle \widetilde{s}_\beta (x',0) \rangle dx', \tag{30}$$

where $D_{\gamma\beta}(x,t)$ is the exact one-dimensional spin diffusion Green's function. In nonmagnetic systems without SO coupling $D_{\gamma\beta}(x,t) = \delta_{\gamma\beta} D(x,t)$ is diagonal in the spin subspace. As a result of the gauge transformation, the uniform density dynamics is determined by a single-momentum Fourier component [130]

$$D(q_{\text{so}},t) = \int_{-\infty}^{\infty} e^{-iq_{\text{so}}x} D(x,t) dx \tag{31}$$

with the wavevector $q_{\text{so}} = 1/L_{\text{so}}$ established by the gauge transformation in Eq.(4). In this case, Eq.(30) simplifies for the physical measurable spins as $\langle s_\gamma (t) \rangle = \langle s_\gamma (0) \rangle D(1/L_{\text{so}}, t)$. Here we use a similar, based on the gauge transformation, however, somewhat different approach based on numerically exact analysis of the direct time evolution of initial spin-polarized state.

In the absence of spin-orbit coupling the eigenstates can be taken in the form $\boldsymbol{\psi}(x) = \psi(x) |1\rangle$ and $\boldsymbol{\psi}(x) = \psi(x) |-1\rangle$, where $|\pm 1\rangle$ are the eigenstates of $\widehat{\sigma}_z$ with the corresponding eigenvalues. The eigenstates of $\hat{H}$, $\boldsymbol{\phi}(x)$ can be obtained by SU(2) rotation of the $\psi(x) |\sigma\rangle$ states. For example, with the spin-up initial state $\psi(x) |1\rangle$ one obtains an entangled state:

$$\boldsymbol{\phi}(x) = \psi(x) \left[\cos\left(\frac{x}{L_{\text{so}}}\right) |1\rangle + \sin\left(\frac{x}{L_{\text{so}}}\right) |-1\rangle \right]. \tag{32}$$

The spin dynamics and spin relaxation in the system, as shown below, is solely due to the entanglement of spin and coordinate in Eq.(32).

To describe the localization and spin dynamics we employ the tight-binding model, similar to that we used to study ballistic spin transport. For simplicity, we assume that the random lattice potential $U_n = U(x_n)$ uniformly spans the range $[-U_0/2, U_0/2]$ with the white noise correlator $\langle U(x_{n_1}) U(x_{n_2}) \rangle = \langle U^2 \rangle \delta_{n_1,n_2}$, where $\langle U^2 \rangle = U_0^2/12$. The effects of 1D disorder at a weak spin-orbit coupling can be approximately characterized through the energy E-dependent momentum relaxation time τ_E, which we define as $\hbar/\tau_E = \langle U^2 \rangle l \nu_E$, where $\nu_E = \sqrt{m^*}/\pi\hbar\sqrt{2E}$ is the density of states per spin component. The resulting mean free path $\ell_E = v_E \tau_E$, where $v_E = \sqrt{2E/m^*}$ is the electron velocity and the corresponding diffusion coefficient $D_E = v_E^2 \tau_E$. However, at any energy the electrons do not propagate freely being localized on the energy-dependent spatial scale. The effect of localization by disorder is seen in the inverse participation ratio [144] (IPR) $\zeta_i = \sum_n \left| \psi_i^4 (x_n) \right|$, which is inversely proportional to the characteristic spread of localized state. The IPR calculated for the low-energy spectrum is presented in Fig. 9. As expected, the degree of localization increases with U_0 and the bottom of the spectrum shifts down. In contrast to the approach of Ref.[140], where the bulk crystals were considered and the spin-orbit coupling could not be gauged out by a SU(2) rotation, the IPR in this system does not depend on the SO coupling. The IPR being a random variable, shows a statistical distribution, too.

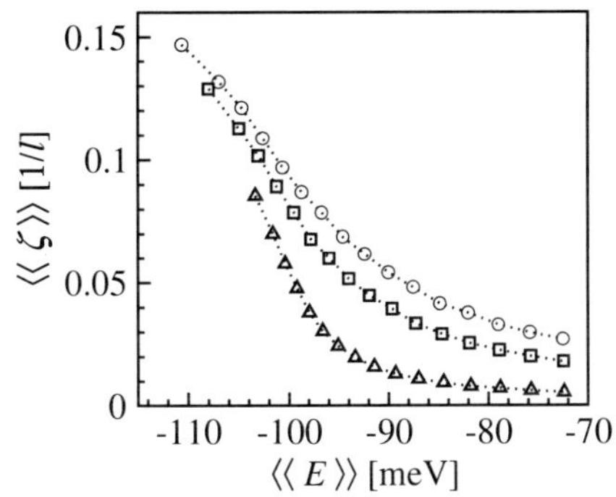

Figure 9. The averaged inverse participation ratio $\langle\langle\zeta\rangle\rangle$ as a function of energy and disorder strength. Triangulars, squares, and circles correspond to $U_0 = 25$, 35, and 55 meV, respectively. The effective hopping matrix element is $t_{\text{is}} = 50$ meV. The lattice constant is $l = \hbar/\sqrt{2m^* t_{\text{is}}} = 3.37$ nm to satisfy $m^* = 0.067\, m_0$ in GaAs. For a free electron with the energy $E \approx 20$ meV the velocity is $v_E \approx 3.5 \times 10^7$ cm/s, the mean free path $\ell_E \sim 2.5 \times 10^{-5}$ cm for $\hbar/\tau_E = 1$ meV, and the corresponding diffusion coefficient $D_E \sim 10^3$ cm^2/s. These parameters correspond to disorder at $U_0 \sim 30$ meV and realistic wires. The double brackets $\langle\langle\ldots\rangle\rangle$ imply averaging over selected bin states and disorder realizations. The calculation is performed for 256 states per energy bin and 8 realizations of disorder.

In this representation the eigenstates of $\hat{\hat{H}}$ and $\hat{H}$ form the basis sets, $\{\boldsymbol{\psi}_i\}$ and $\{\boldsymbol{\phi}_i\}$, respectively, where $1 \leq i \leq 2N$, and N is the total number of states per given spin component. For the same i, these two sets are related by the coordinate-dependent spin rotation $\hat{S}$. We assume that $\boldsymbol{\psi}_i = \psi_i(x_n)\,|1\rangle$ for $1 \leq i \leq N$ and $\boldsymbol{\psi}_i = \psi_{i-N}(x_n)\,|-1\rangle$ for $N < i \leq 2N$.We study the dynamics of initial $\boldsymbol{\psi}_i$ states with $1 \leq i \leq N$, corresponding to the evolution upon instant switching of the SO coupling. The time dependence can be expressed with the spectral decomposition as:

$$\boldsymbol{\psi}_j^{\text{so}}(t) = \sum_{i=1,2N} a_{ij}\boldsymbol{\phi}_i e^{-it\varepsilon_i/\hbar}, \tag{33}$$

where $a_{ij} = \langle\boldsymbol{\phi}_i|\boldsymbol{\psi}_j\rangle$, and ε_i are the corresponding eigenenergies. The spin component expectation value $\langle\widehat{\sigma}_z(t)\rangle_j = \langle\boldsymbol{\psi}_j^{\text{so}}(t)|\widehat{\sigma}_z|\boldsymbol{\psi}_j^{\text{so}}(t)\rangle$ is determined by the spectrum and eigenstates of the system.

For numerical calculations we distribute the eigenstates with respect to the energy over the bins containing a given number of states. The energy width of each bin is much less than the full spectrum of the system, being, however, sufficient to hold a large number of states for a reliable averaging. The calculated bin- and potential realization-averaged spin dynamics is shown in Fig. 10 (left panel) revealing strong influence of the disorder-induced spatial localization of states. Intuitively, the collisions of electrons with impurities force the electron spin to frequently reverse the precession direction. In the classical picture, this leads to a long Dyakonov-Perel' spin relaxation. If the quantum effects are important, the result is the freezing of initial electron spin density. As one can see in Fig. 10 (left panel), the electron spin density relaxes initially and then remains constant. The spin polarization

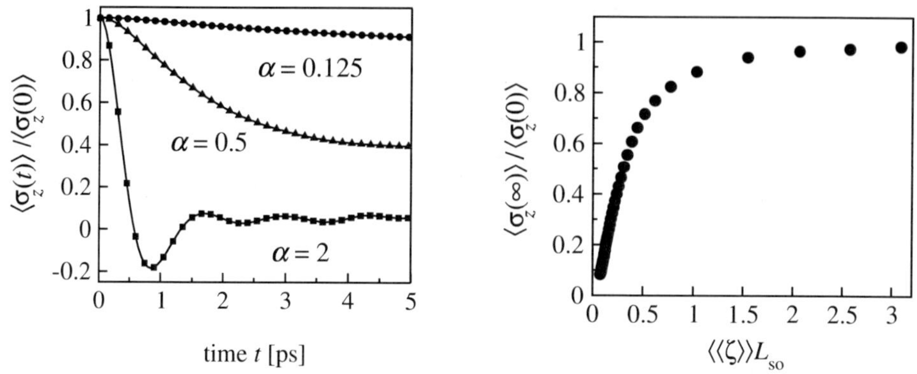

Figure 10. Left panel: Time-dependent spin polarization in the strong-disorder regime with $U_0 = 55$ meV. The energy of electrons is approximately 45 meV counted from the minimum of the numerically calculated spectrum with the IPR given in Fig.9. The spin-orbit coupling α is given in units of 10^{-6} meVcm. Note that at the weak coupling the spin density is almost frozen, while at a strong coupling it relaxes almost to zero. Right panel: numerically universal dependence of the asymptotic spin polarization with three examples demonstrated in the left panel. As in Fig.9, the calculation is performed by averaging over 256 energy bin states and 8 realizations of disorder.

plateau is higher for the localized states and for the weak SO interaction.

To get insight into the physics of spin freezing by disorder, we look at the long-term spin polarization $\langle\sigma_z(\infty)\rangle$ plotted in Fig. 10 (right panel) versus parameter $\langle\langle\zeta\rangle\rangle L_{\rm so}$. This product combines two factors that determine the spin dynamics: SO coupling and spatial localization. The given values of $\langle\sigma_z(\infty)\rangle$ follow a universal dependence indicating a unique trend for the long-term polarization against SO coupling and localization. This trend is a fast increase for $L_{\rm so}\langle\langle\zeta\rangle\rangle < 1$ and a saturation for $L_{\rm so}\langle\langle\zeta\rangle\rangle > 1$. This behavior can be understood as follows. For a spin to show an efficient dynamics, the electron should move the distance of the order of $\pi L_{\rm so}$. Therefore, the spatial spread of the states should be larger than $\pi L_{\rm so}$. In the clean regime, $L_{\rm so}\langle\langle\zeta\rangle\rangle \ll 1$, the spin relaxation is due to the Dyakonov-Perel' mechanism. Here, the time dependence of the spin density is either purely exponential for $\Omega_E\tau_E \ll 1$ or a superposition of oscillations and exponential decay for $\Omega_E\tau_E \geq 1$, where the spin precession rate $\Omega_E = 2v_E/L_{\rm so}$ corresponds to the electron velocity at energy E. For a stronger localization, the spatial spread and the overlap of the orbital states decrease, the electron propagation is suppressed, and the spin polarization remains frozen at the initial value, as shown in Fig.10. This result demonstrates the ability by choosing a Rashba coupling appropriate for a given wire to produce and destroy on demand steady spin states.

3.3. Spin Noise and Spin Memory: Classical Approach

Since the SO coupling is structure- and material-dependent, all sorts of disorder (random doping [145, 146, 147, 148], interface fluctuations [149], random variations in the shape, etc.) which cause electron scattering and finite conductivity, can cause local variations in the coupling. As a result, in addition to the regular Rashba and Dresselhaus SO

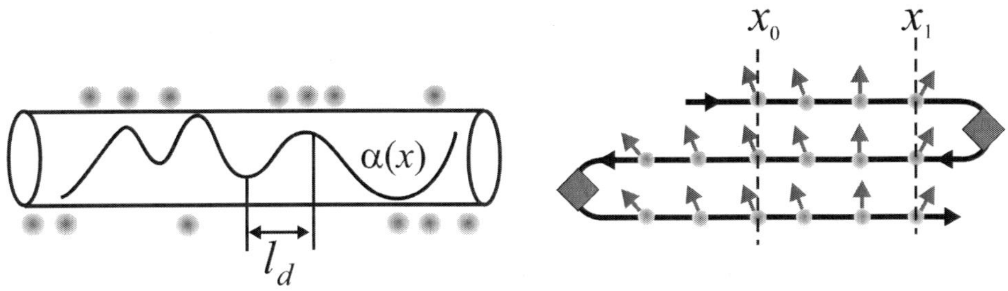

Figure 11. Left panel: Semiconductor nanowire with random spin-orbit coupling $\alpha(x) = \alpha_0 + \alpha_{\rm r}(x)$. Impurities producing the random potential and spin-orbit coupling are shown by filled circles. The correlation length l_d is defined by Eq.(37). Right panel: Diffusive electron path and spin (arrows attached to small circles) dynamics in such a wire. Arrows on the lines correspond to the direction of electron motion. Filled rhombuses correspond to the change in the direction due to impurity or phonon scattering. The spin precession angle for the motion between points x_0 and x_1 does not depend on the history of this motion.

couplings, nanostructures inevitably have a random contribution to the coupling strength. The spatial scale of corresponding random variations is of the order of 10 nm as determined by the characteristic distances in nanostructure. This randomness was shown to give rise to a number of fascinating phenomena [150, 151, 152, 153]. Now we present its role in quantum wires.

We consider a single channel quantum wire extended along the $x-$axis in a random potential and represent the SO Hamiltonian (see Fig.11) as:

$$H_0 = \frac{\hbar^2 k^2}{2m^*} + U_r(x) + H_{\rm so}, \tag{34}$$

where the spin-orbit coupling has the form of anticommutator:

$$H_{\rm so} = \frac{1}{2}\{\alpha(x), k\}(\widehat{\boldsymbol{\sigma}} \cdot \boldsymbol{n}). \tag{35}$$

Here $\alpha(x)$ is the coordinate-dependent SO coupling strength. In Eq.(35) we assumed that the spin quantization axis $\boldsymbol{n}$, is fixed, $(\widehat{\boldsymbol{\sigma}} \cdot \boldsymbol{n})$ is the component of spin along this axis, and the effective field acting at electron spin points randomly either parallel or antiparallel to the axis $\boldsymbol{n}$. This is obvious for a constant $\alpha(x)$ [154, 155, 156, 157], and holds true provided that the microscopic symmetry of the fluctuations forming the SO coupling randomness is the same as the overall symmetry of the system.

The SO coupling is assumed to be the sum of the coordinate-independent contribution, α_0, and the Gaussian random function with zero average, $\alpha_{\rm r}(x)$ such as $\alpha(x) = \alpha_0 + \alpha_{\rm r}(x)$ with the correlation function [152]:

$$\langle \alpha_{\rm r}(x)\alpha_{\rm r}(x')\rangle = \langle \alpha_{\rm r}^2\rangle F_{\alpha\alpha}(|x - x'|), \tag{36}$$

where $\langle \alpha_{\rm r}^2\rangle$ is the mean square of SO coupling fluctuations and $F_{\alpha\alpha}(|x - x'|)$ is the range function. The typical correlation length

$$l_d = \int_0^\infty F_{\alpha\alpha}(x)dx \tag{37}$$

characterizes the size of the domain of the random SO coupling. We assume the short-range correlations, such that the integral in Eq.(37) always converges.

We begin with the semiclassical regime, where SO coupling disorder is smooth on the scale of electron wavelength, that is $l_d \gg \lambda_{\rm F}$. Here the wavelength of the Fermi level electron is $\lambda_{\rm F} = 2\pi/k_{\rm F}$, with $k_{\rm F}$ being the Fermi wave vector for the degenerate electron gas. The Hamiltonian (35) implies that when an electron is displaced from the point x_0 to the point x_1, its spin rotates around the $\boldsymbol{n}$-axis by the angle

$$\theta(x_1, x_0) = \frac{2m^*}{\hbar^2}\int_{x_0}^{x_1} \alpha(x')dx', \tag{38}$$

solely determined by electron initial and final positions whatsoever the motion was: all contributions to $\theta(x_1, x_0)$ of the closed paths, where electron passes the same configuration of $\alpha(x)$ in the opposite directions, cancel each other [58, 77, 130, 156].

The temporal evolution of electron spin is directly related to the electron motion. We consider here the spin projections at a given z axis, perpendicular to the spin quantization axis $\boldsymbol{n}$. Time dependence of the spin z component, averaged over its random spatial motion and over random precession caused by the field $\alpha(x)$ can be most conveniently characterized by the correlator $\langle\sigma_z(t)\sigma_z(0)\rangle = \langle\sigma_z^2(0)\rangle\mathcal{C}_{\sigma\sigma}(t)$ with the normalized correlation function:

$$\mathcal{C}_{\sigma\sigma}(t) = \int_{-\infty}^{\infty} p(x,t)\langle\cos\left[\theta(x,0)\right]\rangle dx, \tag{39}$$

where $p(x,t)$ is the probability that the electron travels the distance x during the time t. Note, that $\mathcal{C}_{\sigma\sigma}(t)$ can be understood as the disorder-averaged z component of spin with the initial condition $\sigma_z(0) = 1$. It results from the linearity of spin dynamics equations: the correlators $\langle\sigma_i(t)\sigma_j(0)\rangle$ satisfy exactly the same equations as the average values $\langle\sigma_i(t)\rangle$ $(i,j = x,y,z)$. In the derivation of Eq.(39) we assumed that the scattering of electrons, which determines $p(x,t)$ is not correlated with the random SO field $\alpha_{\rm r}(x)$. Hence, the averaging over the realizations of $\alpha_{\rm r}(x)$ denoted by the angular brackets and over the trajectories can be considered independently. If the same local disorder determines the electron scattering and random SO fields, then in relatively clean systems the electron mean free path ℓ exceeds by far the disorder correlation length l_d in Eq.(37). As a result, spatial scales of two random processes: ℓ for the electron backward scattering in the random potential and l_d for the spin precession are strongly different, $l_d \ll \ell$. As a result, on the ℓ-scale, the memory of the short-range correlations is lost, and Eq.(39) holds. Although Eq.(39) is presented for a smooth SO coupling disorder, $l_d/\lambda_{\rm F} \gg 1$, the general Green's function approach confirms it for arbitrary $l_d/\lambda_{\rm F}$ values.

To average $\langle\cos\left[\theta(x,0)\right]\rangle$ in Eq.(39) over the random realizations of the $\alpha(x)$-field we use the relation [152]

$$\cos\left[\theta(x,0)\right] = {\rm Re}\left\{\exp\left(2i\frac{x}{L_{\rm so}}\right)\exp\left[i\vartheta_{\rm r}(x)\right]\right\}, \qquad \vartheta_{\rm r}(x) = \frac{2m^*}{\hbar^2}\int_0^x \alpha_{\rm r}(x')dx' \tag{40}$$

where $\vartheta_{\rm r}(x)$ is the contribution of the random SO coupling. As a result, the correlator in Eq.(39) becomes

$$\mathcal{C}_{\sigma\sigma}(t) = \int_{-\infty}^{\infty} p(x,t)\cos\left(2\frac{x}{L_{\rm so}}\right)\exp\left[-\frac{\langle\vartheta_{\rm r}^2(x)\rangle}{2}\right]dx. \tag{41}$$

At large distances where $\langle\vartheta_{\rm r}^2(x)\rangle \gtrsim 1$, the spins are completely dephased. Equation (39) is of crucial importance: if we know the probability $p(x,t)$, we can directly obtain the spin evolution by using Eq.(39). The spin noise power spectrum is given by the transform of $\mathcal{C}_{\sigma\sigma}(t)$ [115]:

$$\left\langle\sigma_z^2\right\rangle_\omega = 2\int_0^\infty \mathcal{C}_{\sigma\sigma}(t)\cos(\omega t)\,dt. \tag{42}$$

To get a better insight into the problem, we begin with the key limits. First, for the ballistic electron dynamics $p(x,t) = \delta(x - v_{\rm F}t)$, where $v_{\rm F} = \hbar k_{\rm F}/m$ is the Fermi velocity. The ballistic motion is realized on the temporal scale $t \ll \tau = \ell/v_{\rm F}$ with τ being the momentum relaxation time. We are interested in the spin dynamics on the time scale $t \gg \tau_d = l_d/v_{\rm F}$, where τ_d is the time during which the electron passes the correlated interval of the SO coupling fluctuations. Using Eq.(39) we obtain the damped oscillations of the spin z-component:

$$\mathcal{C}_{\sigma\sigma}(t) \approx \cos(\Omega_0 t)\exp(-t/\tau_{s,{\rm r}}), \tag{43}$$

with the regular spin precession frequency $\Omega_0 = 2v_{\rm F}/L_{\rm so}$ determined by the averaged SO coupling, and the decay time caused by the fluctuations

$$\frac{1}{\tau_{s,{\rm r}}} = \left(\frac{2m^* v_{\rm F}}{\hbar^2}\right)^2 \langle\alpha_{\rm r}^2\rangle\tau_d. \tag{44}$$

Equation (44) for the spin relaxation time $\tau_{s,{\rm r}}$ is the result of random spin precession [151]. The spin noise calculated using Eqs.(42) and (43) reads:

$$\left\langle\sigma_z^2\right\rangle_\omega = 2\tau_{s,{\rm r}}{\rm Re}\frac{1 - i\omega\tau_{s,{\rm r}}}{\Omega_0^2\tau_{s,{\rm r}}^2 + (1 - i\omega\tau_{s,{\rm r}})^2}. \tag{45}$$

This ballistic regime, however, can be realized only in very clean systems, where $\Omega_0\tau \gg 1$. Although at zero temperature all electron states are localized, at finite temperatures the electron-phonon coupling leads to a temperature-dependent diffusion [158]. At these conditions, the electron spin evolution occurs at the time scale, where electron moves diffusively (Fig.11), i.e.,

$$p(x,t) = \frac{1}{2\sqrt{\pi Dt}}e^{-x^2/4Dt}. \tag{46}$$

Here the diffusion coefficient D depends on the disorder, electron-phonon coupling, and the temperature. In the absence of the SO coupling fluctuations and provided that $\Omega_0\tau \ll 1$ the exponential spin relaxation is due to the Dyakonov-Perel' mechanism [154, 157] with the relaxation time $\tau_{s,{\rm DP}} = 1/\Omega_0^2\tau$. The spin noise spectrum has a Lorentzian form

$$\langle\sigma_z^2\rangle_\omega = 2\frac{\tau_{s,{\rm DP}}}{(1 + \omega^2\tau_{s,{\rm DP}}^2)} \tag{47}$$

with the width determined by the relaxation time.

New physical effects arise when the SO coupling fluctuations dominate over the regular contribution. Now we consider a system where the SO coupling is purely random, $\alpha_0 = 0$, and concentrate on the long-time, $t \gg \tau_d, \tau, \tau_{s,{\rm r}}$, dynamics. At these times, the system in Eq.(39) is characterized by two length parameters. One of them is the time-dependent diffusion propagation distance $\sqrt{Dt}$ in Eq. (46), the other one, time-independent, is

$$L_s = \int_0^\infty \exp\left[-\langle\vartheta_{\rm r}^2(x)\rangle/2\right]dx, \tag{48}$$

and characterizes the spin randomization. Sooner or later $\sqrt{Dt}$ will exceed L_s, and the inequality $\sqrt{Dt} \gg L_s$, will hold at longer times. In this case, in Eq.(41), we are integrating the product of a broad function $p(x,t)$ and a relatively narrow $\exp\left[-\langle\vartheta_{\rm r}^2(x)\rangle/2\right]$. Therefore, in this integration one can take $p(0,t)$ instead of $p(x,t)$ and obtain from Eq.(39) a power-like rather than exponential time dependence:

$$C_{\sigma\sigma}(t) \approx p(0,t)\int_{-\infty}^\infty \exp\left[-\langle\vartheta_{\rm r}^2(x)\rangle/2\right]dx = \frac{L_s}{\sqrt{\pi Dt}}. \tag{49}$$

Equation (49) predicts extremely long spin decoherence described by the inverse square root law: $\langle\sigma_z(t)\rangle \propto 1/\sqrt{t}$. This surprising result has a transparent physical interpretation (see Fig. 11): Indeed, if an electron is displaced from its initial position by a sufficiently large distance, $x \gtrsim L_s$, its spin rotation angle becomes so large, that it does not contribute to the total spin polarization owing to $\exp\left[-\langle\vartheta_{\rm r}^2(x)\rangle/2\right]$ in Eq. (41). As a result, the spin polarization is supported by the electrons located in the vicinity of their initial positions, mainly due to the return after multiple scatterings by the random potential. The fraction of such electrons, in agreement with the diffusion distribution $p(x,t)$ in Eq.(46), decays as $p(0,t) \propto 1/\sqrt{t}$ resulting in the same behavior of spin polarization. It is interesting to mention that this qualitative argument should be modified for the constant SO coupling despite the spin of electron is restored upon the return to the origin. The reason is that due to the oscillations of spin on the spatial scale of the order of $L_{\rm so}$ in Eq.(41), the return of electrons to the origin at a long timescale is insufficient for the formation of algebraic relaxation tail.

Slow non-exponential spin relaxation, described by Eq. (49) manifests itself in the low frequency spin noise spectrum [152]. As follows from Eq. (42), $\left\langle\sigma_z^2\right\rangle_\omega \propto 1/\sqrt{\omega}$, i.e the spin noise diverges at $\omega \to 0$. Such a non-trivial behavior is inherent to the quantum wires with random SO coupling, where the spin is restored upon return to the origin. This observation shows that the low-frequency optical spin noise spectroscopy is an excellent tool for studying the spin phenomena in thin semiconductor nanowires and the characterization of the random potential and the SO coupling there. In multichannel wires for sufficiently fast interchannel scattering and in two-dimensional systems, the spin relaxation is exponential [151] and $\left\langle\sigma_z^2\right\rangle_{\omega=0}$ is finite. Comprehensive reviews on other realizations of a power-like noise, mainly describing the conventional $1/\omega$ behavior can be found in [159, 160, 161].

3.4. Hyperfine Interactions for Disorder-Localized States

In this subsection we address the effects of localization on the spin dynamics due to the contact hyperfine interaction described by the Hamiltonian

$$H = \frac{\hbar^2 k^2}{2m^*} + U_r(x) + \hat{H}_{\rm hf}, \tag{50}$$

where the hyperfine coupling between nuclear magnetic moments and electron spin is given by:

$$\hat{H}_{\rm hf} = \frac{A\nu}{2} \sum_n |\psi(\mathbf{r}_n, t)|^2 \left(I_x^{[n]} \widehat{\sigma}_x + I_y^{[n]} \widehat{\sigma}_y + I_z^{[n]} \widehat{\sigma}_z \right), \tag{51}$$

where the summation runs over all Ga and As nuclei with spin components $I_j^{[n]}$ at sites $\mathbf{r}_n$, A is the effective averaged hyperfine coupling constant for GaAs [17], and ν is the volume per single nucleus. The wave functions are localized due to the random potential $U_r(x)$. The coupling between nuclei and electron depends on the electron cloud density at the position of the nuclei. Hamiltonian (51) can be rewritten in the form

$$\hat{H}_{\rm hf} = \frac{\hbar}{2} \left(\mathbf{\Omega} \cdot \widehat{\boldsymbol{\sigma}} \right), \tag{52}$$

where the precession due to hyperfine interaction $\mathbf{\Omega}$ is an integral that depends on the electron position and can be expressed as

$$\Omega_j = \frac{A\nu}{\hbar} \int |\psi(\mathbf{r}, t)|^2 \rho_j(\mathbf{r}) d^3 r. \tag{53}$$

Here the density of the j-th component of the nuclear magnetic moment $\rho_j(\mathbf{r})$ satisfies the white-noise distribution $\langle\langle \rho_j(\mathbf{r})\rho_j(\mathbf{r}')\rangle\rangle = \rho_0 \delta\left(\mathbf{r} - \mathbf{r}'\right)$, where $\rho_0 = \nu^{-1}\langle \hat{I}_j^2 \rangle$ and $\langle\langle \ldots \rangle\rangle$ stands for the ensemble average. The dependence of the spin dynamics on the qubit spatial width is determined by the number $\mathcal{N}$ of hyperfine interacting nuclei inside the electron cloud: when the size of the cloud decreases, the coupling per single nucleus follows the electron probability density increasing as $1/\mathcal{N}$, and the fluctuation of the nuclear magnetic moments decreases as $\sqrt{\mathcal{N}}$, resulting in $1/\sqrt{\mathcal{N}}$ dephasing rate. In the limit of extended state, $\mathcal{N} \to \infty$, the spin relaxation vanishes.

Here we investigate the localization-induced spin relaxation in nanowires due to the coupling with nuclear spins, as illustrated in the upper panel of Fig.12. We assume that the nanowires have a square cross section $L \times L$. This symmetry leads to the cancellation of the Dresselhaus spin-orbit coupling for the wires grown in the [001]-direction as we consider below. In the absence of the Rashba coupling, the nuclei become the only source of spin dephasing and depolarization. The states of GaAs nuclei are random spinors corresponding to four $z-$components of the $I = 3/2$ total moment. Each of these spinors yields the expectation values $\langle \hat{I}_j \rangle$, which are considered as "frozen" on the time scale of electron precession [17].

Electrons in the wire are characterized by the wavefunctions $\psi(\mathbf{r}_n) = \eta(x_n)\phi(y_n, z_n)$, where the transverse motion is quantized as in a two-dimensional box, $\phi(y_n, z_n) = (2/L)\sin(\pi y_n/L)\sin(\pi z_n/L)$, and $\eta(x_n)$ corresponds to the localization by disorder. We neglect the effect of impurities on the transverse motion assuming that the wavefunction $\phi(y_n, z_n)$ is sufficiently rigid to an external perturbation. For numerical calculations we use the tight-binding model, with the sites separated from each other on the x axis by an effective lattice constant l such that site positions are $x_n = n \times l$. The $U(x_n)$ function is defined for every tight-binding site, similar to the approach applied in the spin freezing problem.

The initially produced fully polarized nonequilibrium spin state, where $|\sigma_z(0)\rangle = |1\rangle$, evolves at $t > 0$ as $|\sigma_z(t)\rangle = e^{-it\hat{H}_{\rm hf}/\hbar}|\sigma_z(0)\rangle$ with $\hat{H}_{\rm hf}$ in Eq.(51). For an ensemble of

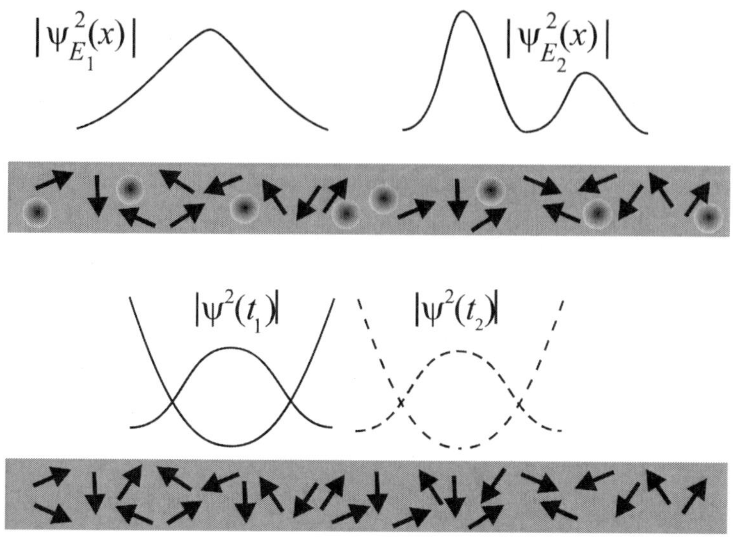

Figure 12. Upper part: Random distribution of nuclear spins shown as arrows. The impurities are shown as filled circles and two examples of density distribution, same as in Fig.8, are presented. This localization leads to the effect of hyperfine interaction on spin precession and dephasing. Lower part: Nuclear spins and electron density distributions (dashed Gaussians) at times t_1 and t_2. Electron is localized by a time-dependent external potential shown as a parabola or by a surface acoustic wave.

localized states, either in quantum dots or in a random potential, one should average this evolution over the ensemble. This can be done by binning the eigenstates, similarly to the procedure described above for the spin-orbit coupling, and calculating averages over the bins and disorder realizations.

The results show that the hyperfine interactions of electron spins in doped nanowires with nuclear magnetic moments yield a characteristic total spin dynamics picture, similar to that in the ensembles of random quantum dots. Correspondingly, the spin relaxation time decreases for a stronger disorder and increases for broader wires. The inverse participation ratio is a reliable measure of the disorder-induced localization leading to this spin dynamics and dephasing.

3.5. Driven Relaxation of Flying Qubits

An interesting and useful for applications feature of nanowires is the ability to produce flying spin-based qubits, where electrons propagate along the wires inside quantum dots produced, e.g., by an external bias. By using the flying qubits, the information encoded in the electron spin can be transferred at relatively large distances inside a spintronics device if the information can be kept intact during the transfer time. Here we investigate the hyperfine mechanism for spin relaxation of flying qubits in semiconductor nanowires and show how new timescales appear in their dynamics. The Hamiltonian of the systems (as shown in Fig.12, lower panel) is:

$$H = \frac{\hbar^2 k^2}{2m^*} + U(x,t) + \hat{H}_{\rm hf}, \tag{54}$$

where $U(x,t)$ is an external time- and coordinate-dependent potential that confines the electrons. For simplicity, we consider $U(x,t) = m\omega^2 (x - \langle x(t)\rangle)^2 /2$ as a moving quantum dot described by the harmonic potential and assume that the electron follows the potential adiabatically, that is always in the time-dependent ground state. Similar approach was applied in Ref.[141] for a 2D quantum dot with spin-orbit coupling in a random potential.

The studied qubits can propagate in two different regimes of motion, demonstrating different spin dynamics. The packet moves straightforwardly with a constant velocity in the first regime and oscillates in the second one. To describe a moving electron, we need to find the characteristic time-dependent correlator which can be obtained assuming Gaussian random fluctuations [138, 150] as:

$$\frac{\hbar^2}{(A\nu)^2} \langle\langle \Omega_j(t_1)\Omega_j(t_2)\rangle\rangle = \rho_0 \int |\psi(\mathbf{r},t_1)\psi(\mathbf{r},t_2)|^2 d^3r. \tag{55}$$

With the course of time, the correlation gradually vanishes, as can be seen in Fig.12. The states characterizing the propagating qubits are defined as $\psi(\mathbf{r}_n,t) = \eta(x_n,t)\varphi(y_n,z_n)$, in a single-mode box, as in the previous subsection. Here $\eta(x_n,t)$ is the Gaussian packet centered at a time-dependent position $\langle x(t)\rangle$,

$$\eta(x_n,t) = \left(\frac{1}{\pi\xi^2}\right)^{1/4} \exp\left[-(x_n - \langle x(t)\rangle)^2/2\xi^2\right]. \tag{56}$$

For this choice of the wavefunction, the correlator in Eq.(55) is expressed as

$$\langle\langle \Omega_j(t_1)\Omega_j(t_2)\rangle\rangle = \langle\langle \Omega_j^2\rangle\rangle \exp\left[-\left(\Delta x(t_2-t_1)\right)^2/2\xi^2\right], \tag{57}$$

where $\langle\langle \Omega_j^2\rangle\rangle = 3\,(A/\hbar)^2\, I(I+1)\nu/8\sqrt{2\pi}L^2\xi$ and $\Delta x(t_2-t_1) \equiv \langle x(t_2)\rangle - \langle x(t_1)\rangle$. This correlator reaches the maximum when the electron comes back to the position occupied at time t. For the motion with constant velocity, $\Delta x(t) = vt$, and the corresponding effective correlation time can be defined as $\tau_\xi \equiv \xi/v$.

We consider three different realizations of the spin dephasing: for static qubits (to have the known system to compare with), and for straightforward and oscillatory motions. We consider the packets with the widths of $\xi = 7$ and 28 nm; their spin evolutions, plotted in Fig. 13, look exactly as expected from the dynamics induced by hyperfine coupling in quantum dots [17, 162]. This steady polarization arises due to the precessional motion of the qubit spin around the time-independent effective field of the nuclear magnetic moments [17]. The spin evolution can be characterized by two timescales originated from the same random position-dependent spin precession: that for the initial polarization to decay down to $1/e$, and that for reaching the 1/3 plateau. The scale for initial decay rate of static electrons is then $\Gamma_{\rm st} = \sqrt{\langle\langle \Omega_j^2\rangle\rangle}$, and the entire time dependence in Fig. 13 (left panel) can be described by using this single parameter [17].

In the next, straightforward motion regime, the packets propagate along the wire with constant velocities $v = 10$ and 20 nm/ns, taken here as examples. A behavior considerably different from the static one appears when $\tau_\xi\Gamma_{\rm st}$ is much less than 1, that is, when the electron moves to a strongly different realization of the hyperfine field before sufficiently changing its spin; this behavior is shown in Fig. 13 (right panel). To get insight into the

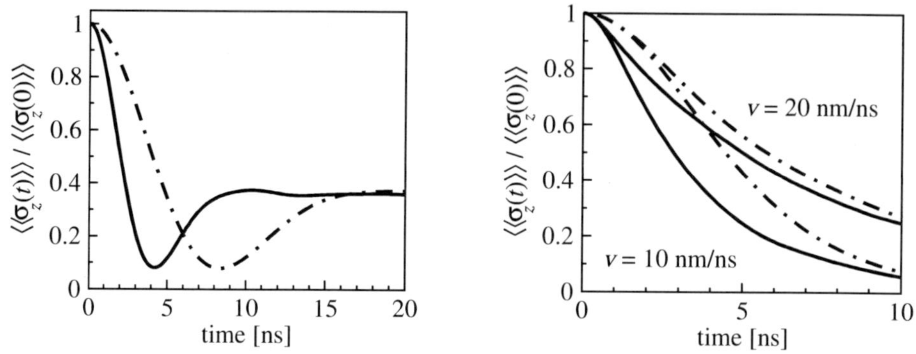

Figure 13. Left panel: Spin relaxation of static qubits due to the hyperfine interaction. Right panel: Average polarization for qubits in the straightforward motion for different velocities v and widths ξ. In both panels solid and dashed-dot lines correspond to $\xi = 7$ nm and $\xi = 28$ nm, respectively.

influence of width and velocity, we have taken into account, again, three distinct packet sizes, $\xi = 7$, and $\xi = 28$ nm. When $v = 10$ nm/ns both polarizations fall below 0.1 after 10 ns and monotonically tend to zero afterwards. Besides, the smaller is the qubit width, the faster is the dephasing. This behavior can be detailed as follows. The averaged spin evolution can be described with [150]

$$\langle\langle\sigma_z(t)\rangle\rangle = \langle\langle\sigma_z(0)\rangle\rangle \exp\left[-\int_0^t dt' \int_0^{t'} \langle\langle\Omega_z(t')\Omega_z(t'-t'')\rangle\rangle dt''\right]. \quad (58)$$

For the exponentially decaying part of the dynamics, this expression has the form

$$\langle\langle\sigma_z(t)\rangle\rangle = \langle\langle\sigma_z(0)\rangle\rangle \exp(-\sqrt{\pi}\Gamma_v t), \quad (59)$$

where the spin relaxation rate is determined by

$$\Gamma_v = \langle\langle\Omega_j^2\rangle\rangle\tau_\xi. \quad (60)$$

Therefore, the fast motion establishes a new and smaller relaxation rate, $\Gamma_v = \Gamma_{\rm st}^2\tau_\xi$, such as $\Gamma_v \ll \Gamma_{\rm st}$. Here Γ_v does not depend on the width of the electron wave packet but only on its velocity, as we explain here with a scaling argument. The square of the spin precession rate being of the order of $\langle\langle\Omega_j^2\rangle\rangle \sim \mathcal{N}^{-1} \sim \xi^{-D}$, while $\tau \sim \xi/v$, is $D-$independent. As a consequence, $\Gamma_v \sim \xi^{1-D}/v$, tends to zero for bulk ($D = 3$) and 2DEG ($D = 2$) with the increase in D, remaining constant in the wire. From Eq.(60), the resulting Γ_v is proportional to $(A/\hbar)^2 \nu/L^2 v$ and, henceforth, is also $\xi-$independent.

A significantly different behavior is obtained when the packet moves back and forth. We analyze oscillations with different velocities and frequencies, which make the amplitudes in the range of several tens of nanometers. We assume that the expectation value of the qubit $z-$coordinate follows the saw-like function,

$$\langle x(t)\rangle = 2\ell_{\rm s}\left|\frac{t}{T} - \left\lfloor\frac{t}{T} + \frac{1}{2}\right\rfloor\right|, \quad (61)$$

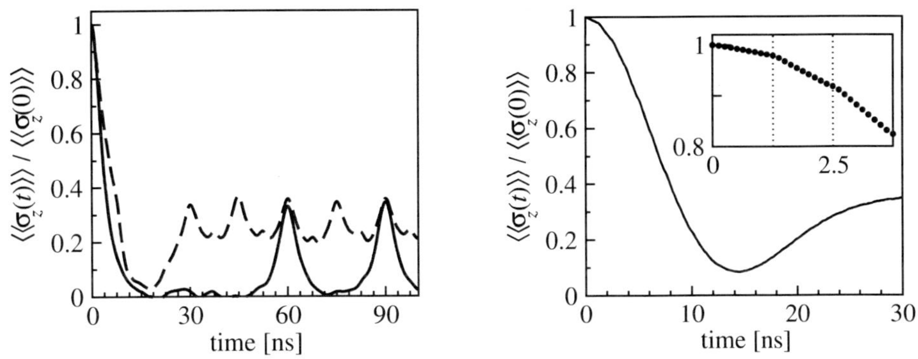

Figure 14. Left panel: Polarization for the qubits moving back and forth along the wires for various velocities and periods, $\xi = 14$ nm. Solid line: $T = 30$ ns and $v = 10$ nm/ns, dashed line: $T = 15$ ns and $v = 20$ nm/ns. Note the sharp revivals in the regime when $T = 30$ ns and $v = 10$ nm/ns. Right panel: memory effect in the spin qubit dynamics for $\xi = 14$ nm, $T = 2.5$ ns, and $v = 160$ nm/ns.

where $\lfloor \ldots \rfloor$ stands for the floor function, T is the period, and $\ell_s = vT/2$ is the maximum displacement from the initial position. Every time the electron returns to the initial position, once per period, it scans exactly the same effective field. As a consequence, the spin experiences strong correlations in this field resulting in the memory effect.

We begin with the regime where $T\Gamma_v$ is of the order of 1 and the typical electron spin strongly changes in a single oscillation. The spin polarization behavior is shown in Fig. 14. For oscillation period $T = 15$ ns and $v = 20$ nm/ns, the calculated dynamics presents clear periodic peaks above the reached plateau. For a larger $T = 30$ ns and smaller $v = 10$ nm/ns, one observes an almost flat stage followed by periodic sharp revivals, see Fig. 14 (left panel). Here, in the first period, the polarization relaxes down to zero. Throughout the second period, the qubits show a revival to 1/3 of the initial spin. Beyond two periods, the relaxation and memory effect both contribute to yield a periodic picture of revivals up to one third and decays down to zero.

Next, we consider very high velocity and frequency, $T\Gamma_v \ll 1$, resulting in a small spin precession angle in one period. When $v = 160$ nm/ns and $T = 2.5$ ns, the average spin polarization decays to a minimum of about 10% of the initial value and monotonically increases up to a constant plateau of 1/3; the ensemble dephasing time roughly is 14 ns, as can be seen in Fig. 14 (left panel). Due to the extremely high velocity, the precession of the spin qubit cannot follow the fast changes in the effective nuclear magnetic field. The characteristic behavior is given in Fig. 14 (left panel), similar to that depicted in Fig. 13 for static qubits. Here, the dependence of polarization with time cannot be modified by tuning the qubit width.

In the oscillatory regime, the relaxation part of the time-dependence is Gaussian rather than exponential [150, 163], as we explain below. At any moment t, the expectation value of spin is $\langle\langle \cos\theta_z \rangle\rangle$; this mean value is $\langle\langle \cos\theta_z \rangle\rangle = \exp(-\sqrt{\pi}\Gamma_v t)$ for $0 < t < T/2$, and is multiplied by an integer factor p, $\langle\langle \cos\theta_z \rangle\rangle = \exp(-\sqrt{\pi}p^2\Gamma_v t)$, at $t = pT/2$. Thus, in this regime, a new spin relaxation rate is established, $\Gamma_{\text{mem}} = \sqrt{\Gamma_v/T} \gg \Gamma_v$, similar to the memory effects in the systems with random spin-orbit coupling subject to external magnetic field [150, 151]. On the other hand, since $\Gamma_{\text{mem}} = \Gamma_{\text{st}}\sqrt{\tau_\xi/T} \ll \Gamma_{\text{st}}$, this relaxation is found even slower than for the static qubits. Moreover, due to the memory effects, the slopes of $\ln\left(\langle\langle\sigma_z(t)\rangle\rangle/\langle\langle\sigma_z(0)\rangle\rangle\right)$ at $t = pT/2$ should change following the ratio $(2p+1)/(2p-1)$, which is indeed very close to that obtained from our numerical data, see inset in the right panel of Fig. 14.

Conclusion

In this Chapter we have discussed examples of unusual dynamics and relaxation of coupled coordinate and spin degrees of freedom in semiconductors where the spin-orbit coupling plays a significant role. Since this coupling is present in various forms in the majority of low dimensional semiconductor structures, these effects can be considered as a general rather than a specific trend stemming from the strong interplay of coordinate and spin degrees of freedom producing the mutual influence on most of the measurable and practically important quantities such as the charge and spin densities, the charge and spin currents, and various relaxation mechanisms. Indeed, during almost a decade that passed after some of the textbooks [92] and reviews [13, 14] on the spin and spin-orbit effects in semiconductors have been published, more and more experimental and theoretical results covering both fundamental and applied aspects of the field emerge continually.

The trends covered here can be viewed generally as deep, and often unexpected modifications of the well-known and fundamental effects such as spin resonance, transport in quantum wires, or spin relaxation produced by the interplay of spin and coordinate evolution highlighted by the reduced dimensionality and/or improved quality of the samples available today. Starting from earlier work by Datta and Das [4] it has been known that the fundamental role played by the spin-orbit coupling can be applied for a novel transistor, although after decades of research it is still has to be found how to fabricate it with industrial standards. To reach them, numerous attempts suggesting new schemes or principles are proposed, and the initially applied goal of engineering new type of transistor paves the way for new ideas, both fundamental and applied. One can argue which type of spin transistor will occupy the place in the major part of future nanodevices, however the variety of such proposals for new generation of spin devices will certainly greatly exceed the initially proposed number of setups.

The same trend can be seen in the study of another well-known fundamental effect such as the spin resonance. The role of spin-orbit coupling here was highlighted more than a decade ago [16] together with potential applications in the qubit engineering [19] which stimulated numerous researchers for deeper understanding of the spin resonance in nanostructures. These studies demonstrated the rich and unusual nature of spin effects in new types of one-dimensional structures such as quantum wires with embedded quantum dots and subject to external driving fields. It became clear that the basic principle of Rabi formula for the spin flip frequency in a simple two-level system needs to be revised carefully and recalculated "ab initio" when a more complicated multilevel system is considered. The results found in such less idealized models demonstrate that the simple predictions for the spin flip mechanisms should be significantly corrected for stronger driving fields or stronger spin-orbit coupling. In particular, the transitions from linear to oscillating dependence of the Rabi frequency on driving field can be seen. No doubt that these non-trivial fundamental findings will have a direct influence on the design of novel applications such as future generations of spin valve devices, spin qubits, and the others.

The modifications of well-known fundamental effects in semiconductor nanostructures by strong or non-uniform spin-orbit coupling have another remarkable manifestation in the field of spin relaxation phenomena [47]. Here the two faces of Janus in Fig.1 stemming from the interplay of the spin-orbit coupling and disorder are both expressed in a non-trivial way. On the one hand, the disorder slows and disturbs the particle motion in the semiconductor channels which provides the conventional difficulties for efficient engineering of transport devices. On the other hand, the disorder slows down the spin relaxation mechanism associated with such ballistic motion, namely, the Dyakonov-Perel' mechanism [92]. A very intriguing consequence of such two-face game is that the spin relaxation can be effectively slower in the presence of disorder and localization, as we have shown, and thus, the better result with longer living spin configurations can be achieved. This result, along with the others, is a good illustration of the continuously emerging opportunities in the field of spintronics where the interplay of separately known and predictable effects can lead to unexpected and promising perspectives with both fundamental and applied values. Future problems and models may create numerous challenges including the effects of the environment, such as spin-dependent effects in open quantum dots, quantum wires and quantum billiards with irregular dynamics [164], the physics of spin-dependent phenomena in low-dimensional structures formed in new classes of materials such as topological insulators and superconductors [165], and many others. There is no doubt that two faces of Janus in the spin-dependent effects in semiconductors will provide us with more fascinating, both fundamentally important and practically promising effects and devices in the forthcoming years of research.

Acknowledgements The work of EYS was supported by the University of Basque Country UPV/EHU under program UFI 11/55, Spanish MEC (FIS2012-36673-C03-01), and "Grupos Consolidados UPV/EHU del Gobierno Vasco" (IT-472-10). DVK is supported by the RFBR (Grants No. 13-02-00717-a and 13-02-00784-a) and by the University of Nizhny Novgorod Program for the development of joint University and Academy of Science RF laboratories. This work is partially supported by the National Science Center in Poland by the Grant Harmonia No. DEC-2012/06/M/ST3/00042. The authors are grateful to C. Echeverria-Arrondo, M. Glazov, A. Sadreev, and I. Tokatly for excellent collaboration. We are grateful to G. Japaridze, J. Siewert, R. Winkler, and L. A. Wu for helpful discussions.

References

[1] Dresselhaus G. *Phys. Rev.* (1955) 100, 580.

[2] Rashba E. I. *Sov. Phys. Solid State* (1960) 2, 1109.

[3] Bychkov Yu. A. and Rashba E. I. *JETP Lett.* (1984) 39, 79.

[4] Datta S. and Das B. *Appl. Phys. Lett.* (1990) 56, 665.

[5] Nitta J., Akazaki T., Takayanagai H., and Enoki T. *Phys. Rev. Lett.* (1997)78, 1335.

[6] Heida J. P., van Wees B. J., Kuipers J. J., Klapwijk T. M., and Borghs G. *Phys. Rev. B* (1998) 57, 11911.

[7] Engels G., Lange J., Schäpers Th., and Lüth H. *Phys. Rev. B* (1997) 55, R1958.

[8] Schäpers Th., Engels G., Lange J., Klocke Th., Hollfelder M., and Lüth H. *J. Appl. Phys.* (1998) 83, 4324.

[9] Hu C.-M., Nitta J., Akazaki T., Takayanagai H., Osaka J., Pfeffer P., and Zawadski W. *Phys. Rev. B* (1999) 60, 7736.

[10] Grundler D. *Phys. Rev. Lett.* (2000) 84, 6074.

[11] Eldridge P. S., Leyland W. J. H., Lagoudakis P. G., Karimov O. Z., Henini M., Taylor D., Phillips R. T., and Harley R. T. *Phys. Rev. B* (2008) 77, 125344.

[12] Matsuyama T., Kürsten R., Messner C., and Merkt U. *Phys.Rev. B* (2000) 61, 15588.

[13] Fabian J., Matos-Abiague A., Ertler C., Stano P., and Zutic I. *Acta Physica Slovaca* (2007) 57, 565.

[14] Fert A. *Rev. Mod. Phys.* (2008) 80, 1517.

[15] Dyakonov (Ed.) M.I. *Spin Physics in Semiconductors*, Springer Series in Solid-State Sciences, Springer-Verlag Berlin Heidelberg (2008).

[16] Rashba E. I. and Efros Al. L. *Phys. Rev. Lett.* (2003) 91, 126405.

[17] Merkulov I. A., Efros Al. L., and Rosen M. *Phys. Rev. B* (2002) 65, 205309.

[18] Kohler S., Lehmann J., and Hänggi P. *Physics Reports* (2005) 406, 379.

[19] Burkard G., Loss D., and DiVincenzo D. P. *Phys. Rev. B* (1999) 59, 2070.

[20] Rashba E. I. and Efros Al. L. *Appl. Phys. Lett.* (2003) 83, 5295.

[21] Nowack K. C., Koppens F. H. L., Nazarov Yu. V., and Vandersypen L. M. K. *Science* (2007) 318, 1430.

[22] Pioro-Ladriere M., Obata T., Tokura Y., Shin Y.-S., Kubo T., Yoshida K., Taniyama T., and Tarucha S. *Nature Physics* (2008) 4, 776.

[23] Semenov Y. G. and Kim K. W. *Phys. Rev. Lett.* (2004) *92*, 026601; Semenov Y. G. and Kim K. W. *Phys. Rev. B* (2007) 75, 195342.

[24] Stano P. and Fabian J. *Phys. Rev. Lett.* (2006) 96, 186602.

[25] Hu X., Liu Y.-X., and Nori F. *Phys. Rev. B* (2012) 86, 035314.

[26] Zhang P., Xiang Z.-L., and Nori F. *Phys. Rev. B* (2014) 89, 115417.

[27] Ban Y., Chen X., Sherman E. Ya., and Muga J.G. *Phys. Rev. Lett.* (2012) 109 206602.

[28] Prabhakar S., Melnik R., and Inomata A. *Appl. Phys. Lett.* (2014) 104, 142411.

[29] Nowak M. P. and Szafran B. *Phys. Rev. B* (2010) 81, 235311; Nowak M. P. and Szafran B. *Phys. Rev. B* (2010) 82, 165316.

[30] Nowak M. P., Szafran B., and Peeters F. M. *Phys. Rev. B* (2012) 86, 125428.

[31] Sánchez D. and Serra L. *Phys. Rev. B* (2006) 74, 153313.

[32] Lü C., Zülicke U., and Wu M.W. *Phys. Rev. B* (2008) 78, 165321.

[33] Romano C. L., Tamborenea P. I., and Ulloa S. E. *Phys. Rev. B* (2006) 74, 155433.

[34] Jiang J. H. and Wu M. W. *Phys. Rev. B* (2007) 75, 035307; Jiang J. H., Wu M. W., and Zhou Y. *Phys. Rev. B* (2008) 78, 125309.

[35] Khomitsky D. V. and Sherman E. Ya. *Phys. Rev. B* (2009) 79, 245321.

[36] Lin W. A. and Ballentine L. E. *Phys. Rev. Lett.* (1990) 65, 2927.

[37] Tameshtit A. and Sipe J.E. *Phys. Rev. E* (1995) 51, 1582.

[38] Murgida G.E., Wisniacki D. A., and Tamborenea P. I. *Journal of Modern Optics* (2009) 56, 799.

[39] Chotorlishvili L., Toklikishvili Z., Komnik A., and Berakdar J., *Phys. Letters A* (2012) 377, 69; Chotorlishvili L., Toklikishvili Z., Komnik A., and Berakdar J., *J. Phys.: Condens. Matter* (2012 24, 255302).

[40] Gómez-León A. and Platero G. *Phys. Rev. B* (2011) 84, 121310(R).

[41] Jiang J. H., Weng M. Q., and Wu M. W. *J. Appl. Phys.* (2006) 100, 063709.

[42] Petta J. R., Johnson A. C., Taylor J. M., Laird E. A., Yacoby A., Lukin M. D., Marcus C. M., Hanson M. P., and Gossard A. C. *Science* (2005) 309, 2180.

[43] Amasha S., MacLean K., Radu I. P., Zumbühl D. M., Kastner M. A., Hanson M. P., and Gossard A. C. *Phys. Rev. B* (2008) 78, 041306.

[44] Stano P. and Jacquod P. *Phys. Rev. B* (2010) 82, 125309.

[45] Shitade A., Ezawa M., and Nagaosa N. *Phys. Rev. B* (2010) 82, 195305.

[46] Quay C. H. L., Hughes T. L., Sulpizio J. A., Pfeiffer L. N., Baldwin K. W., West K. W., Goldhaber-Gordon D., and de Picciotto R. *Nat. Phys.* (2010) 6, 336.

[47] Wu M. W., Jiang J. H., and Weng M. Q. *Physics Reports* (2010) 493, 61.

[48] Nadj-Perge S., Frolov S. M., van Tilburg J. W. W., Danon J., Nazarov Yu. V., Algra R., Bakkers E. P. A. M., and Kouwenhoven L. P. *Phys. Rev. B* (2010) 81, 201305(R).

[49] Nadj-Perge S., Pribiag V. S., van den Berg J. W. G., Zuo K., Plissard S. R., Bakkers E. P. A. M., Frolov S. M., and Kouwenhoven L. P. *Phys. Rev. Lett.* (2012) 108, 166801.

[50] Sánchez R., Cota E., Aguado R. , and Platero G. *Phys.Rev. B* (2006) 74, 035326.

[51] Wang M., Yin Y., and Wu M. W. *J. Appl. Phys.* (2011) 109, 103713.

[52] Zhu Z.-G., Jia C.-L., and Berakdar J. *Phys. Rev. B* (2010) 82, 235304; Wätzel J., Moskalenko A. S. , and Berakdar J. *Appl. Phys. Lett.* (2011) 99, 192101.

[53] Wang L. and Wu M. W. *J. Appl. Phys.* (2011) 110, 043716.

[54] Borhani M. and Hu X. *Phys. Rev. B* (2012) 85, 125132.

[55] Laird E. A., Barthel C., Rashba E. I., Marcus C. M., Hanson M.P., and Gossard A. C. *Phys. Rev. Lett.* (2007) 99, 246601.

[56] Kato Y., Myers R. C., Gossard A. C. and Awschalom D. D., *Nature* (2004) 427, 50.

[57] Khomitsky D. V. and Sherman E. Ya. *Europhys. Lett.* (2010) 90, 27010.

[58] Levitov L. S. and Rashba E. I. *Phys. Rev. B* (2003) 67, 115324.

[59] Tokatly I. V. and Sherman E. Ya. *Annals of Phys.* (2010) 325, 1104.

[60] Romano C. L., Tamborenea P. I., and Ulloa S. E. *Phys. Rev. B* (2006) 74, 155433.

[61] Shirley J. H. *Phys. Rev.* (1965) 138, B979.

[62] Khomitsky D. V., Gulyaev L. V., and Sherman E. Ya., *Phys. Rev. B* (2012) 85, 125312.

[63] Fuchs G. D., Dobrovitski V. V., Toyli D. M., Heremans F. J., and Awschalom D. D. *Science* (2009) 326, 1520.

[64] Rashba E. I. *Phys. Rev. B* (2011) 84, 241305.

[65] Lucero E., Hofheinz M., Ansmann M., Bialczak R. C., Katz N., Neeley M., Connell A. D. O., Wang H., Cleland A. N., and Martinis J. M.,*Phys. Rev. Lett.* (2008) 100, 247001.

[66] Platero G. and Aguado R. *Physics Reports* (2004) 395, 1.

[67] Demikhovskii V. Ya., Izrailev F. M., and Malyshev A. I. *Phys. Rev. Lett.* (2002) 88, 154101.

[68] Streed E. W., Mun J., Boyd M., Campbell G. K., Medley P., Ketterle W., and Pritchard D. E. *Phys. Rev. Lett.* (2006) 97, 260402.

[69] Echanobe J., del Campo A., and Muga J. G. *Phys. Rev. A* (2008) 77, 032112.

[70] Sokolovski D. *Phys. Rev. A* (2011) 84, 062117.

[71] Sokolovski D. and Sherman E. Ya. *Phys. Rev. A* (2011) 84, 030101.

[72] Li R., You J. Q., Sun C. P., and Nori F. *Phys. Rev. Lett.* (2013) 111, 086805.

[73] Sadreev A. F. and Sherman E. Ya. *Phys. Rev. B* (2013) 88, 115302.

[74] Mal'shukov A. G., Tang C. S., Chu C.S., and Chao K.A. *Phys. Rev. B* (2003) 68, 233307.

[75] Wang L. Y., Tang C. S., and Chu C. S. *Phys. Rev. B* (2006) 73, 085304.

[76] Szaszko-Bogar V., Foldi P., and Peeters F. M., *J. Phys.: Condens. Matter* (2014) 26, 135302.

[77] Pershin Y. V., Nesteroff J. A., and Privman V. *Phys. Rev. B* (2004) 69, 121306; Slipko V. A., Savran I., and Pershin Y. V. *Phys. Rev. B* (2011) 83, 193302.

[78] Wang X. F. *Phys. Rev. B* 69, 035302 (2004).

[79] Zhang L., Brusheim P., and Xu H. Q. *Phys. Rev. B* (2005) 72, 045347.

[80] Serra L., Sánchez D., and Lopez R. *Phys. Rev. B* (2005) 72, 235309.

[81] Gong S. J. and Yang Z. Q. *J. Appl. Phys.* (2007) 102, 033706.

[82] Srisongmuang B., Pairor P., and Berciu M. *Phys. Rev. B* (2008) 78, 155317.

[83] Chirolli L., Venturelli D., Taddei F., Fazio R., and Giovannetti V. *Phys. Rev. B* (2012) 85, 155317.

[84] Thorgilsson G., Egues J. C., Loss D., and Erlingsson S. I., *Phys. Rev. B* (2012) 85, 045306.

[85] Ban Y. and Sherman E.Ya. *J. Appl. Phys.* (2013) 113, 043716.

[86] Tang C. S., Tan Y. H., and Chu C. S. *Phys. Rev. B* (2003) 67, 205324.

[87] Wu B. H. and Cao J. C. *Phys. Rev. B* (2006) 73, 245412.

[88] Pikus F. G. and Pikus G. E. *Phys. Rev. B* (1995) 51, 16928.

[89] Tang C. S., Chang S. Y., and Cheng S.J. *Phys. Rev. B* (2012) 86 125321.

[90] Davies J. H., Larkin I. A., and Sukhorukov E. V. *J. Appl. Phys.* (1995) 77, 4504.

[91] Bjorken D. and Drell D. *Relativistic Quantum Mechanics* (McGraw-Hill, New York, 1978), Vol.1.

[92] Winkler R. *Spin-Orbit Coupling Effects in Two-Dimensional Electron and Hole Systems*, Springer Tracts in Modern Physics, Vol. 191 Springer-Verlag Berlin Heidelberg (2003).

[93] Moroz A. V. and Barnes C. H. W. *Phys. Rev. B* (1999) 60, 14272.

[94] Lin C. H., Tang C. S., and Chang Y. C. *Phys. Rev. B* (2008) 78 245312.

[95] Bulgakov E. N. and Sadreev A. F. *Phys. Rev. B* (2002) 66, 075331.

[96] Zhai F. and Xu H. Q. *Phys. Rev. Lett.* (2005) 94, 246601.

[97] Büttiker M. and Landauer R. *Phys. Rev. Lett.* (1982) 49, 1739.

[98] Jauho A. P. *Phys. Rev. B* (1990) 41, 12327.

[99] Wagner M. *Phys. Rev. B* (1998) 57, 11899.

[100] Bulgakov E. N. and Sadreev A. F. *J. Phys.: Cond. Mat.* (1996) 8, 8869.

[101] Li W. and Reichl L. E. *Phys. Rev. B* (1999) 60, 15732.

[102] Sambe H. *Phys. Rev. A* (1973) 7, 2203.

[103] Peskin U. and Moiseyev N. *J. Chem. Phys.* (1993) 99, 4590.

[104] Sadreev A. F. *Phys. Rev. E* 86, (2012) 056211.

[105] Tien P. K. and Gordon J. P. *Phys. Rev.* (1963) 129, 647.

[106] del Valle C. P., Lefebvre R., and Atabek O. *Phys. Rev. A* (1999) 59, 3701.

[107] Lefebvre R. and Moiseyev N. *Phys. Rev. A* (2004) 69, 062105.

[108] de Andrada e Silva E. A. and La Rocca G. C. *Phys. Rev. B* (2003) 67, 165318.

[109] Spezia S., Persano Adorno D., Pizzolato N., and Spagnolo B. *J. Phys.: Condens. Matter* (2012) 24, 052204.

[110] Spezia S., Persano Adorno D., Pizzolato N., and Spagnolo B. *Europhys. Lett.* (2013) 104, 47011.

[111] Ivchenko E. L. *Sov. Phys. Semicond.* (1974) 7, 998.

[112] Müller G. M., Oestreich M., Römer M., and Hübner J. *Physica E* (2010) 43, 569.

[113] Braun M. and König J. *Phys. Rev. B* (2007) 75, 085310.

[114] Starosielec S. and Hägele D. *Appl. Phys. Lett.* (2008) 93, 051116.

[115] Kos S. S., Balatsky A. V., Littlewood P. B., and Smith D. L. *Phys. Rev. B* (2010) 81, 064407.

[116] Li F., Pershin Y. V., Slipko V. A., and Sinitsyn N. A. *Phys. Rev. Lett.* (2013) 111, 067201.

[117] Pershin Y. V., Slipko V. A., Roy D., and Sinitsyn N. A., *Appl. Phys. Lett.* (2013) 102, 202405.

[118] Slipko V. A., Sinitsyn N. A., and Pershin Y. V. *Phys. Rev. B* (2013) 88, 201102(R).

[119] Aleksandrov E. and Zapasskii V. *JETP* (1981) 54, 64.

[120] Ivchenko E. L. and Kavokin A. V. *Sov. Phys. Solid State* (1992) 34, 1815.

[121] Müller G. M., Römer M., Schuh D., Wegscheider W., Hübner J., and Oestreich M. *Phys. Rev. Lett.* (2008) 101, 206601.

[122] Crooker S. A., Brandt J., Sandfort C., Greilich A., Yakovlev D. R., Reuter D., Wieck A. D., and Bayer M. *Phys. Rev. Lett.* (2010) 104, 036601.

[123] Maune B. M., Borselli M. G., Huang B., Ladd T. D., Deelman P. W., Holabird K. S., Kiselev A. A., Alvarado-Rodriguez I., Ross R. S., Schmitz A. E., Sokolich M., Watson C. A., Gyure M. F. and Hunter A. T. *Nature* (2012) 481, 344.

[124] Braun P.-F., Marie X., Lombez L., Urbaszek B., Amand T., Renucci P., Kalevich V. K., Kavokin K. V., Krebs O., Voisin P., and Masumoto Y., *Phys. Rev. Lett.* (2005) 94, 116601.

[125] Glazov M. M. and Ivchenko E. L. *Phys. Rev. B* (2012) 86 115308.

[126] Greilich A., Shabaev A., Yakovlev D. R., Efros Al. L., Yugova I. A., Reuter D., Wieck A. D., and Bayer M. *Science* (2007) 317, 1896.

[127] Coish W. A. and Loss D. *Phys. Rev. B* (2004) 70, 195340.

[128] Dahbashi R., Hübner J., Berski F., Wiegand J., Marie X., Pierz K., Schumacher H. W., and Oestreich M. *Appl. Phys. Lett.* (2012) 100, 031906.

[129] Lyubinskiy I. S. and Kachorovskii V. Yu. *Phys. Rev. B* (2004) 70, 205335.

[130] Tokatly I. V. and Sherman E. Ya. *Phys. Rev. B* (2010) 82, 161305.

[131] Kiselev A. A. and Kim K. W. *Phys. Rev. B* (2000) 61, 13115.

[132] Pramanik S., Bandyopadhyay S., and Cahay M. *Phys. Rev. B* (2003) 68, 075313.

[133] Holleitner A. W., Sih V., Myers R. C., Gossard A. C., and Awschalom D. D. *Phys. Rev. Lett.* (2006) 97, 036805.

[134] Bringer A. and Schäpers T. *Phys. Rev. B* (2011) 83, 115305.

[135] Yamamoto M., Takada S., Bäuerle C., Watanabe K., Wieck A. D., and Tarucha S. *Nature Nanotech.* (2012) 7, 247.

[136] Nadj-Perge S., Frolov S., Bakkers E., and Kouwenhoven L. *Nature* (2010) 468, 1084.

[137] Pareek T. P. and Bruno P. *Phys. Rev. B* (2002) 65, 241305.

[138] Shklovskii B. I. *Phys. Rev. B* (2006) 73, 193201.

[139] Kaneko T., Koshino M., and Ando T. *Physica E* (2007) 40, 383.

[140] Intronati G. A., Tamborenea P. I., Weinmann D., and Jalabert R. A. *Phys. Rev. Lett.* (2012) 108, 016601.

[141] Huang P. and Hu X. *Phys. Rev. B* (2013) 88, 075301.

[142] Anderson P. W., Thouless D. J., Abrahams E., and Fisher D. S. *Phys. Rev. B* (1980) 22, 3519.

[143] Lifshitz I. M., Gredeskul S. A., and Pastur L. A. *Introduction to the Theory of Disordered Systems,* Wiley, NY, (1988); Efros A. L. and Shklovskii B. I. *Electronic Properties of Doped Semiconductors,* Springer, Heidelberg, (1989).

[144] Evers F. and Mirlin A. D. *Phys. Rev. Lett.* (2000) 84, 3690.

[145] Gryncharova E. and Perel' V. I. *Sov. Phys. Semicond.* (1976) 10, 2272.

[146] Sherman E. Ya. *Phys. Rev. B* (2003) 67, 161303.

[147] Sherman E. Ya. *Appl. Phys. Lett.* (2003) 82, 209.

[148] Dugaev V. K., Sherman E. Ya., Ivanov V. I., and Barnaś J. *Phys. Rev. B* (2009) 80, 081301.

[149] Golub L. E. and Ivchenko E. L. *Phys. Rev. B* (2004) 69, 115333.

[150] Glazov M. M. and Sherman E. Ya. *Phys. Rev. B* (2005) 71, 241312.

[151] Glazov M. M., Sherman E. Ya., and Dugaev V. K. *Physica E* 42 2157.

[152] Glazov M. M. and Sherman E. Ya. *Phys. Rev. Lett.* (2011) 107, 156602.

[153] Ström A., Johannesson H., and Japaridze G. I. *Phys. Rev. Lett.* (2010) 104, 256804.

[154] Nishimura T., Wang X.-L., Ogura M., Tackeuchi A., and Wada O. *Japan. Journ. of Appl. Phys.* (1999) 38, L941.

[155] Governale M. and Zülicke U. *Phys. Rev. B* (2002) 66, 073311.

[156] Entin M. V. and Magarill L. I. *Europhys. Lett.* (2004) 68, 853.

[157] Glazov M. M. and Ivchenko E. L. *JETP* (2004) 99, 1279.

[158] Gogolin A. A. *Physics Reports* (1982) 86, 1.

[159] Dutta P. and Horn P. M. *Rev. Mod. Phys.* (1981) 60, 53.

[160] Weissman M. B. *Rev. Mod. Phys.* (1988) 60, 537.

[161] Levinshtein M. E. *Physica Scripta* (1997) T69, 79.

[162] Echeverría-Arrondo C. and Sherman E. Ya. *Phys. Stat. Sol.- RRL* (2012) 6, 343.

[163] Reuther G. M., Hänggi P., and Kohler S. *Phys. Rev. A* (2012) 85, 062123.

[164] Nakamura K. and Harayama T. *Quantum Chaos and Quantum Dots*, Oxford University Press, New York (2004).

[165] Bernevig B. A. *Topological Insulators and Topological Superconductors*, Princeton University Press, Princeton, (2013); Shen S. Q. *Topological insulators. Dirac equation in Condensed Matter*, Springer Series in Solid-State Science, Springer-Verlag Berlin Heidelberg (2012).

In: Advances in Semiconductor Research
Editors: D. Persano Adorno and S. Pokutnyi
ISBN: 978-1-63321-755-3

Chapter 8

ELECTRON SPIN RELAXATION PROCESS IN SILICON CRYSTALS

D. Persano Adorno*[*]*, N. Pizzolato and C. Graceffa
Department of Physics and Chemistry,
Group of Interdisciplinary Physics and CNISM,
University of Palermo, Palermo, Italy

Abstract

Recently, electrical injection of spin polarization in n-type and p-type silicon has been experimentally carried out up to room-temperature. Despite these promising experimental results, a comprehensive theoretical framework concerning the influence of transport conditions on the phonon-induced spin depolarization process in silicon structures, in a wide range of values of temperature, doping concentration and amplitude of external fields, is still in a developing stage. In order to elucidate the charge transport and spin dynamics of conduction electrons in lightly doped n-type Si crystals we have performed semiclassical multiparticle Monte Carlo simulations and numerically calculated the spin lifetimes of drifting electrons heated by the electric field. Spin flipping is taken into account through the Elliot-Yafet mechanism, which is dominant in group IV materials. Our findings are in good agreement with those obtained by using different theoretical approaches and the most recent experimental results achieved in spin transport devices. Moreover, our Monte Carlo estimations, in ranges of temperature and field amplitude yet unexplored, can be used as a guide for future experimental studies oriented towards a more effective optimization of room-temperature silicon-based spintronic devices.

Keywords: Spin relaxation process, Monte Carlo simulation, silicon, electron-phonon interactions

[*] E-mail address: dominique.persanoadorno@unipa.it. (Corresponding Author)

Introduction

The process of spin relaxation in III-V semiconductors structures has been widely investigated in the last decades [1-5]. Researches along this line have been motivated by the possibility to develop electronic devices that use the electron spin rather than charge as a state variable for processing and storing information. This could allow low-power operations and might also have applications in quantum computing. However, the utilization of spin polarization as a carrier of information must face the decay of each initial non-equilibrium orientation during the transport. Hence, in order to open the way to the implementation of spin-based devices, the features of spin relaxation at relatively high temperatures, jointly with the influence of transport conditions, should be firstly fully understood and interpreted in experiment-related terms.

Recently, experiments where low and moderate electric fields (< 1 kV/cm) were applied have shown very long spin lifetimes in ferromagnet-silicon hybrid systems [6-7]. Moreover, electrical injection of spin polarization in n-type and p-type silicon up to room-temperature have been experimentally carried out [8-10]. Despite these promising experimental results, theoretical research is still incomplete and a comprehensive investigation of the influence of transport conditions on the spin depolarization process in silicon structures, in a wide range of values of temperature and amplitude of external fields, is still missing. In particular, the theoretical description of spin relaxation of conduction electrons due to electron-phonon interactions in silicon bulk were studied by using very elaborate analytical approachs, reaching results not always in agreement among each other[11-18]. Yafet derived a general formula for the spin flipping by taking into account the intravalley scattering with acoustic phonon modes only. The intervalley scattering was included in the more recent works [13-18].

In this chapter we present the results obtained by using a semiclassical many-valley anisotropic Ensemble Monte Carlo (EMC) approach to simulate both the charge transport and the spin dynamics of conduction electrons. Spin relaxation is taken into account through the Elliot-Yafet mechanism [11, 12], which is dominant in group IV materials. Here we first set the parameters of our simplied model for spin-flipping dynamics by fitting the experimental data achieved without any external electric field. Moreover, we show that our numerical results are in good agreement with the findings obtained from different analytical calculations. Then, we compare our numerical outcomes with the experimental results obtained by Li et al. by applying low or moderate electric fields, and we find a very good agreement [18]. To our knowledge, this is the first time that full EMC simulations quantitatively reproduce spin-transport experimental results. Furthermore, this chapter provides numerical estimates of the spin lifetime of electrons drifting in n-type Si crystals at different temperatures and under different electron transport conditions, also where experiments are not yet available.

The chapter is organized as follows. Section 2 includes: (i) some details on Monte Carlo simulation and the Silicon band structure; (ii) a brief introduction on the theory of spin relaxation process; (iii) a description of the model used here to calculate the spin lifetime. In Section 3 we discuss the results from the Monte Carlo code, showing a comparison with both experimental and theoretical findings. Few concluding remarks are given in Section 4.

The Theory

The Semiclassical Ensemble Monte Carlo Approach and the Silicon Model

The correct theoretical description of a semiconductor device can be obtained by self-consistently solving the Boltzmann transport equation, or the quantum mechanical equivalent of it, and the Maxwell field equations. In the presence of very strong driving fields or under cyclostationary conditions, no analytical solution of the Boltzmann equation is known. Approximate solutions can be obtained within drift-diffusion or hydrodynamic models, but the validity of these models is very limited. It becomes necessary to perform a numerical simulation of the process and the Monte Carlo approach represents one of the most powerful methods to simulate the transport properties in semiconductor devices, beyond the quasi-equilibrium approximations [19-21]. This technique, representing a space-time continuous solution of the field and transport equations, is suitable for studying both the steady state and the dynamic characteristics of the devices. Such a semiclassical approach is valid in the case that the built-in and applied electromagnetic fields are spatially slowly varying. EMC consists of electrons' classical "free-flights", during which the electrons may be accelerated by classical forces, interrupted by scattering events, which can alter the electrons' momentum. Owing to its flexibility, Monte Carlo method presents the remarkable advantage of giving a detailed description of the particle motion in the semiconductor, by taking into account the main details of band structure, scattering processes and heating effects, specific device design and material parameters. It allows to obtain all the needed electron dynamics information, such as the average velocity, temperature, current density, directly, without the need of calculating first the electron distribution function. The time interval between two collisions (time of free flight), the collisional mechanisms, the scattering angle, and all parameters of the problem are chosen in a stochastic way, making a mapping between the probability density of the given microscopic process and a uniform distribution of stochastically generated random numbers. To avoid the difficulty of solving integral equations for each scattering event, we use the self-scattering technique. If an electron undergoes a "self-scattering" event, its wavevector before and after the event is unchanged [19-21].

The EMC simulation is carried out by: (a) setting up the initial conditions of the system by giving to the electrons spin polarization, random momentum direction and kinetic energies with Maxwellian distribution depending on the lattice temperature; (b) determining the free flight time of each electron and updating the energy, the momentum and the spin polarization vector at the end of the free flight; (c) selecting a scattering process for each electron; (d) calculating the new value of energy (in case of inelastic scattering) and the scattering angles of each electron. The simulation continues from point (b) [22]. At fixed sampling time steps, small enough to properly update the particle motion, the average values of the physical quantities of interest, such as temperature, average electron velocity, spin polarization vector, are calculated.

The band structure of Silicon shows two kinds of minima. The absolute minimum is represented by six equivalent ellipsoidal valleys (X valleys) along the $\langle 100 \rangle$ directions at about 0.85% of the Brillouin zone. The other minima are situated at the limit of the Brillouin zone along the $\langle 111 \rangle$ directions (L valleys). In our simulation the conduction band of silicon is represented by six equivalent X valleys along the $\langle 100 \rangle$ directions, approximated as ellipsoids having rotational symmetry, with two effective, longitudinal m_L=0.9 m_0 and transverse m_T=0.19 m_0 masses, where m_0 is the mass of the free electron. Because the energy gap between the X and L valley is large (1.05 eV), higher valleys and impact ionization are not taken into account, since for the investigated range of electric fields, the electrons do not reach sufficient kinetic energies for these transitions [23-24].

Our numerical code includes both the intervalley and intravalley scattering of electrons in multiple energy valleys. In particular, the scattering by acoustic phonons is taken into account by employing a deformation potential. Ionized impurity scattering is included under the Brooks–Herring approximation and the intervalley scattering is accounted for by considering six types of optical phonon, three concerning transitions between perpendicular valleys (f-type) and three among parallel valleys (g-type). The scattering probabilities are calculated by using the Fermi Golden Rule and the scattering events are considered instantaneous. Scattering probabilities are assumed to be field-independent; accordingly, the influence of the external fields is only indirect through the field-modified electron velocities [22]. At this stage, we neglect the electron–electron interactions and treat electrons as independent particles. Since the applied field is parallel to the crystallographic axis <100>, the different effective mass component in the field direction leads a difference between hot and cold valleys. For all the numerical simulations discussed below, we used the parameters reported in Table I [23-24]. All results were obtained in Si with a free electron concentration $n=10^{13}$ cm^{-3} (nondegenerate low-doped n-type). We assumed that all donors were ionized and that the free electron concentration was equal to the doping concentration.

Electron Spin Relaxation Process in Silicon Crystal

Electron spin depolarization in semiconductor systems may be caused by interactions with local magnetic fields originating from nuclei and spin-orbit interactions or magnetic impurities. The most relevant mechanisms for spin relaxation of electrons under non degenerate regime are: (i) the Elliott-Yafet (EY) mechanism, in which electron spins have a small chance to flip during each scattering, due to the spin mixing in the conduction band [11-12]; (ii) the Dyakonov-Perel (DP) mechanism, based on the spin-orbit splitting of the conduction band in non-centrosymmetric semiconductors, in which the electron spins decay due to their precession around the k-dependent spin-orbit fields (inhomogeneous broadening) during the free flight between two successive scattering events [25]; (iii) the Bir-Aronov-Pikus (BAP) mechanism, in which electrons exchange their spins with holes [26]. Hyperfine interaction is another mechanism, usually important for spin relaxation of localized electrons, but ineffective in metallic regime where most of the carriers are in extended states [27-29].

In a silicon bulk, due to its inversion symmetry, the Dyakonov-Perel spin relaxation mechanism is absent and the spin depolarization process, under non degenerate regime, is mostly related to mechanisms in which electron spins have a small chance to flip during each scattering. Therefore, in n-type bulk Si the main source of spin relaxation is the Elliot–Yafet mechanism, in which spin-orbit interaction causes spin depolarization via spin-flips during the carrier scattering events. Spin-flip mechanisms can be classified as: (i) Yafet processes, involving spin-dependent interactions, whereas the states are viewed as pure spin states [11] and (ii) Elliott processes, governed by spin mixing of the electron states of the conduction band, due to the crystal spin-orbit coupling, whereas the interaction is spin independent [12].

In the presence of spin-orbit coupling, the electronic Bloch states are given by a combination of pure spin-up $|\Uparrow\rangle$ and spin-down $|\Downarrow\rangle$ states [11-12]:

$$|\boldsymbol{k},\Uparrow\rangle = [a_k|\uparrow\rangle + b_k|\downarrow\rangle]e^{i\boldsymbol{k}\cdot\boldsymbol{r}}$$
$$|\boldsymbol{k},\Downarrow\rangle = [a_{-k}^*|\downarrow\rangle - b_{-k}^*|\uparrow\rangle]e^{i\boldsymbol{k}\cdot\boldsymbol{r}}, \quad (1)$$

where k is the momentum, $\Uparrow,\Downarrow$ the effective spins, a_k and b_k are complex periodic functions with the period of the lattice and b_k^2 the spin-mixing parameter [13]. As a consequence of time-reversal symmetry, the two states are degenerate and are chosen to represent electrons with spins polarized along a given direction. Impurity or phonon scattering can induce transitions between the two states, leading to the spin relaxation [30]. For conduction electrons in silicon, spin-flip scattering is described by a matrix element that depends on the initial and final state wave vectors, as well as on the spin orientation, having the form

$$\langle \boldsymbol{k}_2,\Downarrow |H_{SF}|\boldsymbol{k}_1,\Uparrow\rangle, \quad (2)$$

where H_{SF} is a given spin-flip mechanism.

Table I.

Density [kg/m^{-3}]	2329	
Longitudinal sound velocity [m/s]	9180	
Transversal sound velocity [m/s]	4700	
Crystal dielettric permeability	11.70	
	X Valleys	
Electron effective mass: Longitudinal Transversal	 0.9 m_0 0.19 m_0	
Non parabolicity coefficient [eV^{-1}]	0.5	
Number of equivalent valleys	6	
Acustic deformation potential [eV]	9	
Parameters for phonon-electron collisions in silicon		
Phonon	Phonon energy, meV	Electron-phonon coupling parameter 10^{10} eV/cm
f_1	18.1	0.5
f_2	43.1	0.8
f_3	54.3	3.0
g_1	12.1	0.15
g_2	18.1	3.4
g_3	60.3	4.0

Within the Born approximation, the Elliot-Yafet spin lifetime τ_s can be related to the momentum relaxation time τ_m by the following expression [11, 12, 30, 31]:

$$\tau_s = \frac{\tau_m}{4<b^2>}, \quad (3)$$

where $\langle b^2 \rangle$ is the Fermi surface average of the spin-mixing parameter. However, this relationship is not very accurate in silicon. In fact, it has been pointed out that (i) the spin-mixing parameter $|b_k|^2$ varies in a broad range of values, depending on the value of the momentum k in the Brillouin zone [30-31]; (ii) the scattering mechanisms dominating the relaxation of the momentum are different from those leading to spin relaxation [14-15].

In order to elucidate the spin relaxation process it is necessary to calculate the explicit form of the spin-flip matrix elements between electronic states in the lower conduction band. Different theoretical approaches have carried out the calculation of the matrix element expressions for all the phonon-induced spin-flip transitions in the conduction band. Among these, for example, the pseudopotential model reproducing spin-orbit splittings of the relevant electronic states [13-14], the group theory, the k•p perturbation method, the rigid-ion model [15], the parameter-free first-principles method and the density fuctional perturbation theory [16], the adiabatic band charge models and the tight-binding models [17].

Starting from the detailed knowledge of a specific electron distribution, it is possible to calculate the spin relaxation rate Γ_i for each different scattering mechanism. If several independent mechanisms of depolarization are present, the total spin relaxation time τ_s can be calculated by

$$\frac{1}{\tau_s} = \sum_i \frac{1}{\tau_i} \quad (4)$$

Short relaxation times are dominant, while the longer ones can be neglected [32].

Our Model for Spin-Depolarization

In order to describe the electron depolarization process, differently from the seminal work of Yafet [11], and in agreement with more recent works [13-18], we take into account both intra and intervalley scattering mechanisms. Since the rates of transitions of electrons from the $|\Uparrow\rangle$ to the $|\Downarrow\rangle$ states are proportional to the square of the matrix elements of an electron-lattice interaction, which causes a spin flip [12], our EMC code accounts for the electron spin dynamics by making a mapping between the probability density of a given phonon-assisted spin flip transition and a uniform distribution of stochastically generated random numbers.

In our simplified model, spin depolarization due to the electron-impurity interaction is neglected because the electron-phonon interaction is the dominant spin relaxation mechanism in nearly intrinsic samples. We point out that, as shown in Ref [16], in heavily doped Si crystals the additional scattering with the impurities causes a shortening of the spin lifetime. Therefore, we consider the spin-flipping induced by: (i) the intravalley acoustic phonons and (ii) all phonon modes of the intervalley f-processes. Spin-flipping due to g-processes can be neglected because their contribution is not significant in the temperature range considered (60<T<400 K) (see Figure 8 in Ref. [15]). More details and the explicit calculation of the spin-flip matrix elements can be found in Ref. [15].

It has beeen shown that the spin flipping probability is of the order of $10^{-7}\div10^{-6}$, and it depends on the value of **k** [15]. In particular, more energetic electrons strongly drive the average spin relaxation [13]. In our model, the dependence of the spin-flipping probability on the value of the momentum **k** has been considered the same for the two mechanisms here considered, while its dependence on spin orientation has been neglected.

The dependence of spin relaxation times on temperature and/or electric field amplitude, has been investigated by simulating the dynamics of 5×10^3 electrons, all initially spin polarized (P=1) along the z axis of the crystal at the injection plane. We have calculated the polarization P as a function of time by averaging over the ensemble of electrons. Since the mean polarization is found to decrease with time by showing an almost linear trend in a semi-log plot, the spin relaxation time is estimated by considering the spin depolarization as an exponentially process dependent on time [4-5]. Thus, the spin lifetime τ_s corresponds to the time necessary to achieve a reduction of the initial spin polarization by a factor 1/e.

Numerical Results and Comparison with Experiments

In Figure 1 the experimental spin relaxation times τ_s obtained without a driving electric field and reported in Ref. [9, 33-34] are shown as a function of the crystal temperature, together with the solid line which represents the results from our simulations. The agreement with the experimental data is satisfactory in the whole range of lattice temperatures. This accordance was obtained by using a spin-flipping probability equal to 10^{-6} in the case of acoustic phonon assisted transitions, and equal to $2\bullet10^{-6}$ in the case of flipping induced by interactions with optical f-phonons. This is in agreement with the analytical finding that the spin-flip matrix element of an f-process should be larger than that of the other processes [14, 18]. In both cases we have chosen a linear dependence of this probability on the wavevector **k** of the electron suffering the scattering. Furthermore, as shown in Ref. [15], the spin-flip matrix elements, rather than the electron wavevector, explicitally depend also on the phonon polarization, the valley position and spin orientation of the electron. Here we neglect these further dependences.

In Figure 2 the electron spin lifetime, numerically obtained by our simulations, is compared, as a function of the temperature, with the most recent theoretical and analytical findings. Despite our EMC code does not take into account the scattering of electrons with g-phonons, the interactions with the impurities and it uses approximated relations for the spin-flip rates, the good agreement with the analytical theories strengthens the reliability of our EMC algorithm. We point out that the results obtained by very elaborate numerical and analytical methods are not in a full agreement, as demonstrated by the wide range (about one order of magnitude) of possible values of the spin lifetime in the whole range of lattice temperature.

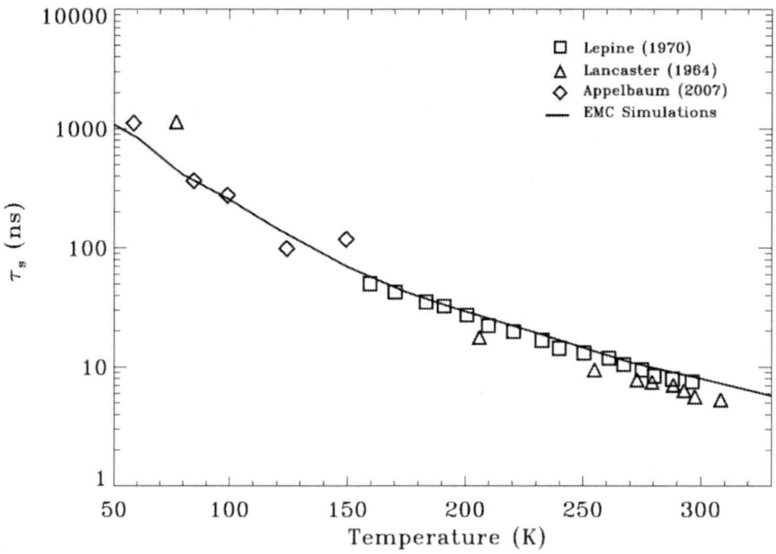

Figure 1. Spin relaxation time in silicon as a function of the temperature. Symbols are electron spin resonance (Lepine and Lancaster) [33, 34] and spin injection (Appelbaum) [9] experimental data; the solid line represents our numerical calculation.

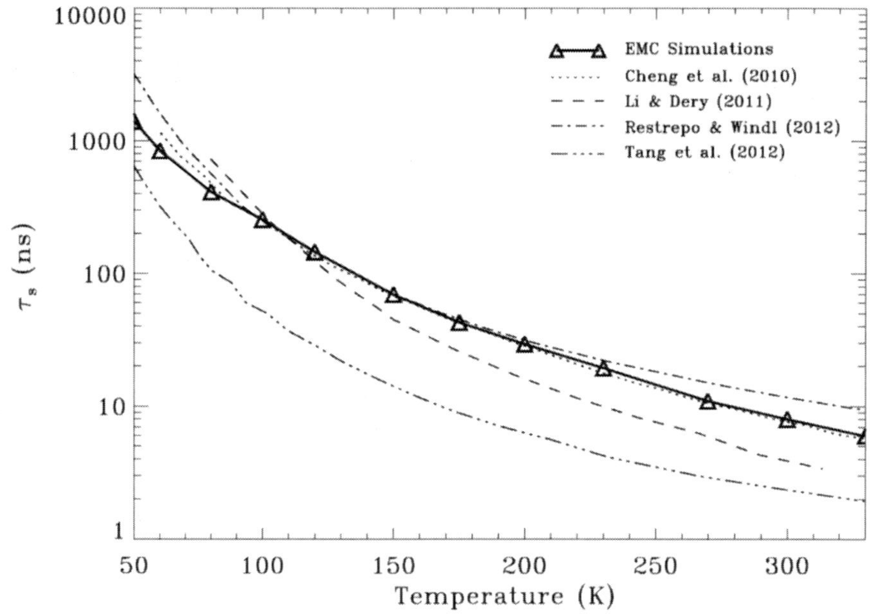

Figure 2. Comparison between our numerical estimates for the spin relaxation time in silicon as a function of the temperature and findings of the most recent theoretical approaches.

A previous comparison between the acoustic and optical phonon contribution to the spin relaxation showed that the AC phonon scattering is dominant in the low temperature range (T<150 K), while OP phonons assisted spin-flip transitions cannot be longer neglected at higher temperatures [35]. This is consistent with the circumstance that long-wavelength acoustic phonon modes are mostly populated at smaller energies (lower temperatures), whereas the growing up of the phonon population with the temperature for optical modes is faster than that for acoustic modes. Therefore there is not a single scattering type governing the whole investigated temperature range: both intra- and inter-valley spin-flip mechanisms are important, making invalid Yafet previsions above T=150 K.

Recently, Li et al. (2012) have performed observations of nonequlibrium spin depolarization by employing all-electrical devices in which spin-polarized hot electrons are tunnel injected through a Schottky metal contact and into a 225 μm-thick wafer of nominally undoped Si <100>, with their spin axis in the device plane. Then the electrons drift across the wafer thickness due to an applied voltage and are collected by a second ferromagnetic film where their spin is analyzed using a ballistic spin detection scheme [18]. Since the electrons are initially spin-polarized, the Elliott-Yafet spin relaxation theory suggests that the spin depolarization during the transport depends only on the time of flight. In that theory the spin and the momentum relaxation times are proportional; thus, the spin polarization should increase with the electron drift velocity. Li et al. (2012) found that, at T>30 K, the spin depolarization of detected electrons first increases with the increasing of the electric field, as expected from the Elliott-Yafet model, but then, unexpectedly, it starts to decrease, showing a Gunn-type behavior [18].

In Figure 3 several values of the electron spin polarization measured after the transport across the Si-channel as a function of the applied electric field are drawn for different temperatures, namely T=60, 90 e 120 K. The experimental data are gathered from the findings reported in Figure 1 of Ref. [18]. Our EMC numerical estimates (lines in Figure 3) well reproduce the experimental negative differential region of spin lifetime. Our curves show that for electric fields strong enough (3-4 kV/cm) the residual polarization reaches a maximum value and then starts to decrease.

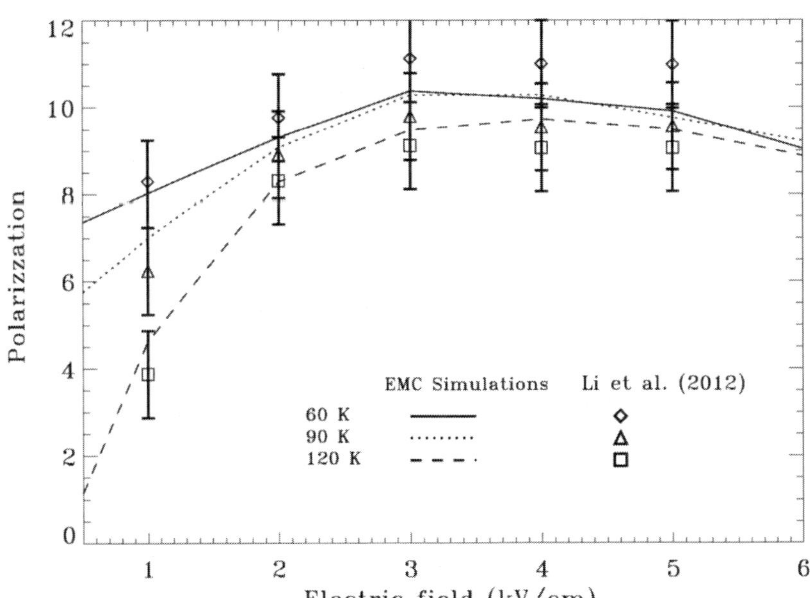

Figure 3. Comparison between the electron spin polarizations measured after the transport across the Si-channel (symbols with error bars) and our EMC numerical estimates (lines) as a function of the applied electric field, for different lattice temperatures, namely T= 60, 90 e 120 K.

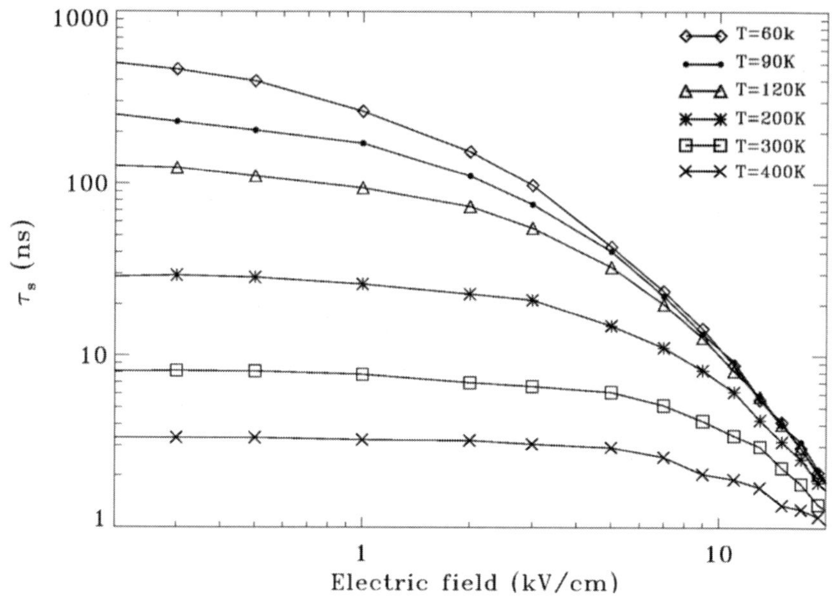

Figure 4. Ensemble Monte Carlo simulation results of the electron spin relaxation time as a function of the driving electric field in a wide range of lattice temperature.

This non-monotonic behavior could be ascribed to a progressive decrease of the spin relaxation time τ_s at the increasing of the applied electric field. Indeed, with the increasing of the electric field, thanks to the switching on of the intervalley scattering, other spin relaxation mechanisms become accessible. In order to verify this hypothesis we studied the dependence of the electron spin polarization on the amplitude of the driving electric field.

Our findings are shown in Figure 4 for different lattice temperatures in the range 60-400 K. The driving electric field significantly affects the electron momentum distribution that deviates from the equilibrium condition even at low temperatures. This fact is responsible for an enhancement of intervalley scattering and explains the rapid drop of spin lifetimes observed at electric fields higher than 3 kV/cm. At room temperature, where experimental data are still missing, our calculation predicts spin lifetimes τ_s of the order of 8 ns up to field amplitudes ~5 kV/cm. When the electric field amplitude is greater than 15 kV/cm, the effect of the temperature becomes nearly negligible; all the curves collapse and the spin relaxation times take values ranging between 1 and 2 ns.

Conclusion

The understanding of spin-related properties and spin transport in Si and its compounds is important for solid state physics and possible applications of these materials in spintronics. In fact, silicon is an ideal material for spintronics due to its long spin relaxation time and central role in semiconductor technology. However, in order to make silicon-based spintronics a viable prospective technology, we need to find out the best conditions to achieve long spin relaxation times (or spin diffusion lengths) in silicon devices.

In this chapter we report preliminary results of numerical calculations of the spin lifetime for conduction electrons drifting in a silicon channel. To model both electronic and spin dynamics in Si and to estimate the spin relaxation time, we employ the ensemble Monte Carlo method. By using a simple model which takes into account the spin-flipping mechanism due to intravalley acoustic and intervalley optical f-type scatterings, we show that the intervalley scattering of hot electrons has a larger role in spin-related phenomena than in charge transport, accordingly with recent experimental findings. This behavior could become critically important in the design of spin-based devices, especially when strong static electric fields are required. Our numerical findings are in agreement with the most recent results obtained by using different theoretical approaches. Moreover, they well describe results experimentally achieved with or without an applied electric field. We also show how spin lifetimes change in a wide range of temperature and electric field amplitude, even where experimental and/or analytical data are not yet available. From this point of view, the results obtained by our Monte Carlo simulations represent a guide for future experimental studies and could be very useful in a more effective design of room-temperature silicon based spintronic-devices.

References

[1] J. M. Kikkawa and D. D. Awschalom, *Phys. Rev. Lett.* 80, 4313 (1998).
[2] J. Fabian et al., *Acta Phys. Slov.* 57, 56 (2007).
[3] M. W. Wu, J. H. Jiang, M. Q. Weng, *Phys. Rep.* 493, 61 (2010).
[4] S. Spezia, D. Persano Adorno, N. Pizzolato and B. Spagnolo, *J. Stat. Mech. Theor. Exp.* P11033 (2010).
[5] S. Spezia, D. Persano Adorno, N. Pizzolato and B. Spagnolo, *Acta Phys. Pol.* B 41, 1171 (2010).
[6] B. Huang, D. J. Monsma, and I. Appelbaum, *Phys. Rev. Lett.* 99, 177209 (2007).
[7] B. Huang and I. Appelbaum, *Phys. Rev. B* 82, 241202 (2010).
[8] B. T. Jonker et al., *Nature Phys.* 3, 542 (2007).
[9] I. Appelbaum, B. Q. Huang, and D. J. Monsma, *Nature* 447, 295 (2007).
[10] S. P. Dash et al., *Nature* 462, 491 (2009).
[11] Y. Yafet, in *Solid State Physics*, edited by F. Seitz and D. Turnbull (Academic New York, 1963), Vol 14 (1963).
[12] R. J. Elliott, *Phys. Rev.* 96, 266 (1954).
[13] J. L. Cheng, M. W. Wu and J. Fabian, *Phys. Rev. Lett.* 104, 016601 (2010).
[14] P. Li and H. Dery, *Phys. Rev. Lett.* 107, 107203 (2011).
[15] Y. Song and H. Dery, *Phys. Rev.* B 86, 085201 (2012).
[16] O. D. Restrepo and W. Windl, *Phys. Rev. Lett.* 109, 166604 (2012).
[17] J. M. Tang, B. T. Collins and M. E. Flattè, *Phys. Rev.* B 85, 045202 (2012).
[18] J. Li, L. Qing, H. Dery and I. Appelbaum, *Phys. Rev. Lett.* 108, 157201 (2012).
[19] C. Jacoboni and L. Reggiani, *Rev. Mod. Phys.* 55, 645 (1983).
[20] K. Tomizawa, *Numerical Simulation of Submicron Semiconductor Devices*, Artech House Publishers (1993).
[21] C. Moglestue, *Monte Carlo Simulation of Semiconductor Devices*, Springer (1993).
[22] D. Persano Adorno, M. Zarcone and G. Ferrante, *Laser Phys.* 10, 310 (2000).
[23] D. Persano Adorno, M. Zarcone and G. Ferrante, *Laser Phys.* 11, 291 (2001).
[24] D. Persano Adorno, M. Zarcone and G. Ferrante, *Laser Phys.* 13, 270 (2003).

[25] M. I. Dyakonov and V. I. Perel, *Sov. Phys. - Solid State* 13, 3023 (1971).

[26] G. L. Bir, A. G. Aronov and G. E. Pikus, *Sov. Phys. - JETP* 42, 705 (1976).

[27] A. Abragam, *The Principles of Nuclear Magnetism*, Clarendon Press, Oxford (1961).

[28] D. Paget, G. Lampel, B. Sapoval and V. I. Safarov, *Phys.Rev.* B 15, 5780 (1977).

[29] G. E Pikus and A. N. Titkov, *Optical Orientation*, edited by F. Meyer, *Nauka* (Leningrad) (1989).

[30] J. Fabian and S. Das Sarma, *Phys. Rev. Lett.* 81, 5624 (1998).

[31] J. Fabian and S. Das Sarma, *Phys. Rev. Lett.* 83, 1211 (1999).

[32] A. W. Overhauser, *Phys. Rev.* 89, 689 (1953).

[33] D. J. Lepine, *Phys. Rev. B* 2, 2429 (1970).

[34] G. Lancaster et al., *Proc. Phys. Soc. London* 84, 19 (1964).

[35] D. Persano Adorno, C. Graceffa, N. Pizzolato and M.A. Lodato, *Lith. J. Phys.* 54, 25 (2014).

SECTION 3.
TECHNOLOGICAL APPLICATIONS

In: Advances in Semiconductor Research
Editors: D. Persano Adorno and S. Pokutnyi ISBN: 978-1-63321-755-3

Chapter 9

MAGNETOSENSITIVE NANOCOMPOSITES WITH FUNCTIONS OF MEDICOBIOLOGICAL NANOROBOTS: SYNTHESIS AND PROPERTIES

P. P. Gorbyk**, *L. B. Lerman, A. L. Petranovska and S. P. Turanska
O. O. Chuiko Institute of Surface Chemistry
of NAS of Ukraine, Kyiv, Ukraine

Abstract

The results of studies are generalized in order to support the concept of chemical design of magnetically sensitive nanocomposites with multi-level hierarchical architecture and functions of medicobiological nanorobots: recognition of microbiological objects in biological media; the targeted drug delivery into target cells and organs, and deposit; complex treatment with chemo-, immuno-, radio-, hyperthermic methods and real-time diagnostics; adsorption of residual cell debris and removing them from the body with the help of an external magnetic field. The data regarding the synthesis, properties, and practical use of nanocomposites is given.

Keywords: Magnetosensitive nanocomposites, multilevel hierarchical nanoarchitecture, medicobiological nanorobots, targeted drug delivery

Introduction

There is no doubt that nowadays malignant tumorous formations represent one of the most difficult medicobiological problems. Positive results of cancer patients therapy depend on the early diagnostics and timely medicinal cure or surgical operation.

* E-mail address: pgorbyk@mail.ru, gorbyk@isc.gov.ua. (Corresponding author)

However, in most cases not only the transformation of a normal cell into malignant form, but also the formation of tumor nidus remains unnoticed until the clinical observation of a disease [1].

Therefore, the actual objective for the researchers all over the world is to develop new methods of early diagnostics of tumors and to elaborate medicines with selective effects upon malignant cells. Great hopes for achievement of these goals are laid upon the development of nanotechnology [1-3].

At present, nanotechnology is the most dynamical project in the field of science and medicine [4]. Its application in oncology would allow diagnostics, prevention and treatment of cancer patients before the clinical observation of disease [1, 3, 5]. For many centuries humanity has been looking for remedies to avoid the tumor diseases. However, the outstanding success of fundamental sciences in the field of cancer biology, chemistry, pharmacology achieved in the last decades did not result in the desired success in clinical oncology. This primarily refers to the selectivity of accumulation and, respectively, to the specificity of the known cytostatic agents. Undoubtedly, it is striking that only a tiny fraction (1:10000–1:100000) of the intraperitoneally introduced dose, even on basis of such modern carriers as monoclonal antibodies, reaches the specified target [3, 6].

In principle, nanotechnology allows delivering the medicine directly to a small clone of malignant cells [3, 7]. This also applies to the contrast agents for visualization [8, 9].

One of the most actual objectives in modern oncology is to design novel antitumor remedies. Nonselectivity of the known medicines, as well as their high general toxicity, stimulates both the further search for new compounds and improvement of the existing medicine forms with the purpose of increasing their selectivity and safety [3]. The modern nanotechnology not only can solve this task, but also is able to realize multivector character of the antitumor action mechanism of nanocomposites [10]. It has been long understood that targeted delivery of the active agents is a fundamental part in the development of medicines. A wide range of systems for drug delivery should improve their stability and therapeutical concentration in a target tissue, as well as directed adsorption and influence on the molecular target [11]. In most cases such nanocomposites include compounds known as agents for pharmaceutical correction or direct cytostatic action [2, 5, 11]. In certain cases the compositional reconstruction of the known preparations involving one or another type of nanoparticles can improve their targeted accumulation upon peroral introduction [12]. The inclusion of nanoparticles with predetermined properties can provide both protection of the mucus cover of gastrointestinal tract and protection of the preparates themselves from degradation in locations with severe pH conditions. The surface modification of nanoparticles allows covalent or noncovalent attachment of a wide variety of different compounds [5, 11].

As active nanosystems for targeted influence one can successfully use folates [13, 14], as well as monoclonal antibodies for the surface proteins of cancer cells and/or integrines [2, 6, 11]. Using such active nanosystems, a composite can react with molecules both at the cell surface and at the level of intercellular cascades of signal transmission, participating in the regulation of broken signal pathways [11]. A number of nanosized systems for medicine delivery have already reached the stage of the first clinical trials [15, 16].

Considering the whole complex of affecting factors in the process of pharmacokinetics directed by a nanocomposite, there are good reasons to hope that a broad range of antitumor action mechanisms can be involved in the achievement of the effect – from antisignal ones to hyperthermia [3, 10, 11]. With the use of multilevel nanocomposites and their electromagnetic maintenance, the real possibility is emerging to modulate pharmacokinetic processes in tumor tissue in the system *in vivo*. The most significant constituent in this process is the possibility of additional creation of uniform and controlled hyperthermic effect using the external magnetic field [17, 18]. The synergism of anticancer effect is reached with the account of cytostatic action of temperature and also owing to additional and accelerated release of the anticancer agent from a nanocomposite. Using a ferromagnetic material composed of iron and nickel alloy and vanadium dioxide, one can create a nanocomposite (dielectric-conductor, conductor-dielectric) that can maintain the temperature at Kurie point (43–45 °C) in an automatic regime on basis of the phase transition.

Ideally, nanocomposites can have longer cycling time not losing their efficiency. Accounting for their selectivity, novel medicinal systems [18–29] can significantly increase not only the efficiency of anticancer drug [22–29], but also largely improve the quality of patient's life [1–5].

In this work the scientific and applied aspects have been analyzed related to the modern approaches for creation of novel medicine forms *via* chemical design of polyfunctional nanocomposites for medicobiological use with multilevel architecture and a programmed complex of functions. Sensations in media regarding the creation of nanorobots capable to destroy cancer cells are not surprising any more.

Unfortunately, these messages contain no data to confirm their authenticity. But certainly timeliness and relevance of the issue raised remain [19]. According to Wikipedia, nanorobots are characterized by functions of movement, processing and transmission of information, the implementation of specific programs. Their sizes are ~ 10 nm.

Modern scientific publications show that the problem of creating nanomachines and nanorobots, in particular for applications in medicine and biology, has acquired relevance [18-21, 23-26, 30-32]. At the initial stages of finding ways to solve it, an important place is given to the task for establishing of promising physical principles and methods for practical fulfillment of controlled mechanical movement of nanoobjects in biological media [18-21, 23-26, 30-34]. For the organization of such movement and for building of nanorobots one can use, for example, the fragments of DNA molecules, muscle nanofibers [20, 21], photoexcited and metastable states of nanoparticles, asymmetric distribution of electric dipoles in the direction of motion, fluctuation processes of molecular and electric nature [33, 34], ring-like biomolecules "strung" on the linear "bridge" of amino acids [31, 32], etc.

However, it should be noted that the practical implementation of such [20, 21, 31-34] ways of nanorobots movement and mechanical manipulation with their components in order to provide, for example, the directed transport of drugs or to realize other useful functions, remains rather problematic perspective.

Nevertheless, the current level of nanotechnology allows one to create unique products for medicine and biology [18, 23, 27, 28]. The implementation of such technologies into practice is the foundation of modern advances in the field of diagnostics and treatment of diseases, particularly on the cellular and genetic levels [19].

In recent years, as an alternative and the most promising trend in terms of practical use, priority was given to the works in the field of creating "nanoclinics" – the multilevel magnetosensitive nanocomposites having the complex of functions that are characteristic of nanorobots, including recognition of specific microbiological objects in biological media, targeted delivery of drugs to organs or cells, diagnostics and treatment of diseases at the cellular and genetic levels, adsorption of cells decomposition products after exposure to the drug or therapeutic hyperthermia, removing them from the body using magnetic fields, etc. [18, 23, 27].

Thus, the authors [18] obtained the magnetosensitive nanocomposite by the multistage sol-gel synthesis in the micellar system. The surface of the nanocomposites was functionalized to ensure recognition and reach for specific cells. Such nanocomposites are the magnetosensitive carriers in silica shells of up to 50 nm size, which can contain different encapsulated optical, magnetic or electrical sensors, therapeutic agents. Nanocomposites are delivered into the target cell and used for diagnosing or change its functions in a predetermined direction, as well as for generation of hyperthermic zones. Application of local hyperthermia in principle allows realizing cancer therapy without using of chemicals.

It should be noted that the interest in magnetic materials for use in medicine appeared long ago and continues unabated today. Attention of researchers to magnetic nanoparticles, especially to biogenic magnetite possessing favorable biodegradability, is, in particular, because they have specific properties that are not typical for a bulk magnetic material. So, the new phenomena are observed in nanomaterials, associated with differences in Curie or Neel temperature up to hundreds of degrees, the emergence of giant magnetoresistance, anomalous magnetocalorimetric effect, etc. [35].

In modern medicine the magnetic drugs based on magnetite are widely used as roentgen contrast agents. For example, the roentgen contrast agent is described in [36], which contains magnetite with a liquid carrier – tetradecane. The efficiency of haemostatic remedy based on the aqueous colloidal solution of magnetite is substantiated in [35]. Another use of the nanosystems is to visualize biological tissues consisting of apoferritine with magnetite crystals to be included in its cavity. Promising trends of using magnetic materials with advanced surface are preparation of sorbents [37-39], magnetosensitive nanocomposites for the targeted drug delivery [40, 41]. The characteristics of magnetite are low toxicity, high level of mutagenic safety, absence of adverse reactions in the organism after intravenous, intraarterial, and intramuscular injections of magnetosensitive colloid.

In addition, therapeutic agents with different mechanisms of action can be immobilized onto the surface of nanodispersed magnetite.

The presence of a drug in the composition of the magnetic therapeutic remedies [42] is not always obligatory. In this case, therapeutic effect is achieved either due to the influence of a constant magnetic field, its source being a magnetic remedy itself, or by hyperthermic treatment when a magnetic remedy is exposed to an external high-frequency magnetic field [18, 27].

By changing the size, shape, composition and structure of nanoparticles one can control to some extent the magnetic characteristics of composites. However, not always these factors are controllable during the synthesis of nanoparticles, therefore the properties of materials of even the same type can vary greatly. In particular, there is a strong dependence of the magnetic properties of ferromagnetic materials on the particle size [25, 43, 44].

Applying the appropriate preliminary treatment of the surfaces of magnetic carriers to increase their sorption capacity (for example, modifying with SiO_2), one can collect the products of cell decomposition and remove them from the body with the use of magnetic field.

While developing magnetically guided pharmaceuticals one needs to solve a number of problems related to the synthesis of efficient nanosized carriers, modification of their surface, immobilization of drugs, encapsulation, etc. It is important to implement theoretical evaluation of the conditions of targeted delivery and calculation of the carrier parameters for their optimization [27, 45-47].

Thus, while the modification of the surface of carriers with biocompatible polymers it is necessary to optimize functions of the polymer component. Polymer molecules can act as a holder of therapeutic or diagnostic remedy and determine the important characteristics of the drugs: solubility, bioavailability, prolonged action by slow desorption of drugs from a polymer matrix, shelf life, etc. The ways of fixing pharmacologically active substances on the polymer surface may also be different, being determined by the destination of drug and the clinical pathology [48].

It should also be taken into account that the anticancer chemotherapeutic drugs *in vivo* disrupt nucleic acid metabolism, prevent the processes of biosynthesis and function of DNA, inhibit hematopoiesis, digestion, being cardio-, neuro- and nephrotoxic [42]. Therefore, the problem of targeted drug delivery in tumor cells using magnetic carriers is extremely important. Their fixation and deposition in tumor nidus by magnetic field makes it possible to reduce significantly the drug dose and, consequently, to minimize the toxic and allergic reactions in an organism.

The use of monoclonal antibodies in the oncology practice refers to methods of passive immunotherapy [49, 50]. Clinical data suggest that passive immunotherapy is definitely effective and its results equate with chemotherapy ones, but the toxicity level is much lower. The combined use of these methods is considered as the most promising, since it greatly enhances the possibilities of treatment with modern drugs and favors to increase in their effectiveness.

One of the varieties of malignant tumors radiotherapy is a neutron capture therapy (NCT) [51]. NCT is a cancer therapy that uses radiation from the reaction of capturing thermal neutrons by ^{10}B or ^{157}Gd isotopes, previously accumulated in tumor cells [52, 53]. The result of thermal neutrons interaction with isotopes is a significant cytotoxic effect caused by the influence of the products of nuclear reaction on cancer cells.

Despite the much larger capture section of neutron by ^{157}Gd isotope, compared to ^{10}B, the use of gadolinium neutron capture therapy (GNCT) was prevented by the toxicity of free, i.e. unbound gadolinium ions. Interest in GNCT was escalated in late 1980 with the introduction into practice the universal gadolinium-containing agents for magnetic resonance contrast imaging diagnostics (MRI).

NCT agents proposed till the present day are in a molecular form and thus have a number of shortcomings. Among them one can distinguish the inability of targeted delivery to the tumor and deposit, the combined T_1/T_2 diagnostics and therapy in real time, a small content of boron or gadolinium, and so on.

These drawbacks can be avoided by using nanocomposites based on the magnetosensitive magnetite nanoparticles as well as boron and gadolinium containing preparations [29, 54]. The nanocomposites synthesized can be used to create the new types of highly efficient remedies for neutron capture therapy with additional functions of magnetically directed targeted delivery to specific organs or cells, selective accumulation and deposition, hyperthermia, combined T_1- , T_2- MRI diagnostics and real-time therapy.

In the developed countries the priority has been given to researches aimed at the development of magnetically guided medicinal preparations for complex chemo-, immuno-, radiotherapeutic, hyperthermia and diagnostic actions. The trends associated with early diagnostics and cancer treatment at the cellular level are particularly relevant. The mentioned circumstances have led to the choice of scientific issues described in this review.

It summarizes the results of studies in order to support both concept and methodology of chemical design of magnetically sensitive nanocomposites with multilevel hierarchical architecture and functions of medicobiological nanorobots [22]: recognition of microbiological objects in biological media; the targeted drug delivery into specific cells and organs, and deposit; complex treatment with chemo-, immunotherapeutic, radiation neutron capture, hyperthermic methods and real-time diagnostics; adsorption of residual cell remains and removing them from the body with the aid of an external magnetic field.

1. Analysis of Delivery and Retention Conditions of Magnetically Sensitive Drug Carriers with the Help of Magnetic Field

The analysis of delivery and retention conditions of magnetically sensitive drug carriers with the help of magnetic field was carried out in [25, 55].

Realization of function of the targeted delivery and deposition of drugs into specific cells and organs by medicobiological magnetically sensitive nanocomposites (nanorobots) is to create the therapeutic concentration of drug in the corresponding place of the body, and prolong its action. Magnetically guided targeted delivery of drug and its retention over time might be the way to solve this problem. Therefore, in recent years the study on the possibility of using magnetic carriers as the means to deliver drugs have attracted considerable attention. First of all, this is due to the expected significant extension of the scope of chemotherapy, particularly by minimizing the undesirable toxic and allergic effects. Since the treatment is carried out in a localized area of the body, the total dose of the drug may be minimal. The external static magnetic field should affect the motion of the particles and prevent leaching the carriers from a target organ. However, widespread use of this method has not yet been found because of the difficulties arising from its practical implementation.

The requirements for magnetic carriers are rather rigid. Their size should be less than 1 micron to prevent aggregation in the magnetic field and occlusion of capillary vessels. At the same time, the force caused by the applied external magnetic field to keep the carrier in the blood flow is proportional to the characteristic size of the particle in the third degree and to the quantity of magnetic component in it.

Opposing hydrodynamic force trying to leach the carrier from the target organ is proportional to the size of the carrier only in the second degree. Thus, for efficient retention of the carrier, the size of particles and the content of magnetic material should be increased.

Besides resolving this contradiction, when creating magnetic carriers, one should take into account the factors such as reducing residual time of carriers (as foreign bodies), diminishing the influence of external magnetic field on the organism, etc.

To keep the preparation in a given limited area, one must create a zone of effective action of inhomogeneous magnetic field, under the influence of which the significant portion of (or all) the carriers should remain in the target organ. These circumstances lead to the need for solvation of the following tasks:

- calculation of the force of inhomogeneous magnetic field acting on separate carriers;
- definition of the conditions (the magnetic field gradient value, magnetization degree of the carrier, size distribution, etc.) necessary for obtaining quantitative estimates of the parameters in order to determine the possibility of keeping containers with medicinal preparation at the vessel wall;
- analysis of magnetic carriers movement in a viscous fluid (blood) under the influence of inhomogeneous magnetic field to obtain criterion of the possibility of their delivery to the walls of blood vessels.

Some aspects of these issues are outlined in [55-58]. The technological aspects of development of magnetically guided drug carriers were discussed. Analysis of these papers shows the complexity of constructing of both physico-mathematical models of the given phenomena and related computational methods and algorithms.

Therefore, special attention was paid to the improvement of models and construction of numerical methods that adequately describe the experimental results [25].

Here are the formulas for deriving the magnetic force acting on the separate carriers with taking into account their size distribution. Calculation of the magnetic field distribution in space is associated with the decision of specific tasks of magnetostatics. The most appropriate sources are these having an axis of symmetry [55]. They create axially symmetric fields, and in this case the vector potential has only one other than zero component, A_φ, that must satisfy the scalar Poisson's equation

$$\Delta A_\varphi = -4\pi \mu_0 j_\varphi \tag{1}$$

where Δ - the Laplace operator; j_φ - the apparent density of ring current, $\mu_0 = 4\pi \cdot 10^{-7}$ Hn / m – the magnetic constant.

Solution of the equation (1) can be represented as a volumetric potential

$$A\varphi = \frac{\mu_0}{4\pi} \iiint_V \frac{j_\varphi}{R} dv \quad , \tag{2}$$

where $R = \sqrt{r^2 + \rho^2 - 2r\rho\cos(\varphi - \theta) + (z - \zeta)^2}$.

Integration in (2) is performed over V area occupied by the current sources. As $(r,\ \varphi,\ z)$ one can mark cylindrical coordinates of a free point in space, and as $(\rho,\ \theta,\ \zeta)$ – cylindrical coordinates of a point that belongs to V area). Magnetic induction vector is determined from the equality $\mathbf{B} = rot\mathbf{A}$. Magnetic carriers (separate nanoparticles or their aggregates) can be considered as magnetic dipoles. In general, the dipole is influenced by the force $\mathbf{F}$ and mechanical moment $\mathbf{M}$ which are defined by the known formulas:

$$\mathbf{F} = -\nabla(\mathbf{m} \cdot \mathbf{B}) \tag{3}$$

where $\mathbf{m}$ – vector of the effective magnetic moment of a single carrier or nanoparticle aggregates; $\mathbf{B}$ – vector of magnetic induction, and ∇ symbol marks the gradient operator.

Due to the mechanical moment, free dipoles orient along the field, and then $\mathbf{M} = \mathbf{0}$. Dipoles turning quite quickly, we can assume that they are always focused along the field. The force of magnetic interaction depends not only on change of magnetic field, but also variations of magnetic moment of the carrier. If these values are constant in space then $\mathbf{F} = 0$. Hence, to obtain the required values of the forces, magnetic field should vary in space.

Note that

$$\nabla(\mathbf{m} \cdot \mathbf{B}) = |\nabla(\mathbf{m} \cdot \mathbf{B})| \cdot \mathbf{n} \tag{4}$$

where $\mathbf{n}$ – the unit vector of outer normal to the field lines $\varphi = \nabla(\mathbf{m} \cdot \mathbf{B})$. Then derivative for a free direction $\mathbf{e}$ looks as

$$\frac{\partial \varphi}{\partial e} = \mathbf{e} \cdot \nabla \varphi = |\nabla(\mathbf{m} \cdot \mathbf{B})| \cos(\mathbf{e}, \nabla \varphi),$$

where

$$|\nabla(\mathbf{m} \cdot \mathbf{B})| = \sqrt{\left(\frac{\partial(\mathbf{m} \cdot \mathbf{B})}{\partial x}\right)^2 + \left(\frac{\partial(\mathbf{m} \cdot \mathbf{B})}{\partial y}\right)^2 + \left(\frac{\partial(\mathbf{m} \cdot \mathbf{B})}{\partial z}\right)^2}$$

Consider the one-dimensional inhomogeneous field with induction $B = B(y)$, acting along the axis. If you ignore the change of magnetic moment in space, the expression for the magnetic force according to equation (4) can be written as

$$F_y = V \cdot \frac{d(m(H)B)}{dy} \tag{5}$$

where V – the volume of the particle; – the field intensity associated with the magnetic induction by the ratio $B = \mu_0 H$; – the average magnetic moment corresponding to the unit volume of the particle.

The value of the effective magnetic moment of nanoparticles and aggregates depends on their size, shape, magnetization, temperature and some other physical factors, and it is unknown in advance. For ferromagnetic materials, magnetization is a function of module of the magnetic field intensity vector, and for small values of intensity of the fields, the average magnetic moment per unit volume, $m(H)$, is proportional to the module of the magnetic field intensity vector, H, i.e. $m(H) = \kappa H$, where κ – the proportional coefficient. From the formula (5) it follows that in this case the magnetic field force to act on a particle of free shape is proportional to its volume and the product of intensity module and induction module gradient. Such regularity is fulfilled only for the intensities, while magnetization of the particles is far from saturation.

Then formula (5) can be rewritten as

$$F_y = \kappa V \frac{d(HB)}{dy} \quad (6)$$

The value of the proportional coefficient, κ, depends on the particle shape. For continuous rotation ellipsoids with different ratios of length to diameter D_p, the values for are given in Table 1, which reveals that the ratio (the shape factor) affects substantially the value of magnetic moment of the particle.

In the case of particles with a magnetic core and shell in the form of an ellipsoid, due to the existence of analogy between the tasks of electrostatics and magnetostatics, in order to calculate the magnetic moment, one can use the method [40] proposed for the calculation of laminate ellipsoid polarization.

At high levels of field intensity for ferromagnetic materials the magnetization reaches saturation, i.e. $m(H) \rightarrow m_{sat}$, where – the magnetic moment per unit volume of the material in reaching the saturation magnetization. In this case, from (6) we have

$$F_y = \kappa V m_{sat} \frac{dB}{dy}, \quad (7)$$

and thus the level of field intensity ceases to influence the force, the latter depending only on the induction module gradient. These regularities will be fulfilled also for a free field distribution in space, if we assume that the y axis coincides with the direction of magnetic induction vector.

Since the force depends strongly on the geometrical parameters of the particle, it is necessary to take into account the size distribution of nanoparticles or their aggregates.

The density of particle size distribution (by the diameter, volume) is largely determined by the preparation technology. The study was carried out [56] on the size distribution of nanoparticles using the data obtained by three different methods. It has been found that the most suitable for magnetic carriers is logarithmically normal distribution with its probability density $p(D_p)$ for spherical particles given by

$$p(D_p) = \frac{1}{D_m \sigma \sqrt{2\pi}} \exp\left\{-\ln^2(D_p/D_m)/(2\sigma^2)\right\} \quad (8)$$

where D_p and – the diameter and the most probable diameter (the mathematical expectation) of the particle, respectively; σ – standard deviation from the mean value.

Table 1. Proportional coefficient for calculation of the magnetic moment of rotation ellipsoids with different ratio of semiaxes values

L/D_p	0.0	0.1	0.25	1.0	2.0	3.0	4.0	5.0
$\kappa = m/H$	1.0	1.19	1.45	1.9	3.0	5.75	12.5	18.0

In a view of the size distribution law it is not difficult to calculate the average value (mathematical expectation M) of the force acting on the spherical particles with their volume $V=\pi D^3/6$ according to the formula

$$F_m = M[F_y] = \frac{\pi}{6}\kappa m_{sut}\frac{dH}{dy}\int_{-\infty}^{\infty} D_p^4 p(D_p)dD_p \quad (9)$$

To calculate the motion characteristics of both individual magnetic carriers and their aggregates under the influence of magnetic field, we give some theoretical estimates of the parameters in the task of carriers transport in blood vessels. The branched structure of blood vessels in target organ is shown schematically in Figure 1, *a*, and also the scheme for calculation of motion for the carriers in the section of horizontal vessel of finite length (Figure 1, *b*; the dotted lines designate the area of active action of the magnetic field). Note that each j-th branch of the main vessel of the organ forms a certain α_j angle (Figure 1).

Consider the motion of a single carrier (magnetic particle or aggregate) in a vessel section that we regard as a cylindrical tube with rigid walls. Since the carrier motion is onward, the particle can be considered as a physical point. Let's describe the particle movement in the Cartesian *xyz* rectangular coordinate system, directing axis along the axis of the vessel, and *z* axis – perpendicularly to axis so that they would form the right three, moreover directing y axis along the line of the maximal field growth. Let us bind the coordinate origin with the axis of the vessel. The particle is influenced by external forces: gravity $G = mg$ (here m is the particle mass, g – acceleration of gravity), $\mathbf{F}_A$ – Archimedes force, $\mathbf{F}_m$ – magnetic force, $\mathbf{F}_S$ – the movement resistance force. Further assume that the force of gravity, Archimedes force and inertial force (ma; a – acceleration of the particle) can be neglected because of small mass of the particles (10^{-11}–10^{-10} g).

During the detailed study of blood flow, it was revealed experimentally and theoretically that movement is laminar both in thin vessels and in areas of large curvature.

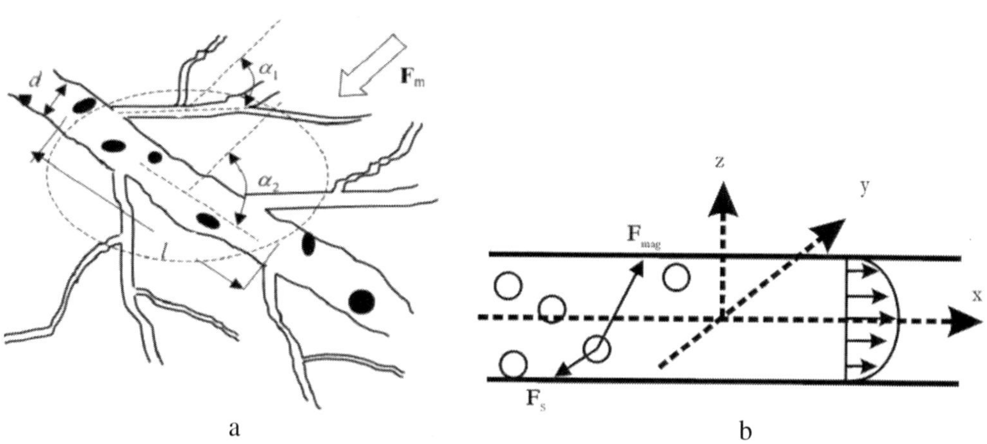

Figure 1. A scheme of vascular environment (*a*) and calculation plan of the magnetic carriers movement in the vessel section (*b*).

Due to such character of movement, one can assume that movement of the carrier is virtually stopped in a free section of the vessel, when reaching the vessel wall. Note that the particles can slide on the walls of blood vessels, but in this case the speed of their movement is much less than in the middle part.

At first consider the conditions of carriers stopping at the inner wall of the vessel by the magnetic field. The carrier to be holding in the blood flow by the magnetic field is influenced by the Stokes force:

$$F_s = 3\pi D_g v \eta_1,$$

where D_g is the hydrodynamic diameter of the carrier (including the shell); v – the rate of blood flow; η_1 – its viscosity. As blood flow in the vessel is laminar, it is much easier to stop the carrier near the wall where the flow rate is minimal and proportional to γ (γ parameter has the dimension of angular velocity) by the order of magnitude.

Let the magnetic field with the value of intensity gradient be localized in the area of size within the vessel of diameter (Figure 1). In the Poiseuille (parabolic) approximation the velocity distribution in cross-section of the vessel taken as a circular cylindrical tube, there is the following velocity distribution law on radial coordinate r:

$$v = \frac{p_1 - p_2}{4\eta_1 l}(d^2/4 - r^2),$$

where – the pressure at the inlet and outlet area of l length of the magnetic field active action. The average rate of fluid flow $\bar{U}$ is given by

$$\bar{U} = \frac{p_1 - p_2}{\eta_1 l}\frac{d^2}{128}.$$

Then the volume of blood (blood flow) Q in terms of the mean velocity can be found from the formula $Q = \pi d^2 \bar{U}$.

Equating the Stokes force to the value of axial component of magnetic force [58]

$$F = m\nabla H \cos\alpha,$$

the condition of the carriers retention at the vessel wall by the magnetic field was obtained:

$$m\nabla H \cos\alpha \geq 6\pi r_g^2 \gamma_1 \eta_2 \qquad (10)$$

where m – the magnetic moment of the carrier; – the angle between direction and the axis of the vessel; $r_g = D_g/2$ – the hydrodynamic radius; – blood viscosity in the near-wall layer, and is given by [58]

$$\gamma_1 = \frac{32Q}{\pi d^3} = \frac{8\bar{U}}{d}.$$

If the volume content of ferromagnetic material with a specific magnetic moment in the carrier is equal to k, and the volume of the carrier is $(4/3)\pi r^3$ (here and below $r = r_g$), then

$$m = \frac{4}{3}\pi r^3 \cdot m_0 \cdot \chi \cdot k \tag{11}$$

where χ is the magnetization [58]. With an accuracy sufficient for estimative calculations, we assume that for the magnetic fields with intensity $H < H^*$, the saturation for the ferromagnetic $\chi \sim H/H^*$, and under the condition of $H > H^*$ we have $\chi = 1$. Now the condition for carrier stopping at the inner wall of the vessel will take the form [58]:

$$k_1 = \frac{9}{2} \cdot \frac{\gamma \eta_2}{\mu_0 \chi r \nabla H \cdot \cos\alpha}. \tag{12}$$

Equation (12) defines such a content of ferromagnetic material in the carrier k_1, at which for the average velocity of blood flow in the vessel of d diameter the carrier can be stopped near the vessel wall by the magnetic field with the parameters and ∇H.

Later we proceed taking into account the characteristics of human blood flow [58], given in Table 2.

Let us consider now the effect of inhomogeneous magnetic field on the particles in a free vessel and evaluate the conditions necessary for delivery of the carriers to the vessel wall in the area of effective action of the magnetic field with a characteristic length l (Figure 1). We consider the field level such as for magnetization of the carriers to reach saturation.

Table 2. Some characteristics of human blood flow [58]

Type of vessel	Vessel diameter d, cm	Average velocity of blood flow in the vessels $\bar{U}$, cm/s	Average value of the parameter $\gamma_1 = 8\bar{U}/d$, s^{-1}
Aorta	1.6 – 3.2	30 – 60	300 – 600
Large arteries	0.6 – 0.1	20	270 – 1600
Arterioles	0.1 – 0.02	10 – 0.2	800
Capillaries	0.0005 – 0.001	0.05 – 0,07	500 – 800
Venules and small veins	0.02 – 0.2	0.1 – 1	40 – 200
Large veins	0.5 – 1.0	10 – 20	160
Hollow veins	2.0	50	200 – 300

Analysis of the equations of particle motion in a viscous fluid with some simplifications allows one to estimate the time required for the particle to reach the vessel wall [57]. For a free particle with radius vector, which moves in the area opposite to location of the magnet, one can estimate the particle motion velocity module in direction.

From the formula (7) we determine the value of particle velocity in this direction [55]:

$$v_r = \frac{\kappa m_{sat} G}{\rho f(D_g)}\left[1 - e^{-f(D_g)t}\right], \tag{13}$$

where is the function of the effective particle diameter D_g; – particle density, κ – coefficient depending on particle shape; – the magnetic induction module gradient that can be considered constant in a section of thin vessel.

From (13) it follows that under the influence of magnetic field, the particle velocity becomes rapidly constant due to the small value of the mass. Then the motion velocity v_r relative to blood flow at certain values of magnetic carriers concentration and the field gradient along y axis one can find from the formula

$$v_r = \frac{\kappa m_{sat} G}{\rho f(D_g)}. \tag{14}$$

We estimate time of the carriers motion to the vessel wall. If we denote as the time of the carriers residence in the organ vessel and as the time required for all the carriers to reach the vessel wall, then the condition of successful drug delivery to target organ can be written in the form of inequality $t \le \tau$. The fulfillment of this inequality is the main criterion for successful drug delivery.

For further estimative calculations it is reasonable, as above, to use the average flow velocity $\bar{U}$, with its value for different vessels to be shown in Table 2. Then the time of the carriers residence in the organ vessel in the absence of the field is $\tau = l / \bar{U}$, where – the area of active action of the magnetic field (Figure 1).

In general, the magnetic force vector and the axis of the vessel may form some angle ($0 < \alpha < \pi$). The time for the carrier to be in the vessel depends on size of this angle, because the projection of the relative velocity to this direction is added to the average velocity of the motion along the axis. The times and can be calculated by the formulas:

$$t = \frac{d}{v_r \sin\alpha}, \quad \tau = \frac{l}{\bar{U} + v_r \cos\alpha}. \tag{15}$$

From these correlations it follows that on condition $0 < \alpha < \pi/2$, the time of the carrier residence in the vessel is reduced, while increasing as long as $\pi/2 < \alpha < \pi$. In making so, the minimal residence time of the drug in the vessel, obviously, is $\tau_1 = l/(\bar{U} + v_r)$ and the maximal one is $\tau_2 = l/(\bar{U} - v_r)$.

The most unfavorable case is where the particle is located at the opposite wall, and therefore the maximum possible distance that it had to overcome is equal to the diameter of the vessel. Taking into account (15), the criterion of successful drug delivery can be written in the form of inequality

$$v_r(\sin\alpha - \frac{d}{l}\cos\alpha) \geq \frac{d}{l}U, \quad (0 \leq \alpha \leq \pi) \tag{16}$$

from which it follows that for certain values of the angles the expression in brackets can be negative. So, fulfillment of the condition (16) for some limit angle $\alpha_{\lim} = \text{arctg}(d/l)$ becomes impossible. Thus, part of the vessels for which the angles between the axes and the direction of magnetic force enter the interval $0 \leq \alpha \leq \alpha_{\lim}$, does not satisfy the criterion of successful drug delivery.

In general, the formal solution of the inequality (16) has the form $\alpha_1 \leq \alpha \leq \alpha_2$, where

$$\alpha_1 = \text{arctg}\frac{d}{l} + \arcsin\frac{d}{l} + \arcsin\frac{d}{\sqrt{d^2+l^2}}\frac{U}{v_r},$$

$$\alpha_2 = \pi + \text{arctg}\frac{d}{l} + \arcsin\frac{d}{l} - \arcsin\frac{d}{\sqrt{d^2+l^2}}\frac{U}{v_r}. \tag{17}$$

This solution exists under the condition

$$\frac{d}{\sqrt{d^2+l^2}}\frac{U}{v_r} \leq 1,$$

and otherwise – there are no solutions.

The graphical illustration of the inequality (16) is represented in Figure 2 which schematically shows the dependencies $t(\alpha)$ and $\tau(\alpha)$ at the fixed values of length and diameter of vessels and average speeds.

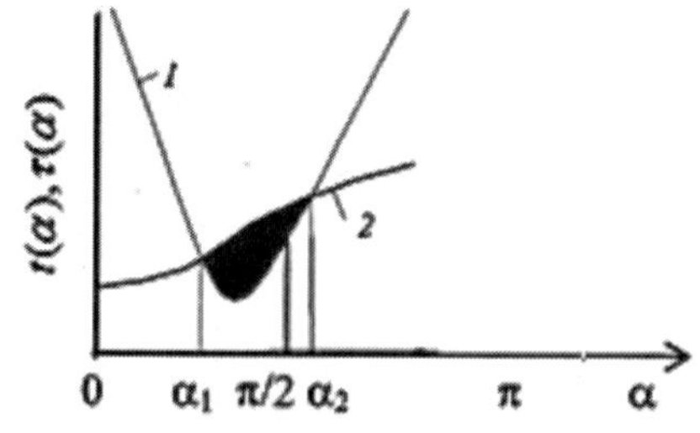

Figure 2. Graphical illustration of the inequality (16) solution: 1 – the dependence $t(\alpha) = d/(v_r \sin\alpha)$, 2 – the dependence $\tau(\alpha) = d/(U + v_r \cos\alpha)$.

It should be kept in mind that for most blood vessels, the ratios of the diameters to the length of the area of the field active action are small, and therefore the value of limit angle is small. For example, for the vessel with ratio $d/l = 0.1$ the limit angle $\alpha_{\lim} \approx 6^{\circ}$. Since α angle can be considered as a random value, uniformly distributed in the specified range of angles, the probability of getting vessels in this interval is small, and in the example considered, it is equal to only 0.03.

For thinner vessels this probability is even lower. Therefore, with very high probability it is permissible to believe that the inequality $\alpha > \alpha_{\lim}$ holds.

In making so for small values of the ratio d/l one can use the approximation $\alpha_{\lim} \approx d/l$. The range of solutions of the inequality is limited by the angle values $\alpha_1 < \alpha < \alpha_2$; in Figure 2 it is drawn for clarity reasons.

For different values of length and diameter of the vessel and speeds, this interval may be different (including blank if the curves do not intersect). The interval length $A = \alpha_2 - \alpha_1$ has the following meaning: for all the vessels with their axes to form the angles with the magnetic force vector, which enter this interval, there drug delivery to their walls is guaranteed. Since the angles can be free values within the interval $[0, \pi]$, and upon this they are distributed uniformly, the ratio $\lambda = A/\pi$ will show in what part of the organ vessels, randomly oriented in space, the drug can be delivered.

Taking into account the notations introduced one comes to the inequality looking as

$$v_r \geq \overline{U} \frac{\sin \alpha_{\lim}}{\sin(1-\lambda)\pi/2} \approx \overline{U} \frac{d}{l} \frac{1}{\cos(\lambda\pi/2)} \tag{18}$$

which allows one to solve fully the problem of determining the required values of velocity and, consequently, the characteristics of the field. The inequality binds the rate of relative motion, which the applied magnetic field should provide, with the average speed of blood flow in the vessel, geometric parameters of the vessels and the part of the target organ vessels used.

Table 3 shows the values of the parameter $\lambda = (\alpha_2 - \alpha_1)/\pi$ at different values $v_r / v_r^{\min}$, where $v_r^{\min} = \overline{U}d/l$.

From these data it turns out that even at a slight increase in the minimal critical velocity $v_{cr}^{\min} = \overline{U}d/l$, that is a fixed characteristic of a particular vessel, one can reach 50% of the success in delivering the drug to the vessel wall.

Table 3 gives a possibility to estimate the required speeds for any vessel when the corresponding characteristics are available.

Table 3. Critical velocities for different values of λ parameter

λ, %	30	50	70	90
$v_r / v_{cr}^{\min}$	1.12	1.41	2.20	6.39

Taking into account the above (in the framework of simplified model of one-dimensional field) expressions, one obtains the inequality to estimate the required average levels of gradients in the vessels

$$G \geq \frac{\bar{U}d}{l\cos(\pi\lambda/2)}\frac{\rho f(D_g)}{\kappa m_{sat}}. \quad (19)$$

This criterion allows one to define the required values of motion velocities of the magnetic carriers for a specified organ. In turn, it provides an opportunity to assess the required levels of fields and gradients. The more exact solution of these problems requires further studies and improvement of models to describe the motion of carriers in the vessels. We note that from the general criterion obtained, the result presented in [58] appears – the formula (5).

One can keep all the carriers that get into a large vessel only when the characteristic size l of localization area of the magnetic field is equal $l = b\bar{U}\tau - d\text{ctg}\alpha$, where b – the coefficient of an order of one, $\tau = d/V_\perp$ – the time for the carrier to cross the vessel under the influence of transverse component of magnetic force. The transverse component $V_\perp$ of the carrier velocity has the form

$$V_\perp = \frac{\mu \cdot \nabla H \cdot \sin\alpha}{6\pi r\eta_1}.$$

As before, expressing μ through k, we find the condition of 100 % carriers delivery to the vessel wall at the desired area in action of the magnetic field (here and below r – the hydrodynamic radius):

$$k_2 = \bar{U}d\eta_1[br^2\mu_0\nabla H\sin\alpha(l + d\text{ctg}\alpha)]^{-1}. \quad (20)$$

Comparing this expression with the condition of the carriers stopping at the vessel wall when $\alpha = 45^\circ$, b = 4/3 and provided that $l >> d$ and $\eta_1/\eta_2 \sim 2$ we obtain

$$\frac{k_1}{k_2} = \frac{1}{18}\frac{d^2}{rl}. \quad (21)$$

For 100 % carriers delivery to the wall on condition $d > \sqrt{18 \cdot l \cdot r}$, it is necessary to have the larger content of ferromagnetic material in the carriers than to stop them at the wall; for the case $d < \sqrt{18 \cdot l \cdot r}$ – on the contrary. If the content of ferromagnetic substance is k_1/k_2 =1%, then for realization of 100% carriers delivery to the vessel wall and stopping them in the vessels of different diameter, the magnetic field is necessary with $H \geq 20$ kOe (the saturating field) and $\nabla H \geq 7$ kOe/cm for veins, where $\overline{U} = 0.1$ cm/s, or $\nabla H \geq 28$ kOe/cm for arteries, where $\overline{U} = 0.4$ cm/s.

In the vessels with a diameter of more than 3 cm, it is difficult to reach 100% deposition of the carriers, because the necessary gradient of magnetic field intensity is 10^4 -10^5 kOe/cm at $k=1$. Nonetheless, this does not exclude the possibility of stopping that part of magnetic carriers at the inner wall of large vessels, which has reach the surface for the time of "flying" the external magnetic field localization area.

In such a way, in the thin vessels ($d < \sqrt{18rl}$) one can realize both 100% delivery of the carriers to the vessel wall and their stopping. At the same time, in the large vessels ($d > \sqrt{18rl}$) it is impossible in practice to reach 100% delivery, because it is technically difficult to create the values of gradients necessary for this, but some part of the carriers that have got into these vessels can be still kept by the magnetic field.

From the estimates obtained in this paper it turns out that when using the optimally selected magnetic systems, one can achieve retention of the containers (nanocomposites) with medicine preparations even in large blood vessels. Further studies will show the effectiveness of the approaches developed, but it should be particularly noted that the calculations carried out and the found levels of gradients indicate the real possibility of delivery and retention of the magnetic carriers in a target organ. At the same time, taking into account the complexity of the drug delivery problem, from the estimations carried out above one can draw a conclusion that to solve the specific therapeutic tasks with the help of magnetic carriers, one needs a comprehensive implementation of both theoretical studies on model systems and the necessary preliminary experiments, including the use of cell cultures and animals.

2. Synthesis of Magnetosensitive Polyfunctional Nanocomposites for Medicobiological Purposes and Investigation of Their Properties

2.1. The Scheme of Chemical Construction of Nanocomposites with Functions of Nanorobots

Figure 3 shows a scheme of chemical design of multifunctional nanocomposites nanoarchitecture with properties of nanorobots [24, 27] according to the formulated concept worked experimentally for all the basic stages *via* successive synthesis of nanoarchitecture elements and comprehensive checking their functionality.

At the first level of the hierarchical nanostructures, monodomain magnetite with the size of 8-50 nm was selected as source material for chemical design of the nanocomposites. The magnetite is known to be of biogenic nature and to be eliminated satisfactorily by the body. In this scheme it can act as a magnetosensitive drug carrier; a converter of the energy of high-frequency magnetic field of an external source into the heat in order to create hyperthermal zones; a nanoparticle with reactive surface to allow the further design of a given hierarchical architecture of multifunctional nanocomposite.

The second hierarchical level of the nanostructure (1-4 nm) can be rather complicated structure.

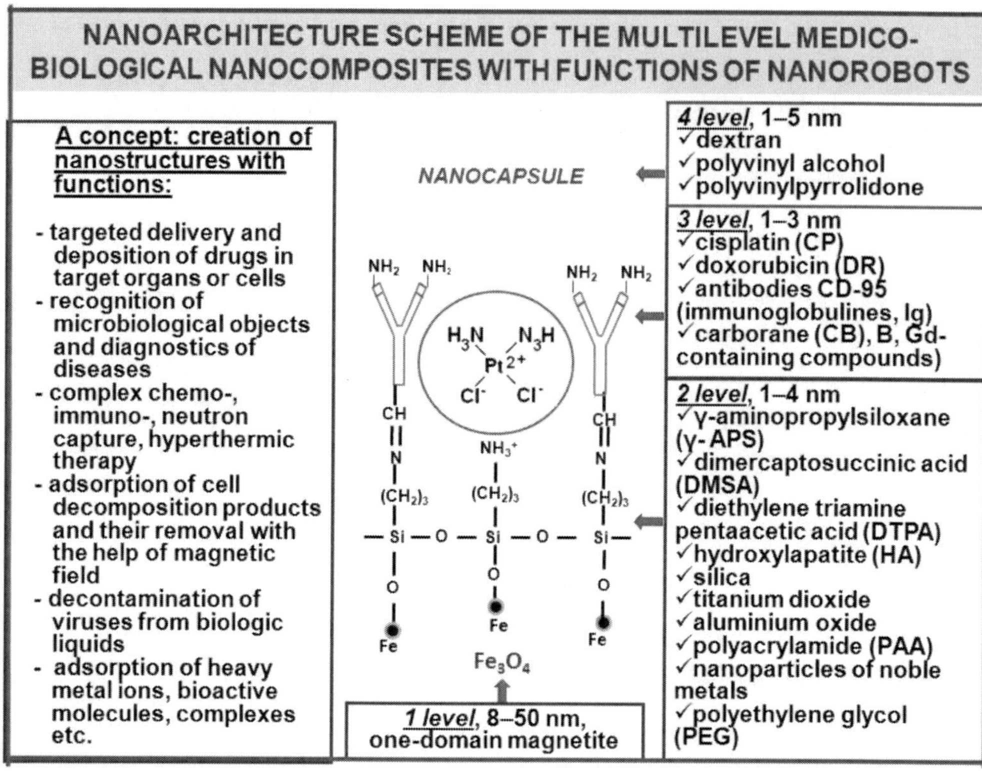

Figure 3. Chemical construction scheme of multilevel nanocomposites nanoarchitecture with functions of nanorobots.

It stabilizes the nanosized carrier by modifying the surface with biocompatible coverage (polyacrylamide, silica, hydroxyapatite, titanium or aluminium oxide, etc.). The presence of a modifying layer retains a high specific surface of the nanoparticles and allows the necessary chemical functionalization, for example, with hydroxyl, carboxyl, thiol and amino groups.

Biofunctionalization of the nanocomposite is carried out at the third level (1-3 nm) by immobilization of chemo- (cisplatin, doxorubicin), immuno- (antibodies), radiotherapy (boron and gadolinium containing neutron capture agents) and diagnostic (T_1-, T_2-contrast agents, nanoparticles of noble metals) drugs, and also sensors that provide recognition of specific microbiological objects. It should be noted that nanoparticles of noble metals, for example, due to the phenomenon of plasmon resonance, can act as sensors (optical marks) and therapeutic agents (thermal and photodynamic therapy). The functions of the fourth hierarchical level are associated with capsulation of nanocomposites to maintain their properties and prolong the action of drugs. Nanocapsulation is carried out with dextrane, gelatin, polyvinyl alcohol (PVA), polyvinyl pyrrolidone (PVP). The results of experimens at all the main stages of chemical construction of polyfunctional nanocomposites with properties of nanorobots, as well as investigation of their properties and examination of functions according the shown scheme are cited below [22–29]. Synthesis nanotechnology promising for practical application was developed, physical and chemical properties of magnetosensitive nanocomposites and magnetic fluids on their basis were studied including colloidal systems stable in water, phosphate buffer and physiological solution, as well as biofunctionalizated magnetic nanocomposites and fluids for pre-clinical studies (Table 4).

Table 4. List of synthesized and investigated experimental samples

Magnetosensitive polyfunctional nanocomposites	Magnetic fluids	Biofunctionalizated magnetosensitive nanocomposites	Biofunctionalizated magnetic fluids
Fe_3O_4/PAA/CP Fe_3O_4/γ-APS/CP Fe_3O_4/HA/CP Fe_3O_4/HA-Ag/CP Fe_3O_4/HA/DR Fe_3O_4/Gd_2O_3 $Gd_xFe_{3-x}O_4$	Fe_3O_4/Na ol. Fe_3O_4/Na ol./CP Fe_3O_4/Na ol./PEG Fe_3O_4/Na ol./PEG/CP Fe_3O_4/ DMSA Fe_3O_4/Gd_2O_3/Na ol. Fe_3O_4/Gd_2O_3/Na ol./PEG Fe_3O_4/DMSA/CP	Fe_3O_4/PAA/Ig Fe_3O_4/γ-APS/Ig Fe_3O_4/HA/Ig Fe_3O_4/HA-Ag/Ig Fe_3O_4/γ-APS/CD-95 Fe_3O_4/γ-APS/CP/CD-95 Fe_3O_4/PAA/CD-95 Fe_3O_4/PAA/CP/CD-95 Fe_3O_4/HA/CD-95 Fe_3O_4/HA/CP/CD-95	Fe_3O_4/DMSA/SMCC/Ig Fe_3O_4/Na ol./CP Fe_3O_4/Na ol./PEG/CP Fe_3O_4/HA/DR

The list of shortenings in table 4: PAA – polyacrylamide, γ-APS – γ- aminopropylsiloxane, DMSA – *meso*-2,3-dimercaptosuccinic acid, SMCC – sulfosuccinimidyl-4-(N-maleimidomethyl)-cyclohexane-1-carboxylate, HA – hydroxylapatite, CB – carborane, Na ol. – sodium oleate, CP – cisplatin, DR – doxorubicin, PEG – polyethylene glycol.

Let us consider in more detail the basic stages of synthesis and functionalization of the samples of polyfunctional nanocomposites, as well as results of studying their properties and possibilities of practical use.

The Synthesis and Properties of Magnetite

The synthesis of magnetite was carried out by the coprecipitation of iron salts according to the reaction:

$$Fe^{+2} + 2Fe^{+3} + 8NH_4OH \rightarrow Fe_3O_4 + 4H_2O + 8NH_4^+.$$

In order to obtain the nanoproduct yield not less that 85 % (mass) of monodomain particles Fe_3O_4 with the assigned size distribution, the cryochemical technique was used for heterogenic synthesis of magnetite on the phase boundary: one solid – frozen solution of iron(II) sulfate and iron(III) chloride salts, and the second liquid – the ammonia solution of fixed concentration. As distinct from homogenic synthesis, it is possible to change the growth conditions of nanoparticles in the synthesis process.

The samples of nanocrystalline magnetite were obtained, with a specific surface area change depending on the parameters of initial solution within ~ 40 – 180 m^2g^{-1}. The average particle size was dependent on the synthesis conditions, with values of 6 – 50 nm; the interval of size distribution can be technologically controlled, being sufficiently narrow (Figure 4).

The advantage of the technique proposed is the possibility of obtaining nanomaterials with high content of monodomain particles with the assigned size distribution, and is simplification of the ultradisperse product separation process.

Nanocrystalline one-domain magnetite (fraction of 20–50 nm) was obtained also by the method of solid phase synthesis in the inert atmosphere. As initial compounds, metal-organic compounds of bi- and trivalent iron were used.

Magnetite obtained by the solid phase method showed the highest values of coercitive force (150–240 Oe), with the specific surface of ~ 40 m^2g^{-1}.

The Investigation of Hyperthermal Effect

In order to clarify the possibilities for creation of hyperthermal zones, the nonresonance (thermal) influence of electromagnetic radiation (λ = 3 cm) was investigated on the model animal muscular tissues *in vitro* injected by nanodisperse magnetite (Figure 5).

The mass of the tissue samples was 5 g. The samples 2 and 3 were differed by the content of small quantity of grease layer. The area of radiation was ~ 2 cm^2.

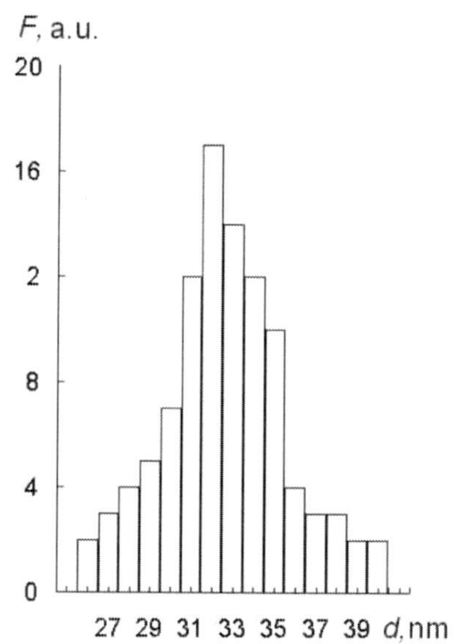

Figure 4. Size distribution of magnetite particles (*F* is given in arbitrary units, *d* is particle diameter, nm).

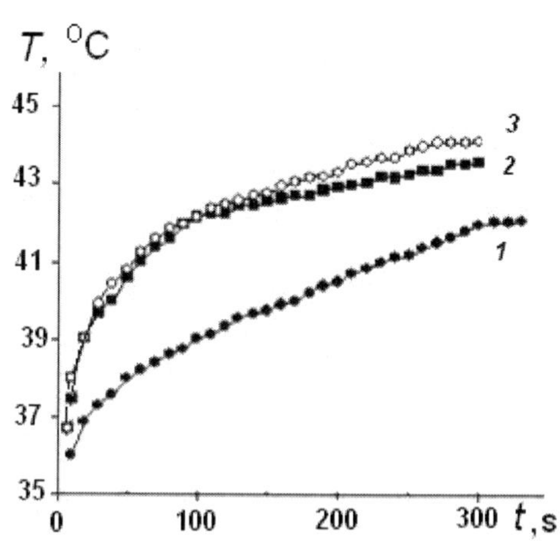

Figure 5. Temperature of muscular tissue samples as function of treatment time: 1 – initial, 2, 3 – with magnetite.

It was discovered that in the case of injection of Fe_3O_4 nanoparticles into muscular tissues (0.1 mass %) the absorption of electromagnetic radiation increased, and the rate of their heating increased by 1.5–2 times (in comparison with the control samples without magnetite) and reached 4 grad/min at the irradiation power value of 0.5 Wt.

Creation of a Biocompatible Covering on the Surface of Nanosized Magnetite

In order to stabilize the properties of magnetosensitive carriers and to increase their specific surface area, they were modified with biocompatible coverings.

The nanotechnology of magnetite surface modification with polyacrylamide consists of the formation of covering on the surface of nanosized magnetite by polymerization of acrylamide with sewing agent N,N'-methylene-bis-acrylamide in high-frequency glow discharge plasma.

It was shown from the analysis of kinetic curves of titration of double carbon (C=C) bonds that the full polymerization of acrylamide layer was realized in 2 minutes.

The samples of Fe_3O_4/PAA nanocomposites were investigated by the atomic force and magnetic force nanoscopy. Figure 6 (*a* - *d*) shows the atomic force (*a, c*) and magnetic force (*b, d*) images of Fe_3O_4/PAA nanocomposite arrays (*a* - *d*) in the formats 2*d* (*a, b*) and 3*d* (*c, d*).

The dried powder-like samples with a broad size distribution of the particles and their aggregates (from 8 nm to 1 μm) were used for the investigations. The images of the nanocomposites obtained by the atomic force and magnetic force methods are apparently in good accordance, showing coordination of the magnetic and "force" phases, as well as small thickness and sufficiently uniform distribution of cross-linked polyacrylamide around the magnetite particles.

Modification of the magnetite surface with γ-aminopropylsiloxane was carried out by the liquid phase method in toluene. As a result of polycondensation reaction the surface of magnetite acquired base properties because of the grafting of γ-aminopropyl groups at their concentration of 24 $\mu mol/m^2$ when the specific surface area was 90 m^2g^{-1}. The atomic force (*e, g*) and magnetic force (*f, h*) images of Fe_3O_4/γ-APS nanocomposite arrays (e - h) in the formats 2d (*e, f*) and 3d (*g, h*) are shown in Figure 6. As in the previous case, one can draw a conclusion about the coordination of magnetic and "force" phases and small thickness and sufficiently uniform distribution of γ-aminopropylsiloxane layer around the magnetite particles.

Modification of the magnetite surface with hydroxylapatite was carried out to give high biocompatibility to Fe_3O_4 magnetosensitive carriers. It was determined that hydroxylapatite phase on the surface of the carrier was characterized by the correlation Ca/P=1.67, which corresponds to the stoichiometry of the reaction of its creation.

The average size of magnetite and hydroxylapatite crystallites was calculated in accordance to (311) and (002) X-ray diffraction peaks, respectively, with the use of Sherrer formula. The thickness of hydroxylapatite layer on the surface of magnetite nanoparticles was ~ 4 nm, which was evaluated by the area ratio of Fe2p- and Fe3p-lines (studied by X-ray photoelectronic spectroscopy) and the increase in the Fe_3O_4/HA nanocomposite mass (~ 30 %). Modification of the magnetite surface with meso-2,3-dimercaptosuccinic acid (DMSA) is one of the variants to solve the problems of colloidal system stability in aqueous medium, the material biocompatibility, the immobilization of necessary compounds through thiol and carboxyl functional surface groups.

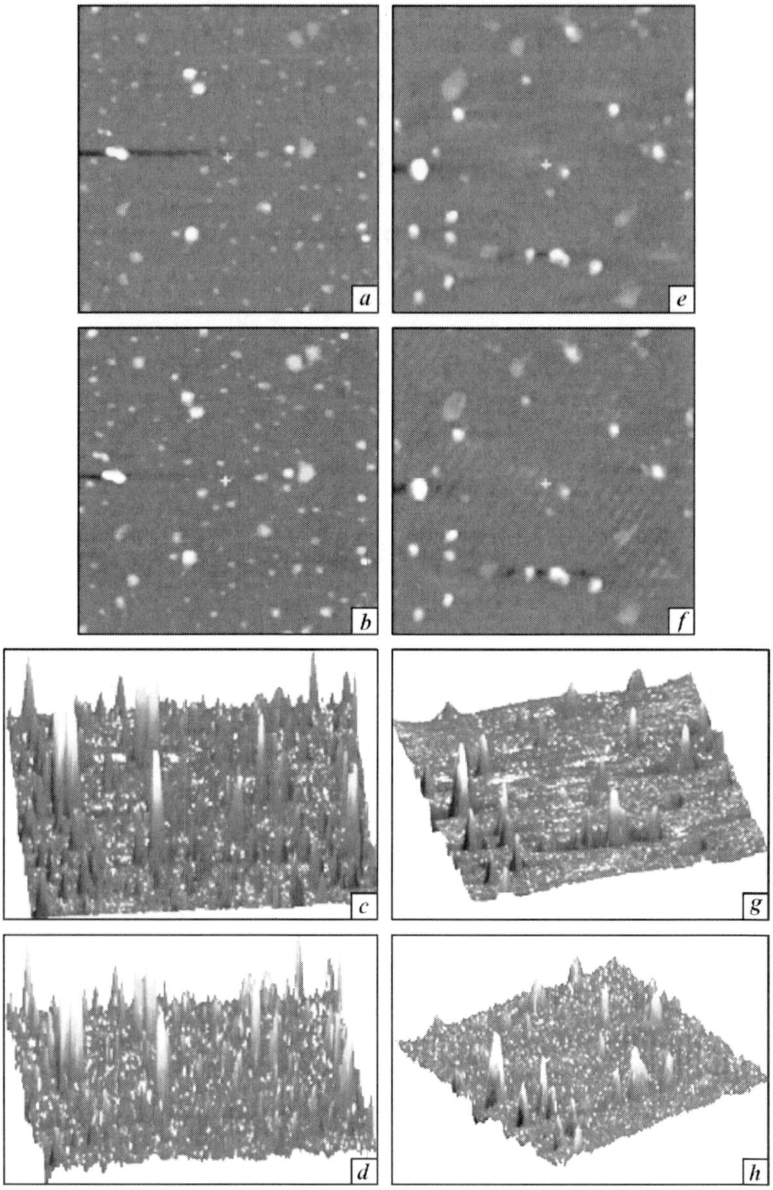

Figure 6. Atomic force (*a, c, e, g*) and magnetic force (*b, d, f, h*) images of nanocomposite arrays Fe_3O_4/PAA (*a - d*) and Fe_3O_4/γ-APS (*e - h*) in formats 2*d* (*a, b, e, f*) and 3*d* (*c, d, g, h*).

Intermolecular disulfide bonds of DMSA bound with the surface increase the stability of the cover. The aqueous sol obtained from the magnetite nanoparticles covered with DMSA is stable in the broad range of pH values (3-11) in aqueous and phosphate buffered systems. It was found that the concentration of SH groups was 0.019 mmol/m^2 when the specific surface area being 130 m^2g^{-1}.

The results of measuring the distribution of the particles of Fe_3O_4/DMSA/water magnetic fluids by the method of dynamic light scattering on zeta potential and the average size of nanocomposite particles are shown in Table 5.

Table 5. Zeta potential and average particle size for magnetic fluids Fe_3O_4/DMSA/water

Zeta potential, mV	- 40.3
Zeta potential distribution width, mV	9.84
Sample conductivity, mS/cm	0.129
Average size of particles, nm	40.5

The negative potential of the particles in magnetic fluids prevents their aggregation. With increasing pH, the zeta potential diminishes because of the increase in the ionization degree of COOH and SH functional groups.

Biofunctionalization of the Nanocomposites

Biofunctionalization is an important stage in the preparation of medicobiological nanocomposites, because it provides them with functions of recognition of microbiological objects in biological media and targeted delivery of medical preparations to them.

Immobilization of maleimide activated immunoglobulines on the surface of Fe_3O_4/DMSA was carried out to obtain the immunomagnetic nanosorbents and nanocomposites with recognition functions. A technique was developed for magnetic fluid creation with simultaneous modification of the surface of magnetite particles with DMSA and immobilization of antibodies with the help of spacer molecules of sulfosuccinimidyl-4-(N-maleimidomethyl)-cyclohexan-1-carboxylate (sulfo-SMCC). It was shown that the chemical immobilization of antibodies on the surface of Fe_3O_4/DMSA nanocomposites may be realized by the interaction of the reactive sulfo-groups of the carrier with the maleimide functional groups of immunoglobuline (Ig) activated previously with sulfo-SMCC.

The two stage reaction of the immobilization of immunoglobuline on the surface of Fe_3O_4/DMSA nanocomposites by means of sulfo-SMCC spacer leads to the formation of specific complexes with antibody:

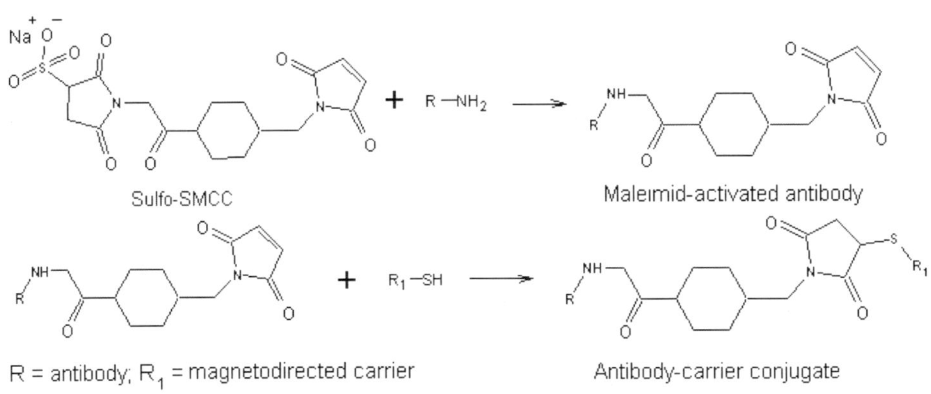

The isoterms of adsorption of maleimide activated immunoglobulines on the surface of Fe_3O_4/DMSA were investigated. The adsorption capacity value A = 92.12 mg/g, with the specific surface area $S_{specific}$ = 130 m^2g^{-1}. The release of immunoglobuline, immobilized by means of sulfo-SMCC spacer, in the model medium was 8–12 % for 24 h.

The covalent immobilization of human normal immunoglobuline at the surface of Fe_3O_4/PAA and Fe_3O_4/γ-APS is based on the idea of binding immunoglobuline with the surface of the nanocomposites by means of using the interaction of aldehyde groups (created on Ig molecules as a result of periodate oxidation of side carbohydrate chains) with aminogroups of the modifying agent at the carrier surface. This leads to formation of Schiff-bases (imines).

In the case of Fe_3O_4/PAA nanocomposites, covalent binding of Ig was carried out after activation of polyacrylamide surface with ethylenediamine to form reactive $-NH_2$ groups:

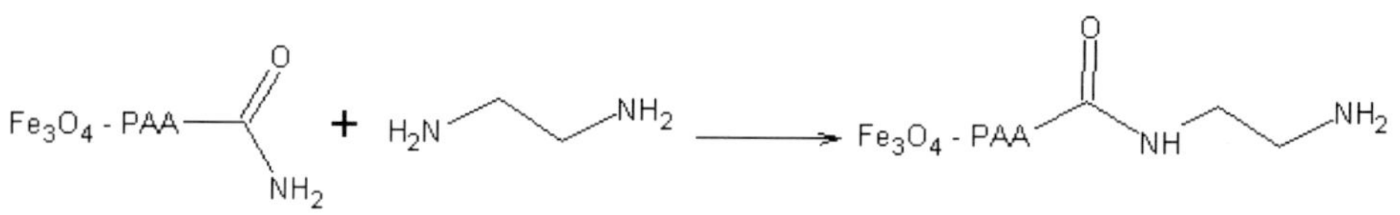

The oxidation of human normal Ig was carried out in acetate buffer-based solution of $NaIO_4$ according to the reaction:

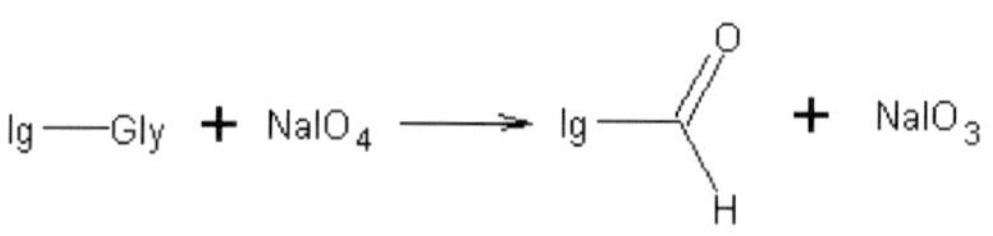

The covalent binding of oxidized Ig with the surface of Fe_3O_4/PAA nanocomposites (activated with ethylenediamine) and Fe_3O_4/γ-APS was carried out in carbonate buffer based on physiologic solution according to the reaction:

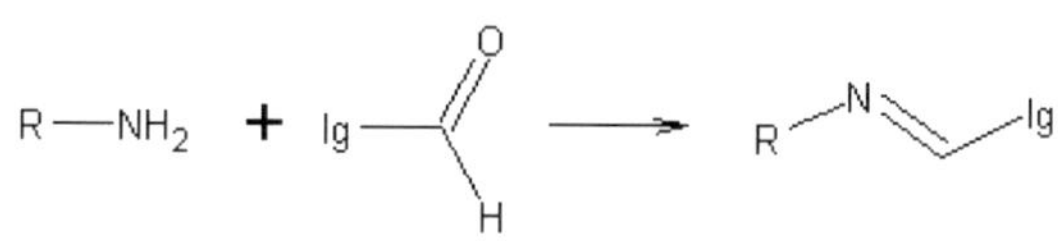

The comparative studies of isotherms (Figure 7, 8) of nonspecific (physical) adsorption of human normal Ig (Figure 7, *a*, *b*, curves *1*, Figure 8) and covalent binding of the oxidized Ig (Figure 7, curves *2*) onto the surface of nanocomposites were carried out. The distribution coefficients (E) were calculated for immunoglobuline between the surface of Fe_3O_4/PAA, Fe_3O_4/γ-APS, Fe_3O_4/HA, Fe_3O_4/HA/Ag nanocomposites and the solution, as well as quantities of Ig immobilized at maximal concentration of its initial solution C = 1.4 mg/g (Table 6).

In the case of covalent binding of Ig the distribution coefficients of Fe_3O_4/PAA and Fe_3O_4/γ-APS nanocomposites are higher than the respective coefficients for nonspecific adsorption by more than an order and reflect the equilibrium shift towards surface immobilization of the substance.

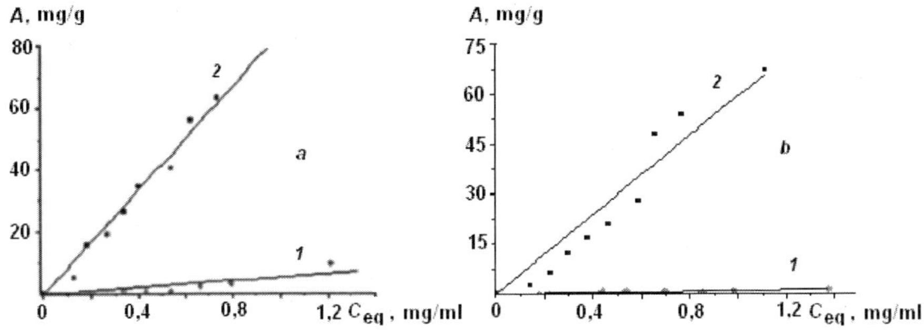

Figure 7. Isotherms of nonspecific adsorption of human Ig (*1*) and covalent binding of the oxidized human Ig (*2*) onto the surface of the nanocomposites Fe_3O_4/PAA (*a*) and Fe_3O_4/γ-APS (*b*).

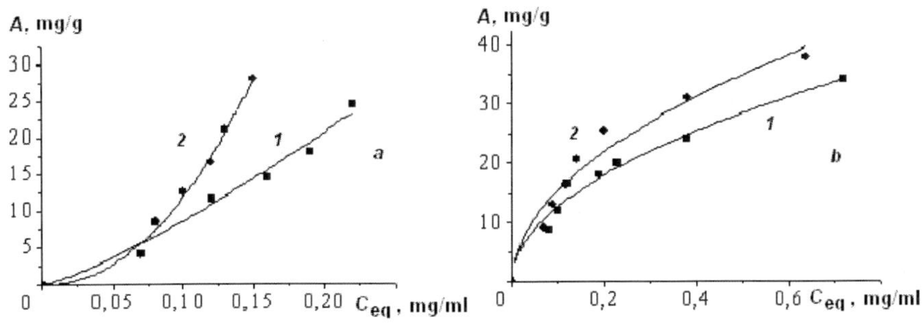

Figure 8. Adsorption isotherms of human Ig on the surface of the nanocomposites Fe_3O_4/HA and Fe_3O_4/HA/Ag from phosphate buffer (*a*) and physiological solution (*b*).

Table 6. The quantity of human Ig immobilized onto the nanocomposite surface

Nanocomposite	A_{phys}, mg/g	E_{phys}, ml/g	A_{cov}, mg/g	E_{cov}, ml/g
Fe_3O_4/PAA	9.48	6.1	116.00	83.53
Fe_3O_4/γ-APS	1.18	0.92	67.41	59.51
Fe_3O_4/HA (physiological solution)	34,0	47.2		
Fe_3O_4/HA (phosphate buffer)	25	111.36		
Fe_3O_4/HA/Ag (physiological solution)	38.0	59.4		
Fe_3O_4/HA/Ag (phosphate buffer)	25.5	186.67		

It was shown by experiment that covalent binding of Ig according to the reaction proposed has essential advantages in comparison with nonspecific adsorption: the oriented fastening of antibody molecules and restriction of Ig desorption by hydrolysis velocity of Schiff bases.

The adsorption of Ig on the surface of Fe_3O_4/HA/Ag exceeds that on the surface of Fe_3O_4/HA in two investigated buffer systems. It may testify for silver nanoparticles at the composite surface to be additional adsorption centers. Considering high adsorption of immunoglobuline by Fe_3O_4/HA and Fe_3O_4/HA/Ag nanocomposites (A = 34–38 mg/g) in comparison with adsorption by the nanocomposites functionalized with aminogroups (A =

1.18–9.48 mg/g), one can suppose the formation of compact oriented packing of adsorbed antibodies on the surfaces of the nanocomposites modified with HA and silver.

Immobilization of antibodies at the surface of nanocomposites magnetite–polyacrylamide, magnetite–γ-aminopropylsiloxane and magnetite–hydroxylapatite was carried out by the techniques described above. The nonspecific (physical) adsorption and covalent binding of CD-95 monoclonal antibodies on the surface of Fe_3O_4/HA, Fe_3O_4/PAA, and Fe_3O_4/γ-APS nanocomposites, respectively, were investigated.

As sensor molecules we used the monoclonal mouse antibody CD-95 against the human Fas-antigen of the isotype Ig1, kappa, DX2 clone produced by DakoCytomation (Denmark). The quantity of antibodies immobilized was determined by calibrated graph and the adsorption value A was calculated to be 165 μg/g for Fe_3O_4/PAA nanocomposites, 137.7 μg/g – for Fe_3O_4/γ-APS and 590 μg/g – for Fe_3O_4/HA. One may remark that a magnetosensitive sensor carrier, when recognizing a specific microbiological object in a biological medium, for example a cell, is in direct proximity from it, or penetrates through a membrane into intracellular space.

Then there is a possibility to determine its space location and to observe its time change by magnetic resonance methods. Using computer tomography, one can visualize this process and obtain three-dimensional images. The latter is used in practice, being especially important for early diagnostics of diseases and determination of exact sizes of tumours.

Immobilization of Cisplatin at the Surface of the Nanocomposites

Cisplatin (CP) is a platinum-based antitumour preparation. The mechanism of its antitumour action is related to bifunctional alkylating of DNA chaines, which suppresses biosynthesis of nucleic acids and induces cell apoptosis.

It was shown that cytotoxic activity of cisplatin in the samples studied was preserved for one month.

Adsorption kinetics of CP at surfaces of the nanocomposites was investigated (Figure 9).

It was shown that the adsorbed amounts of cisplatin (calculated for Pt^{2+}) at the surfaces of the nanocomposites Fe_3O_4/PAA, Fe_3O_4/γ-APS and Fe_3O_4/HA were 128 mg/g, 98.3 mg/g and 60.1 mg/g, respectively. The major part of the drug is adsorbed during the first 2–3 hours.

Release of the Preparations from the Nanocomposite Surface

Desorption kinetics for Ig immobilized on the surface of the nanocomposites was studied with the use of physiological solution as a model medium. It has been shown that the release of immunoglobuline immobilized at the surface of Fe_3O_4/PAA and Fe_3O_4/γ-APS nanocomposites *via* covalent binding is going slowly and to lesser extent than in the case of nonspecific immobilization. The antibodies immobilized at the surface of Fe_3O_4/HA and Fe_3O_4/HA/Ag nanocomposites showed weak desorption in a model medium.

By investigating kinetics of cysplatine release from the surface of Fe_3O_4/PAA nanocomposites into the physiological solution it was found that about 50 % of cytotoxic preparation was washed out in 80 min.

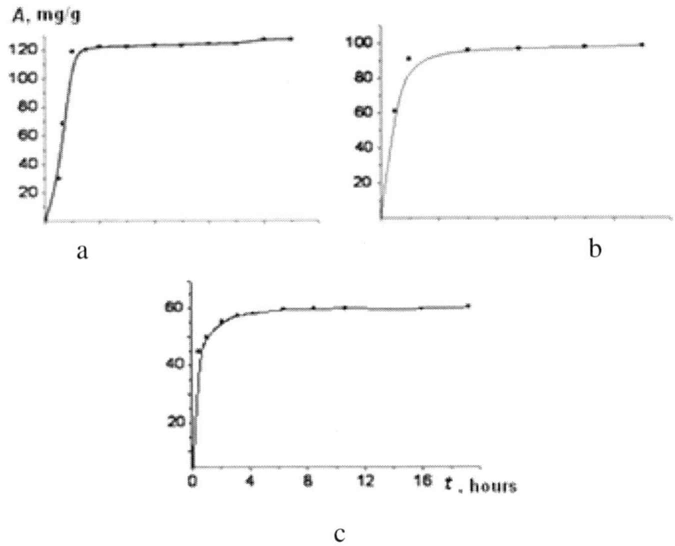

Figure 9. Adsorption kinetics of cisplatin on the surface of the nanocomposites Fe_3O_4/PAA (a), Fe_3O_4/γ-APS (b), Fe_3O_4/HA (c).

Investigation of Cytotoxic Effect of the Magnetosensitive Nanocomposites

Cytotoxic influence was studied for the magnetosensitive nanocomposites with the adsorbed cytostatic preparation conjugated with monoclonal antibody (nanorobot models) upon viability of the cell line of human breast cancer MCF-7 (Table 7). For comparison, the influence of control samples on cell viability with the separately immobilized monoclonal antibodies and cytostatic drug was investigated.

As a reference, we used: the pure nutritious medium; cisplatin in a concentration $C = 2.5$ μg/ml, corresponding the fourth dose of biological equivalent efficacy IC_{25}; monoclonal antibody CD-95 in $C = 0.2$ μg/ml (the dose used in curing is 10 – 30 μg/ml).

The biocompatibility of the initial magnetite and Fe_3O_4/PAA, Fe_3O_4/γ-APS and Fe_3O_4/HA nanocomposites (modified carriers) to the said cell line was shown experimentally. The use of magnetic nanocomposites that comprise antitumour drug and monoclonal antibody CD-95 was accompanied by significant synergic effect of cytotoxic action. Their efficacy exceeds the combined action of the respective control doses of cisplatin preparations and antibody. So, it was shown that the efficacy of Fe_3O_4/PAA/CP/CD-95 and Fe_3O_4/HA/CP/CD-95 model nanorobots that comprise antitumour drug CP and monoclonal antibody CD-95 exceeds by 1.5–2 times the combined cytotoxic action of the respective control doses.

The synergic effect of action of the nanocomposite biologically active components may be explained as follows.

Firstly, the targeted delivery of the complex "cytostatic drug – oriented monoclonal antibody" is carried out by the magnetosensitive carrier to tumour cells due to recognition of the respective receptors on their surface. The cytotoxic effect of cisplatin is realized through the formation of covalent bonds of preparation molecules with DNA.

Table 7. Results of the nanocomposite cytotoxicity studies

References	Number of dead cells, %									
	Effect	Fe_3O_4/γ-APS+ CP	Fe_3O_4/γ-APS+CD-95	Fe_3O_4/γ-APS+ CP+CD-95	Fe_3O_4/PAA+CP	Fe_3O_4/ PAA+CD-95	Fe_3O_4/ PAA+CP+CD-95	Fe_3O_4/HA+ CP	Fe_3O_4/ HA+CD-95	Fe_3O_4/ HA+CP+ CD-95
Cisplatin (CP)	25	31			38			48		
Antibody CD-95	10		20			21			27	
CP+CD-95	40			46			57			57

With this, cell death is caused by traumatic action of nanosized carrier onto cellular membrane which essentially enhances penetration of the medicine preparation through the membrane barrier. Secondly, the optimized system ligand-receptor plays an important role in apoptosis of malignant cells. Binding with its receptor, the antibody launches the immunotherapeutical mechanism that leads to apoptosis.

So, the use of magnetosensitive polyfunctional nanocomposites that include antitumour drug and monoclonal antibody CD-95 allows one to realize recognition of specific cells and to reach cytotoxic effect with lower concentrations of medicine remedies, diminishing toxico-allergical influence of chemicotherapeutical preparation onto the organism as a whole.

Thus, it was shown that the synthesized nanocomposites can fulfill such functions as magnetodirected delivery of drugs, recognition of microbiological objects in biological media, diagnostics, local chemical and immunotherapy, as well as hyperthermia, including the cellular level. Realization of these functions is a characteristic property of medicobiological "nanoclinics" [18] and nanorobots [22–29].

Nanocomposites in the Composition of Magnetic Fluids

To make specimens of experimental magnetic fluids with cytotoxic properties we used cisplatin and doxorubicin cytostatics.

The ensembles of Fe_3O_4 particles stabilized with sodium oleate and modified with polyethylene glycol were investigated. Modification of the surface of nanoparticles with PEG was carried out in order to increase in stability of magnetic fluid and decrease in aggregation of the particles. Besides, in using magnetic fluid in biological medium, polymer chains prevent adsorption of different biomolecules onto the surface of nanocomposite effectively, including their agglutination by active agents in composition of blood. The synthesized magnetic fluids were purified by dialysis against 3l of water. Immobilization of cisplatin was carried in the systems: Fe_3O_4/Na ol. and Fe_3O_4/Na ol./PEG.

The obtained specimens of the synthesized magnetic fluids were investigated by FTIR spectroscopy and X-ray diffraction analysis. Preservation of Fe_3O_4 phase and formation of stabilizing phase of sodium oleate and polyethylene glycol in the composition of magnetic fluid was confirmed.

The dependence of magnetic characteristics of specimens on nanocomposite architecture was studied by the methods of vibration magnetometry. The influence of PEG, CP concentration and dialysis on magnetic and size characteristics of the colloidal systems Fe_3O_4/ Na ol. and Fe_3O_4/Na ol./PEG was investigated. The field dependencies of specific magnetization $\sigma = M/\rho$, (M and ρ – magnetization and density, relatively) of the specimens were measured depending on the chemical nature of the magnetite surface modifier.

It was determined that coercitive force value (H_c) depends on sizes and shape of the particles, and also on magnetic interaction force between them. In most cases, the interaction between particles caused a decrease in H_c.

The availability of hysteresis for the specimens is explained by that certain volume of the ensemble of magnetite particles is occupied by the particles of size larger than superparamagnetic boundary for Fe_3O_4 particles at room temperature.

One can suppose that after chemical modification, the sizes of stabilized magnetite particles and their size distribution regularity do not change.

Figure 10 shows the hysteresis loops of Fe_3O_4/Na ol./PEG (*a*) and Fe_3O_4/Na ol./PEG/CP (CP immobilized by adsorption (b)) nanocomposites.

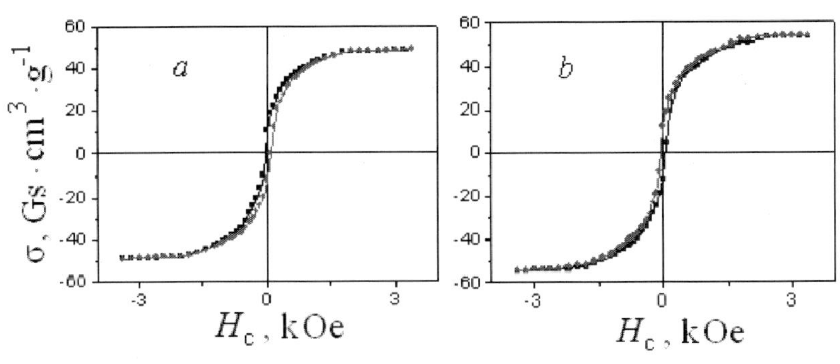

Figure 10. Hysteresis loops of the specimens: *a* - Fe_3O_4/Na ol./PEG, *b* - Fe_3O_4/Na ol./CP.

The mass concentration of the magnetite particles stabilized with sodium oleate in ensemble (C_{Fe3O4}^{mass}) was found as relation of specific saturation magnetization of the stabilized (σ_s^{stab}) and initial (σ_s^{Fe3O4}) particles: $C_{Fe3O4}^{mass} = \sigma_s^{stab} / \sigma_s^{Fe3O4}$. Calculation based on the geometrical models shows that thickness of sodium oleate layer on the surface of most particles investigated is ~ 1 nm.

The mass concentration of the magnetite particles stabilized with sodium oleate in ensemble of the particles modified with PEG ($C^{mass}{}_{Fe3O4/Na\ ol.}$) was found as a relation of specific saturation magnetization of the modified (σ_s^{mod}) and stabilized (σ_s^{stab}) particles: $C^{mass}{}_{Fe3O4/Na\ ol.} = \sigma_s^{mod} / \sigma_s^{stab}$.

Tables 8 and 9 show the values H_c, σ_s, relative residual magnetization M_r/M_s, and also the mass concentrations of the initial magnetite and the magnetite stabilized with sodium oleate in the synthesized ensembles of nanoparticles in Fe_3O_4/Na ol. and Fe_3O_4/Na ol./PEG systems; the average particle size values in specimens calculated using Sherrer formula (D_{XRD}). The characteristics of the initial magnetite are cited for room and liquid nitrogen temperatures. The values C_{mass} Fe_3O_4 (%) in Fe_3O_4/Na ol. system, C_{mass} Fe_3O_4/Na ol. (%) in Fe_3O_4/Na ol./PEG system, C_{mass} Fe_3O_4 (%) in Fe_3O_4/Na ol./PEG system are also shown.

Table 8. Influence of PEG, CP concentration, and dialysis on the magnetic and size characteristics of Fe_3O_4/Na ol. and Fe_3O_4/Na ol./PEG systems

№	Nanostructure	T, K	H_c, Oe	σ_s, Gs·cm^3/g	M_r/M_s	C^{mass}_{Fe3O4}, %	D_{XRD}, nm
1	Fe_3O_4	90	88.8±0.5	72.0±0,2	0.20	100	7.5
		293	54.0±0.5	54.2±0.2	0.19	100	11.0
2	Fe_3O_4/Na ol.	293	56.0	50.7	0.16	94	10.5
3	Fe_3O_4/Na ol. + dialysis	-	50.0	51.6	0.17	95	10.5
4	Fe_3O_4/Na ol. + 18mg PEG	-	50.0	48.9	0.08	90	10.7
5	Fe_3O_4/Na ol. + 18mg PEG + dialysis	-	50.0	49.5	0.09	91	10.7
6	Fe_3O_4/Na ol. + 36mg PEG	-	50.0	45.2	0.11	84	10.7
7	Fe_3O_4/Na ol. + 180mg PEG	-	50.0	25.5	0.11	47	12.5
8	Fe_3O_4/Na ol. + 360mg PEG	-	50.0	24.8	0.11	46	13.7
9	Fe_3O_4/Na ol. + 0.5mg CP	-	50.0	54.1	0.11	~100	10.5
10	Fe_3O_4/Na ol. + 0.5mg CP + dialysis	-	50.0	54.3	0.13	~100	10.6
11	Fe_3O_4/Na ol. + 1.5mg CP	-	50.0	43.9	0.13	81	10.5
12	Fe_3O_4/Na ol./ PEG + 0.5mg CP	-	50.0	51.8	0.09	96	10.7
13	Fe_3O_4/Na ol./PEG + 0.5mg CP + dialysis	-	50.0	52.7	0.11	97	10.8

The data shown testify that increase in PEG concentration in the composites causes decrease in their specific saturation magnetization and lower concentration of magnetite in the composite. Dialysis promotes increase in magnetite concentration and, accordingly, specific magnetization in Fe_3O_4/Na ol. and Fe_3O_4/Na ol./PEG system (Table 8).

According to the experimental data obtained, an increase in saturation magnetization (σ_s) and relative residual magnetization (M_r/M_s) is observed for the specimens with adsorption of small quantities of CP (0.5 mg in calculation onto platinum). Upon growing CP concentration

(1.5 mg), σ_s decreases (Table 8). This effect is displayed in both systems – Fe_3O_4/Na ol. and Fe_3O_4/Na ol./PEG. One can suppose that small ferromagnetic particles have two spin undersystems formed by the atoms to be found on the surface and in bulk [59]. In this case one can not exclude the influence of Pt atoms on the spin undersystem connected with magnetite surface atoms in such a way that magnetization of the particle grows.

The nanocomposites containing Gd – magnetite/gadolinium oxide with core/shell structure – were also synthesized and investigated. Table 9 shows magnetic characteristics of the colloidal systems of magnetite doped with gadolinium ions, as well as gadolinium ferrite obtained by specimens annealing at 1000 °C. The magnetic characteristics and mass concentrations are cited for two systems: Fe_3O_4/Gd_2O_3/Na ol. and Fe_3O_4/ Gd_2O_3/Na ol./PEG.

Table 9. Influence of Gd^{3+} ions concentration on the magnetic and size characteristics of Fe_3O_4/Gd_2O_3/Na ol. systems

№	Nanostructure	H_c, Oe	σ_s, Gs·cm^{3}g^{-1}	C^{mass}_{Fe3O4}, %	C^{vol}_{Fe3O4}, %	δ, nm	M_r/M	D_{XRD}, nm (T_{syn}=323 K)
1	Fe_3O_4	5.0±0.5	4.2±0.2	100	100	0	0.19	11.0
2	Fe_3O_4/Gd_2O_3	106.5	33.0	61	4	1.9	0.47	13.8
4	Fe_3O_4/Gd_2O_3/ Na ol.	106.5	33.7	62			0.41	13.9
5	$Fe_3O_4/2Gd_2O_3$ /Na ol.	227.9	15.8	29			0.23	16.0
6	Fe_3O_4/Gd_2O_3/ Na ol./PEG	106.5	33.7	90			0.37	13.9
7	$Fe_3O_4/2Gd_2O_3$/ Na ol./PEG	227.5	15.9	89			0.18	16.1
8	$GdFeO_3$	144.00	2.6	-			0.44	10.7

Table 10. Density (ρ), specific saturation magnetization (σ_s), mass fraction of magnetite in (m_{Fe3O4}/m_{comp}) structure, saturation magnetization (M_s), coercitive force (H_c), specific residual magnetization M_r/M_s of the nanocomposites

№	Nanostructure	ρ, g/cm^3	σ_s, Gs cm^{3}g^{-1}	m_{Fe3O4}/m_{comp}	M_s, Gs	H_c, Oe	M_r/M_s
1	Fe_3O_4/Na ol. + H_2O (MF)	1.127	7.3±0.2	0.106±0.6	8.2	~ 0.0	0.00
2	Fe_3O_4/Na ol.	2.290	25.0±0.8	0.363±0.8	57.1	69.0	0.08
2a	Fe_3O_4/Na ol. + H_2O (MF)	1.130	5.5±0.2	0.080±0.05	6.2	22.0	0.02
3	Fe_3O_4/Na ol./CP	3.320	29.8±0.9	0.433±0.06	98.9	56.0	0.11
3a	Fe_3O_4/Na ol./CP + H_2O (MF)	1.560	15.1±0.5	0.219±0.013	23.6	19.0	0.13
4	Fe_3O_4/HA/DR	1.143	49.4±1.6	0.718±0.043	56.5	81.0	0.13
4a	Fe_3O_4/HA/DR + H_2O (MF)	1.057	16.4±0.5	0.238±0.014	17.3	50.0	0.09

So, the main properties of colloidal systems based on the synthesized nanocomposites were experimentally investigated, and their size parameters were calculated.

It was determined that dialysis promotes increase in specific magnetization in Fe_3O_4/Na ol. and Fe_3O_4/Na ol./PEG systems.

It was ascertained by experiment that DMSA, HA, APS, PAA layers available on the magnetite surface practically did not affect the magnetic properties of the initial magnetite (the nanocomposite core).

The magnetization curves of ensembles of the magnetite nanoparticles modified with sodium oleate; the magnetite-based nanocomposites containing hydroxylapatite, sodium oleate, cisplatin (0.5 mg), or doxorubicine (0.5 mg); magnetic fluids were measured by a magnetometer with a vibrating specimen (283 Hz, t_{chem} ~ 100 s) at the temperature of 290 K. The results, as an example, are shown in Table 10 and Figure 11.

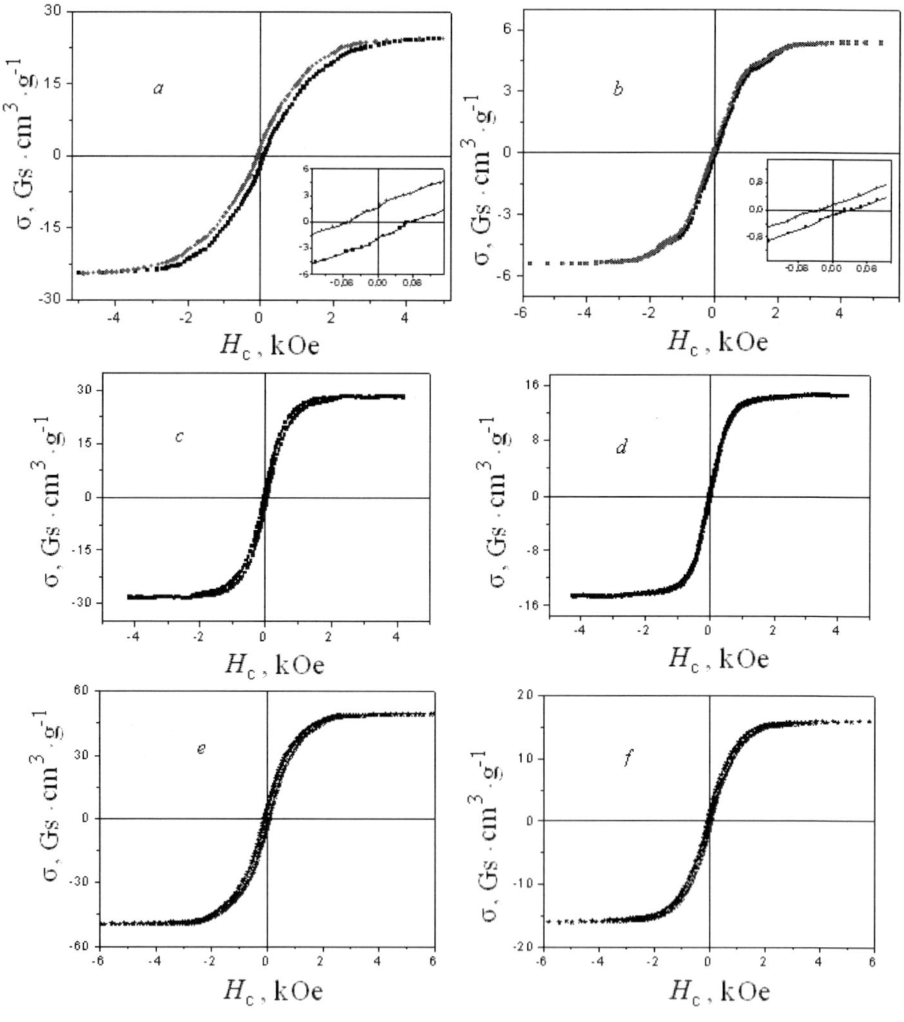

Figure 11. Hysteresis loops (on insertions – the initial areas) of the modified nanocomposites Fe_3O_4/Na ol. (*a*), Fe_3O_4/HA/Na ol./CP (*c*), Fe_3O_4/HA/Na ol./DR (*e*), and their MF (*b*), (*d*), (*f*), respectively.

In such a way, the theory on magnetism of ferromagnetic particles ensemble has been applied to describe the magnetic properties of the synthesized ferrimagnetic Fe_3O_4 nanoparticles, nanocomposites of new types with functions of nanorobots, and magnetic fluids on their base for biomedical application.

One can determine with high authenticity the size parameters of particle ensembles by means of magnetic measurements.

The results obtained will be useful in practical application of nanocomposites and magnetic fluids in oncotherapy.

Besides the magnetic fluids pointed in Table 10, water-based magnetic fluids (physiologic solution) and nanocomposites containing B and Gd were synthesized and investigated.

Study on the Influence of the Magnetic Fluids In vitro and In vivo

In order to estimate the prospects of the nanosystems developed for practical use, pre-clinical investigations were started. It was shown (MTT-test) that the magnetic fluids Fe_3O_4/ Na ol. and Fe_3O_4/Na ol./PEG (Figure 12) with magnetite content C = 10–50 μg/ml show practically no cytotoxic activity *in vitro* upon the cells of breast cancer line MCF-7.

Cytotoxic activity of the samples of magnetic fluids Fe_3O_4/Na ol./CP and Fe_3O_4/Na ol./ PEG/CP exceeded cytotoxic activity of the free form of cisplatin in IC_{25} dose for the mentioned cell line.

The investigations were carried out *in vivo* with male mouse hybrids of the line C57B1/ 6DVA/21, which were inoculated intraperitoneally with Erlich ascite carcinoma at $6 \cdot 10^6$ cells per animal.

It was shown that the use of Fe_3O_4/Na ol./PEG/CP nanocomposites as the magnetic fluid components increases essentially (by ~40 %) the therapeutic effect of cytostatic preparation.

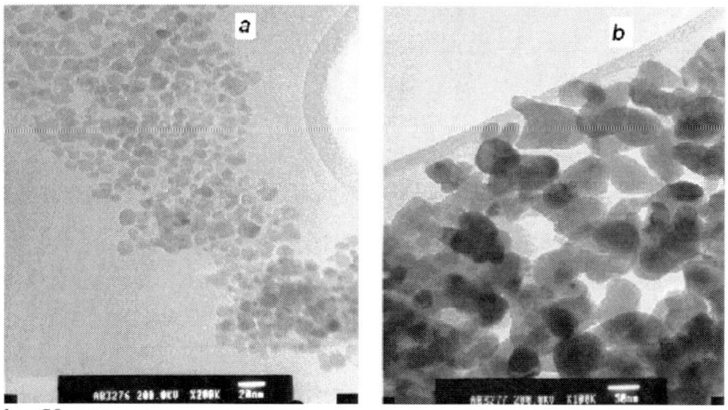

Scales: *a* – 20 nm, *b* – 50 nm.

Figure 12. Scanning electron microscopy images of dried magnetic fluids Fe_3O_4/Na ol.

Conclusion

The analysis of the ways for preparation of polyfunctional medicobiological nanocomposites with multilevel hierarchic architecture and functions of nanorobots determining their ability for recognition of microbiological objects in biologic media, the targeted delivery and deposition of drugs in target organs, the diagnostics and therapy of diseases at the cellular level, as well as their physicochemical and medicobiologic properties, shows that such structures can be promising for preparation of novel medicinal forms with high efficiency and complex (cytotoxic, immunotherapeutic and hyperthermic) activity.

Further work is planned aimed at optimization of the full technological cycle of medicobiological nanocomposites production for their use in oncology, completion of pre-clinical investigations, clarifying the problems related to toxicological aspects, standardization, possibilities of manufacturing, etc.

Acknowledgments

The authors are sincerely grateful to all the collaborators and colleagues participated in the investigations for their creative cooperation.

The work was carried out with a support of goal complex programmes on Fundamental Studies of the National Academy of Sciences of Ukraine "Fundamental problems of nanostructured systems, nanomaterials, nanotechnologies" (projects 53/10-H and 57/10-H) and "Fine Chemicals" (project 31/14).

About the Authors

Gorbyk Petro Petrovych, doctor of physicomathematical sciences, professor, head of the department of nanomaterials at Chuiko Institute of Surface Chemistry of NAS of Ukraine, E-mail: pgorbyk@mail.ru, gorbyk@isc.gov.ua.

Lerman Leonid Borysovych, candidate of physicomathematical sciences, senior researcher from the department of theory of nanosystems, E-mail: llerman@yandex.ua.

Petranovska Alla Leonidivna, researcher from the department of nanomaterials, E-mail: petranovska@ukr.net.

Turanska Svitlana Petrivna, candidate of chemical sciences, researcher from the department of biomedical problems of surface, E-mail: turanska@isc.gov.ua

References

[1] Chekhun, V. F.: Rol innovatsiinych technologiy u rozvyazanni problem onkologiyi. *Visn. NAN Ukrayiny*. 9, 38 (2008).

[2] Nie, S., Xing, Y., Kim, G. J., Simons, J. W.: Nanotechnology applications in cancer. *Annu. Rev. Biomed. Eng*. 9, 257 (2007).

[3] Chekhun, V. F.: Nanotechnology in oncology: present and future. *Exp. Oncology*. 30, 2 (2008).

[4] Lakhtin, V. M., Afanasyev, S. S., Lakhtin, M. V., et al.: Nanotechnologies and perspectives of their application in medicine and biotechnology. *Vestn. Ross. akad. med. nauk*. 4, 50 (2008).

[5] Lammers, T., Hennink, W. E., Storm, G.: Tumour-targeted nanomedicines: principles and practice. *Br. J. Cancer*. 99, 392 (2008).

[6] Okamoto, O. K., Perez, J. F.: Targeting cancer stem cells with monoclonal antibodies: a new perspective in cancer therapy and diagnosis. *Expert Rev. Mol. Diagn*. 8, 387 (2008).

[7] Zuo, L., Wei, W., Morris, M., et al.: New technology and clinical applications of nanomedicine. *Med. Clin. North Am*. 91, 845 (2007).

[8] Hartman, K. B., Wilson, L. J., Rosenblum, M. G.: Detecting and treating cancer with nanotechnology. *Mol. Diagn. Ther*. 12, 1 (2008).

[9] Rhyner, M. N., Smith, A. M., Gao, X., et al.: Quantum dots and multifunctional nanoparticles: new contrast agents for tumor imaging. *Nanomed*. 1, 209 (2006).

[10] Kawasaki, E. S., Player, A.: Nanotechnology, nanomedicine, and the development of new, effective therapies for cancer. *Nanomedicine*. 1, 101 (2005).

[11] Wang, M. D., Shin, D. M., Simons, J. W., Nie, S.: Nanotechnology for targeted cancer therapy. *Expert Rev. Anticancer Ther*. 7, 833 (2007).

[12] Mystakidou, K., Tsilika, E., Tsiatas, M., Vlahos, L.: Oral transmucosal fentanyl citrate in cancer pain management: a practical application of nanotechnology. *Int. J. Nanomedicine*. 2, 49 (2007).

[13] Pan, J., Feng, S. S.: Targeted delivery of paclitaxel using folate-decorated poly(lactide)-vitamin E TPGS nanoparticles. *Biomaterials*. 29, 2663 (2008).

[14] Schroeder, J. E., Shweky, I., Shmeeda, H., et al.: Folate-mediated tumor cell uptake of quantum dots entrapped in lipid nanoparticles. *J. Control Release*. 124, 28 (2007).

[15] Peer, D., Karp, J. M., Hong, S., Farokhzad, O. C., et al.: Nanocarriers as an emerging platform for cancer therapy. *Nat. Nanotechnol*. 2, 751 (2007).

[16] Retel, V. P., Hummel, M. J, van Harten, W. H.: Early phase Technology Assessment of nanotechnology in oncology. *Tumori*. 94, 284 (2008).

[17] Colbert, A. P., Wahbeh, H., Harling, N., et al.: Static Magnetic Field Therapy: A Critical Review of Treatment Parameters. *Evid. Based Complement Alternat. Med*. (2007) /http://ecam. oxfordjournals.org/cgi/reprint/neml31vl.

[18] Levy, L., Sahoo, Y., Kim, K.-S., Bergey, J., E. J., Prasad, P.: Synthesis and characterization of multifunctional nanoclinics for biological applications. *Chem. Mater*. 14, 3715 (2002).

[19] Roco, M. C., Williams, R. S., Alivisatos, P.: *Nanotechnology research directions. Vision for Nanotechnology R and D in the Next Decade*. Dordrecht: Kluwer Academic Publishers. 156, 171 (2002).

[20] Gu, H., Chao, J., Xiao, S.-J., Seeman, N. C.: A proximity-based programmable DNA nanoskale line. *Nature*. 465, 202 (2010).

[21] Lund, K., Manzo, A. J., Dabby, N., Michelotti, N., Johnson-Buck, A., Nangreave, J., Taylor, S., Pci, R., Stojanovic, M. N., Walter, N. G., Winfree, E., Yan, H.: Molecular robots guided by prescriptive landscapes. *Nature*. 465, 206 (2010).

[22] Gorbyk, P. P., Petranovska, A. L., Usov, D. G., Storozhuk, L. P.: Nanokapsula z funkziyamy nanorobota. *Patent na vynachid № 86322*. Ukraine. 2009.

[23] Nanomaterials and Supramolecular Structures. *Physics, Chemistry, and Applications*. Shpak, A. P., Gorbyk, P. P. (eds.) Springer. 63 (2009).

[24] Gorbyk, P. P., Chekhun, V. F., Shpak, A. P.: Fiziko-khimicheskiye i mediko-biologicheskiye aspekty sozdaniya polifunktsionalnykh nanokompozitov i nanorobotov. Nanostrukturnyye materialy – polucheniye, svoistva, primeneniye. Minsk, *Belorusskaya nauka*. 131 (2009).

[25] Fiziko-khimiya nanomaterialov i supramolekularnykh struktur. Shpak, A. P., Gorbyk, P. P. (eds.) Kiyev, *Naukova dumka*. 1, 45 (2007).

[26] Gorbyk, P. P., Petranovska, A. L., Turelyk, M. P., Abramov, N. V., Chekhun, V. F., Lukyanova, N. Yu.: Construction of magnetocarried nanocomposites for medicobiological applications. *Chem. Phys. Tech. Surf.* 1, 360 (2010).

[27] Gorbyk, P. P., Turov, V. V.: *Nanomaterialy i nanokompozity v meditsyne, biologii, ekologii*. Shpak, A. P., Chekhun, V. F. (eds.) Kiyev, Naukova dumka. 444 (2011).

[28] Gorbyk, P. P., Gorobez', S. V., Turelyk, M. P., Chekhun, V. F., Shpak, A. P.: Biofunktsionalizatsiya nanomaterialiv i nanokompozytiv. *Navchal'nyi posibnyk*. Kyiv, Naukova dumka. 283 (2011).

[29] Gorbyk, P. P., Petranovska, A. L., Turelyk, M. P., Pylypchuk, Ye. V., Chekhun, V. F., Lukyanova, N. Yu., Shpak, A. P., Karbovskyi, V. L., Korduban, A. M., Trishin, V. V., Shevchenko, Yu. B.: Sintez, fiziko-khimicheskiye svoistva i bioaktivnost' polifunktsional'nych nanokompozitov dlia biologii i medizyny. *Nanosystemy, nanomaterialy, nanotekhnologiyi*. 8, 749 (2010).

[30] Gorbyk, P. P., Chekhun, V. F.: Nanocomposites of medicobiologic destination: reality and perspectives for oncology. *Functional materials*. 19, 145 (2012).

[31] *Muscat, R. A., Bath, J., Turberfield,* A. J.: A programmable molecular robot. *Nano Lett.* 11, 982 (2011).

[32] Lewandowski, B., De Bo, G., Ward, J. W., Papmeyer, M., Kuschel, S., Aldegunde, M. J., Gramlich, P. M. E., Heckmann, D., Goldup, S. M., D'Souza, D. M., Fernandes, A. E., Leigh, D. A: Sequence-Specific Peptide Synthesis by an Artificial Small-Molecule Machine. *Science*. 339, 189 (2013).

[33] Korochkova, T. Ye., Dekhtyar, M. L., Rosenbaum, V. M.: Osobennosti kinematiki dipol'nykh fotomotorov. *Khimiya, fizika i tekhnologiya poverkhnosti*. Kiyev, Naukova dumka. 14, 52 (2008).

[34] Korochkova, T. Ye., Rosenbaum, V. M.: Molekulyarnyi nasos, upravlyayemyi fluktuaziyami elektricheskogo polya. *Khimiya, fizika i tekhnologiya poverkhnosti*. Kiyev, Naukova dumka. 11-12, 29 (2006).

[35] Gubin, S. P., Koksharov, Yu. A., Khomutov, G. B., Yurkov, G. Yu.: Magnitnyye nanochastitsy: metody polucheniya, stroyeniye i svoistva. *Uspekhi khimii*. 74, 539 (2005).

[36] Belikov, V. G., Kuregyan, A. G.: Polucheniye i medico-biologicheskoye ispol'zovaniye magnitnykh poley i nositeley. *Khim-farm jurn*. 35, 27 (2001).

[37] Vasilevskiy, Yu. A., Volodina, A. P., Pan'kov, V. V.: Magnitnyi poroshok ferrita bariya. Moskwa, *NIITEkhim*. 2 (1995).

[38] Mishchenko, V. M., Kartel', M. T., Lutsenko, V. A., Nikolaichuk, A. D., Kusyak, N. V., Korduban, O. M., Gorbyk, P. P.: Magnitochutlyvi adsorbenty na osnovi aktyvovanogo vugil'lya: syntez ta vlastyvosti. Poverkhnost'. Mizhvid. zb. nauk. prats'. *Kyiv*. 2 (17), 276 (2010).

[39] Semko, L. S., Khutornyi, S. V., Storozhuk, L. P., Dzyubenko, L. S., Abramov, N. V., Gorbyk, P. P.: Khimichne konstruyuvannya ta doslidzhennya vlastyvostey magnitokerovanykh adsorbentiv dlya ekstraktsiyi nukleyinovykh kyslot. Poverkhnost'. Mizhvid. zb. nauk. prats'. *Kyiv*. 2 (17), 330 (2010).

[40] Gorbyk, P. P., Petranovska, A. L., Turelyk, M. P., Abramov, N. V., Turanska, S. P., Pylypchuk, Ye. V., Chekhun, V. F., Lukyanova, N. Yu., Shpak, A. P., Korduban, A. M.: Problema napravlennogo transporta lekarstvennykh preparatov: sostoyaniye i perspektivy. *Khimiya, fizyka ta tekhnologiya poverkhni*. 2, 433 (2011).

[41] Gorbyk, P. P., Dubrovin, I. V., Petranovska, A. L., Turelyk, M. P., Storozhuk, L. P., Mishchenko, V. M., Abramov, N. V., Turanska, S. P., Machno, S. N., Pylypchuk, Ye. V., Chekhun, V. F., Lukyanova, N. Yu., Shpak, A. P., Korduban, A. M.: Magnitoupravlyayemyi transport lekarstvennykh preparatov: sovremennoye sostoyaniye razrabotki i perspektivy. Poverkhnost'. Mizhvid. zb. nauk. prats'. *Kyiv*. 2 (17), 286 (2010).

[42] Oborotova, N. A.: Napravlennaya dostavka protivoopukholevykh preparatov. *Antibiotiki i khimioterapiya*. 36, 47 (1991).

[43] Hafely, U., Schut, W., Teller, Y., Zbororsky, N.: *Scientific and clinical application of magnetic carriers*. London: Plenum. 628 (1997).

[44] Rymarchuk, V. I., Radkevich, L. A., Sarbash, V. I.: Fizicheskiye osnovy primeneniya ferromagnetikov, vvedennykh v organism. *Biofizika*. 35, 145 (1990).

[45] Fertman, V. Ye.: *Magnitnyye zhydkosti*. Minsk, Vysshaya shkola. 179 (1988).

[46] Alyautdin, R. N., Kroyter, Y., Kharkevich, D. A.: Dostavka lekarstvennykh preparatov v mozg s pomoschyu nanochastiz. Eksperim. i klinich. *farmakologiya*. 66, 65 (2003).

[47] Gorbyk, P. P., Petranovska, A. L., Storozhuk, L. P., Lukyanova, N. Yu., Korduban, O. M., Machno, S. M., Chuiko, O. O., Chekhun, V. F., Shpak, A. P., Nanokompozyty na osnovi magnetytu. *Ukrayins'kyi khimichnyi zhurnal*. 73, 24 (2007).

[48] Berezov, T. T., Yaglova, N. V., Dmitriyeva, T. B., Zhyrkov, Yu. A., Chekhonin, V. P.: Napravlennyi transport lekarstvennykh sredstv s pomoshchyu liposom. *Vestn. Ros. akad. med. nauk*. 42 (2004).

[49] Herrmann, J. E., Wang, S., Zhang, C., Panchal, R. G., et al.: Passive immunotherapy of Bacillus anthracis pulmonary infection in mice with antisera produced by DNA immunization. *Vaccine*. 24, 5872 (2006).

[50] Mitra, S., Li, G., Harsh, G. R.: Passive antibody-mediated immunotherapy for the treatment of malignant gliomas. *Neurosurg. Clinics of North America*. 21, 67 (2010).

[51] Bregadze, V. I., Sivayev, I. B.: Lekarstvo-snayper, ili Khimicheskiy pritsel dlya neytrona. *Priroda*. 1 (2004).

[52] De Stasio, G., Casalbore, P., Pallini, R., Gilbert, B., Sanità, F., Ciotti, M. T., Rosi, G., Festinesi, A., Larocca, L. M., Rinelli, A., Perret, D., Mogk, D. W., Perfetti, P., Mehta, M. P., Mercanti, D.: Gadolinium in Human Glioblastoma Cells for Gadolinium Neutron Capture Therapy. *Cancer Research*. 61, 4272 (2001).

[53] Detta, A., Cruickshank, G. S.: L-Amino Acid Transporter-1 and Boronophenylalanine-Based Boron Neutron Capture Therapy of Human Brain Tumors. *Cancer Res*. 69, 2126 (2009).

[54] Turanska, S. P., Turelyk, M. P., Petranovska, A. L., Turov, V. V., Gorbyk, P. P.: Nanokompozity v neytronozakhvatnoy terapii. *Poverkhnost'. Mizhvid. zb. nauk. prats'*. Kyiv. 2 (17), 356 (2010).

[55] Grechko, L. G., Gorbyk, P. P., Lerman, L. B., Chuiko, O. O.: Mozhlyvist' vykorystannya magnitnykh nanochastynok yak zasobiv transportu i utrymannya likars'kykh preparativ v organi-misheni. *Dop. NAN Ukrayiny*. 181 (2006).

[56] Coper, S.: Nanoparticle sizing: a comparative study using atomic force microscopy, transmission electron microscopy, and ferromagnetic resonance. *J. Magn. and Magn. Mater.* 225, 79 (2001).

[57] Zubarev, A. Yu., Iskakova, L. Yu.: Strukturnyye prevrashcheniya v polidispersnykh ferrozhydkostyakh. *Kolloid. zhurn*. 63, 778 (2003).

[58] Rusetskiy, A. N., Ruuge, E. K.: Gemodinamicheskiye aspekty magnito-napravlennogo transporta lekarstv. *Byul. Vsesoyuz. kardiolog. nauch. zentra AMN SSSR*. 7, 85 (1984).

[59] Gorbyk, P. P.: Nanokompozyty z funktsiyamy medyko-biologichnykh nanorobotiv: syntez, vlastyvosti, zastosuvannya. *Nanosystems, nanomaterials, nanotechnologies*. 11, 323 (2013).

In: Advances in Semiconductor Research
Editors: D. Persano Adorno and S. Pokutnyi

ISBN: 978-1-63321-755-3

Chapter 10

Titanium Dioxide Coupled with Small Band-Gap Metal Oxides Sputtered Films for Bacterial Inactivation on Textiles

Sami Rtimi[1,•], Cesar Pulgarin[1], Rosendo Sanjines[2] and John Kiwi[1]

[1]EPFL-SB-ISIC-GPAO, Ecole Polytechnique Fédérale de Lausanne, Lausanne, Switzerland

[2]EPFL-SB-IPMC-LNNME Ecole Polytechnique Fédérale de Lausanne, Lausanne, Switzerland

Abstract

This chapter addresses the design, preparation, evaluation and characterization of the surfaces properties of TiO_2, a wide band-gap semiconductor coupled with thin layers of metal nanoparticulate oxides films presenting narrow band gaps. The TiO_2, In_2O_3, Cu(I)Cu(II)O and Ag_2O have been sputtered in the form of nanoparticulate films on polyester the most widely synthetic textile fabric presenting high chemical resistance, durability and available at a low cost. The sputtering of the highly oxidative nanoparticulate narrow band gap oxide films was carried out by direct current magnetron sputtering (DC), pulsed magnetron sputtering (DCP) and high power pulsed magnetron sputtering (HIPIMS). The sputtering approach to deposit uniform, adhesive thin films on low heat resistant material flexible substrates is shown to lead to films inducing fast bacterial inactivation. *E. coli* was taken as a probe to monitor the surface species generation, oxidative capacity and bacterial inactivation induced by light and in the dark by these composite films in the minute range. The amounts of Cu and Ag needed to inactivate *E. coli* by HIPIMS were far lower compared to films loaded with DC. This leads to a significant saving of metal. Polyester surfaces were pretreated by RF-plasma or UVC to introduce organic-polar groups containing oxygen functionalities binding the metal-clusters to promote the interfacial adhesion to the substrates of TiO_2 and In-, Cu- and Ag-oxides. The coupling of the wide band-gap TiO_2 with In-, Cu- and Ag-oxides led to composite semiconductor-metal oxide films increasing the light absorption in the visible region. The light excitation leads to interfacial charge transfer (IFCT) involving electron (e^-) and hole (h^+) transfer processes. A model is suggested for the charge injection based on the relative position of the conduction band and valence band of TiO_2/In_2O_3, TiO_2/Cu(I)Cu(II)O and TiO_2/Ag_2O.

• E-mail address: sami.rtimi@epfl.ch.

The surface science techniques necessary to characterize the properties of the composites was done out by up-to-date methods involving mainly spectroscopy, X-ray diffraction (XRD) and X-ray fluorescence (XRF) and transmission electron microscopy (TEM).

Keywords: Titanium dioxide, metal oxides, sputtering, thin films, antibacterial activity, oxidation, optical absorption

Introduction

Antibacterial films research is an actual area of research directed to find surface microstructure leading to an accelerated bacterial inactivation. This involves the release of metal-ions producing toxicity below the accepted limits set for its use in medical applications. The regular, periodic and continuous elution of antibiotics and antiseptics (or both) added on medical devices and implants over long period of times is the central problem to avoid infections leading to the replacement of the implants.

TiO_2 has been reported for the last five decades to mediate photocatalytic destruction of organics in all their forms when activated with band gap irradiation [1-3]. TiO_2 has been used in bactericide surfaces for the last 30 years and some reviews have detailed the efficiency and mechanism of the charges in this semiconductor leading to bacterial inactivation. The addition of metal and metal/oxide to TiO_2 increases the TiO_2 reaction kinetics of these destructive processes occurring under mild ecological conditions. Metal/oxide gives rise to Schottky barriers on contact with TiO_2. The TiO_2-metal/metal oxide nanoparticulate reactions involve interfacial electron transfer accelerating the bacterial inactivation kinetics compared to TiO_2 or metal/metal-oxides applied separately.

The field of antibacterial films preparation and evaluation has gained much attention during the last decade since they have been reported to be effective in reducing hospital-acquired infections (HAI) [4-7]. The microbes causing HAI have two main properties: most of them are toxic and resistant to antibiotics and they survive for long times on surfaces present in hospital environment. In addition the increasing bacterial resistance to antibiotics in the last decade and the widespread biofilm formation by toxic pathogens spreading bacteria in schools, hospitals and public places motivates the present work fixing nanoparticles at relatively low temperatures within the heat resistance level limit of textile or polymer films. The effect of nanoparticulate metal-ions is known to be extremely toxic to bacteria at very low concentrations (ppb range) due to their high oxidation potential and easy penetration through the cell envelope porins. Many formulations of TiO_2 surfaces by being antibacterial under light have been reported and the particular interest in finding more effective antibacterial textiles is based on the fact that their microstructure provides a suitable environment for bacterial growth. This can be decreased/suppressed by the addition of bactericide agents like Ag and Cu.

We address in this chapter three coupled nanocomposite thin films prepared by magnetron sputtering: TiO_2/In_2O_3, TiO_2/Cu(I)Cu(II)O and TiO_2/Ag_2O. These composites are tested to destroy organics using *E. coli* as a probe to evaluate the reaction kinetics under similar experimental conditions.

The first system TiO_2/In_2O_3 focuses on the preparation, bacterial inactivation kinetics and characterization of TiO_2-In_2O_3 films sputtered on polyester fabrics. We intend to evaluate the increase of the visible light absorption of TiO_2-In_2O_3 up to 600 nm with respect to TiO_2 alone and the concomitant degree of inhibition of the recombination of the e^-/h^+ generated upon irradiation of TiO_2 particles, thus enhancing the formation of highly oxidative radicals on the TiO_2 surface. These effects should accelerate the bacterial inactivation kinetics [8]. TiO_2/In_2O_3 has been reported as a catalyst for isopropanol decomposition under solar simulated radiation [9-10]. Photo-electrochemical measurements have yielded an indirect band-gap for In_2O_3 with values between 2.3 and 3.0 depending on the size and physical state of In_2O_3 [11].

The second system TiO_2/Cu(I)Cu(II)O focuses on the preparation, bacterial inactivation kinetics and characterization of the films sputtered on polyester fabrics. More recently Cu/CuO has been reported as an effective bactericide, fungicide, algaecide and antiviral agent [12-20]. Cu-bactericide properties have been known for decades due to the release of metal-ions into the surrounding fabrics [21]. Cu-films would avoid the use of chemicals to clean/disinfect surfaces leaving residues after use [22]. The particular interest in Cu-textiles is based on the fact that the porous hydrophilic structure of textiles provides a suitable environment for bacterial growth. For some years a technology is being developed to introduce Cu into natural and artificial fibers, latex and polymer films. The application of these Cu-fibers was directed towards the treatment of viral transmissions, nosocomial infections and antibiotic resistant bacteria.

Plasma-based film deposition processes have lately evolved toward the higher ionization of Cu-ions/metal in the plasma flux. This development was necessary since metal colloids do not lead to uniform, reproducible and adhesive films. These films could be wiped off easily by hand or cloth contact and could not be annealed on low resistant substrates like polymer films and textile fabrics [23]. Plasma deposition methods like magnetron sputtering allow metal and metal-oxide deposition at low temperatures (<130°C) of TiO_2, Cu and their combinations which was not possible when using CVD requiring a higher operating temperature. The deposition of uniform thin solid films was observed to be reproducible and show high adherence and compactness [24-26]. Our laboratory has recently reported Cu-films prepared by direct-curent magnetron sputtering (from now abbreviated as "DC") and direct-curent pulsed magnetron sputtering (DCP) [27-29]. Hashimoto et al., prepared the TiO_2/Cu thin films by the dip-coating technique to extend the light absorption of TiO_2 into the visible light region and reduce due to the surface Cu the electron-hole recombination upon band-gap recombination [30-32]. Metal nanoparticulates possess the most favorable ratio of bacterial inactivation and cell toxicity without decreasing the biocompatibility since the in vitro cytocompatibility was retained [33].

Cu addition has shown to improve effectively the antibacterial properties of textile and other substrates when added in extremely low concentrations. In recent years PVD has been applied to modify fibers and surfaces promoting antimicrobial effects compared to chemical vapor depositions. Recent detailed and systematic work on TiO_2-Cu and TiO_2-Ag thin films prepared by CVD deposition on glass, silica has been reported [34-37]. Stranak et al., [38] prepared Ti-Cu films on Ti-alloys and demonstrated that these films presented strong antibacterial activities without cytotoxic effects to osteoblast cells. Dong et al., [39] deposited Cu-alloys on stainless steel by sputtering (PVD) and Zhang et al., incorporated Cu on polyethylene obtaining films reporting long-term antibacterial properties [40].

The last meta-material presented in this study TiO_2/Ag_2O presents many common properties with the two previous coupled semiconductor/metal oxides. The interest in using bactericide, anti-viral and fungicide textiles to treat infections has significantly increased during the last several years using Ag/Ag-ions as the active component to inactivate bacteria [41-42]. The Ag original use was to extend the optical spectrum of TiO_2 into visible region [43-44]. But Ag-particles in contact with the cell wall of bacteria release Ag-ions that due to their small size penetrate into the cells through the wall porins [5, 45]. Ag-nanoparticles affect the bacteria respiratory enzyme due to the production of ROS and associated DNA damage leading to death. The antimicrobial effect of Ag-solutions have been recently reported. The release of Ag-ions from Ag-grafted surfaces involves Ag-oxidation reactions due to the oxygen available to the Ag-clusters. This release is directly dependent on the Ag-particle size that is effective only when the Ag-ions concentration is in a sufficient concentration and available in a sustained way during relatively long periods. Uniform sputtered textiles with effective Ag bactericide action have been prepared recently by sputtering methods [46-48]. When applying solar or visible light Ag_2O (bg 1.5-1.7) has been observed to act as a semiconductor enhancing bacterial inactivation [49].

Bacteria have a low resistance to Ag, yet Ag is known to be benign to human cells. The Ag-deposition on cotton by sputtering is important since cotton has the property to adsorb a large amount of moisture making this textile prone to microbial attack. With ambient temperature and humidity, cotton becomes a nutrient for bacterial and fungal growth. Ag-nanoparticles on textiles expose a significant amount of Ag-clusters on the surface of textiles that can interact with pathogen without hindrance with the surrounding medium. In the past decade textiles were pretreated by RF-plasma and/or UVC and then colloidal Ag-suspension were contacted with the textile surface to graft Ag-nanoparticles [41-42]. But smearing and Ag-loses occur when this approach was used. Later work, magnetron sputtering provided a more robust bonding of the Ag-nanoparticles revealing to be effective in the removal of airborne bacteria [43].

The addition of Ag, Cu and In-oxides to TiO_2 gives rise to plasmonic nanoparticulate interacting with UV or visible electromagnetic radiation. This type of interaction induces oscillations of the free meta/metal oxide electrons localized on these surfaces giving rise to plasmon resonances. Electron scattering and electron-phonon coupling associated with the localized plasmon resonance (LPR) the strong electric field generated around the metal/metal-oxides nanoparticles is converted into heat. During this work the strong electric field generated around Ag, Cu and In/TiO_2 due to the large nanoparticle absorption cross section is converted into heat leading to the inactivation of pathogens.

Another aspect that has not being fully explored is the effect of the doping of TiO_2 during the sputtering process. Doping the TiO_2 topmost layers with In, Cu and Ag induce a higher fraction of Ti^{4+} species on the TiO_2 making it more p-type. This pushes the Fermi level of the TiO_2 up. No substitution of the Ti^{4+}-cation is possible due large radii difference between the Ag, Cu and In with respect to Ti^{4+} [50]. The porosity of the sputtered film will also be altered due to the metal/metal oxides nanoparticles interaction affecting the composite photocatalyst microstructure.

This study is directed to present antibacterial coatings showing the interdisciplinary nature of this work involving: microbiology and surface design which includes the preparation of films, photocatalysis, catalysis, material science and surface characterization. The sterilization of air and of solid-air and solid–solution interfaces makes disinfection processes a three phase problem. This reality has been recognized by the most important laboratories working in this area [4-7, 25, 33-38].

This work describes thin TiO_2 films increasing their reaction kinetics in the presence of metal-metal oxides In_2O_3; Cu(I)Cu(II)O; Ag_2O. The most important features related to the film preparation, kinetic evaluation of their performance, surface characterization and the modeling of the interfacial charge transfer (IFTC) between the TiO_2 and the metal-oxides are investigated in this chapter. Electronic energy levels available in the coupled semiconductors were respected.

Methods

RF-Plasma and UVC Pre-Treatment Increasing the Number of Sites for the Attachment of TiO_2/In_2O_3; TiO_2/Cu(I)Cu(II)O and TiO_2/Ag_2O

RF-Plasma Pretreatment

RF-plasma activation of textiles in implied the application of electromagnetic radiation at low P (0.1- 1 torr) since vacuum enhances the capture length of the electron avoiding collision deactivation [51]. Residual O_2 generates cations anions (O^-, O^+), radicals, excited states, molecule-ions and electrons with high energy in a system where the local generated heat has a non–homogeneous distribution in a non-equilibrium situation. But recently pressures up to 1 atm have been used in conjunction with RF-plasma where the RF-plasma treatment has been used to generate localized heat effect [43]. In the RF-cavity intense local heat develops breaking H^- and intermolecular bonds I textile fabrics. Intermolecular C-C bond scissions occur and the O_2 containing hydrophilic functional groups increase for polyester leading to an improved wettability of these textiles. In the presence of O_2, -COO-, percarboxylic species, -O-O-, phenolic and lactam species develop as detected by XPS.

UVC Pretreatment to Increase the Binding of TiO_2

The light excitation at 185 nm does not have the energy content to produce cationic O^+ and anionic oxygen O^- species, as it is the case for RF-plasma pretreatment. This light only leads to excited O*: O_2 + UVC → 2O*. The optical absorption between the light and the textile positioned at 3 mm of the light source at 1 atm air is very low since ε_{O2} (185 nm) = 2.6 $M^{-1}cm^{-1}$ and ε_{N2} (185 nm) = 0.3 $M^{-1}cm^{-1}$ at 185 nm/6.70 eV.

Practically, no Vacuum-UVC light is lost in the optical pathway between the light source and the cotton sample even at atmospheric pressures. Fabrics obtained by UVC pretreatment introduced surface activation with more evenly distributed than when activation by RF-plasma was used. An UVC lamp emitting at 185 nm (6W) from a 25W (254 nm + 185 nm light) low-pressure Hg-lamp (Ebara Corp. Tokyo, Japan) was used for UVC pretreatment. Vacuum-UVC light radiant energy at 185 nm is higher than 241 nm (495 kJ/mole), the energy required for the reaction O_2 → 2O*. The absence of cationic or anionic oxygen leads to more uniform distribution in the activated sites on the polyester. This in turn leads more uniform sites on the textile surface.

Sputtering of TiO_2/In_2O_3; TiO_2/Cu(I)Cu(II)O and TiO_2/Ag_2O Nanoparticulate Films

DC sputtering was carried out in the sputtering chamber by means of a 5 cm diameter target plate bombarded by Ar-ions [52]. The diagram of the magnetron chamber used for the DC and DCP magnetron sputtering is shown in Figure 1. Applying a current of 300 mA needed a bias voltage of -400 V for DC-sputtering. DC pulse-magnetron sputtering (DCP) was operated at 50 kHz with 15 % reversed voltage. A negative voltage was applied of 430 V and then the voltage is switched to 65 V (15% of 430 V) to accelerate the sputtered particles towards the substrate. DCP pulses of 10 microseconds were applied, but with time the target gets overcharged and when this occurs, the unit tries three times to clear the charging arc with additional three pulses before the power supply shuts down for 7 ms and turns on automatically after this recovery time. The mode of operation is neither: a) unipolar pulsed sputtering, where the target voltage is pulsed between the normal operating voltage and ground nor b) bipolar pulsed sputtering where the target voltage is reversed and becomes more positive during the pulse-off period. The DCP operation is bipolar asymmetric. The Ar gas used in the sputtering chamber at 0.4 Pa and the power densities of about 6.6 W/cm^2 were used.

The HIPIMS deposition of metals was carried out in a CMS-18 Vacuum system from Kurt Lesker Ltd. evacuated to 10^{-5} Pa by a turbomolecular pump [53]. Figure 2 shows the scheme of the HIPIMS chamber.

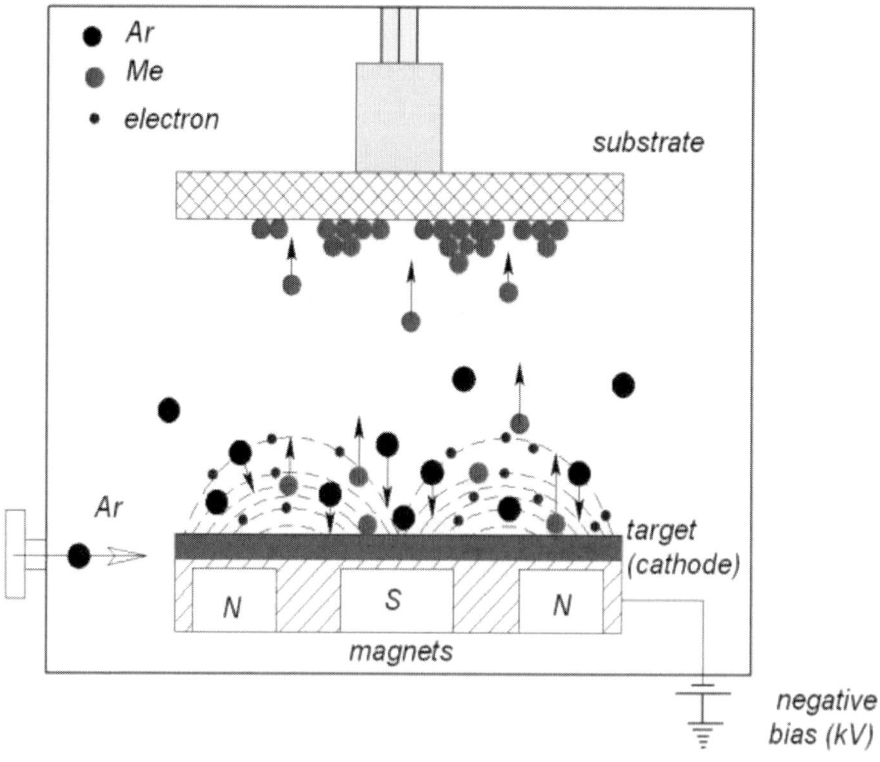

Figure 1. Schematic diagram of the direct current magnetron sputtering (DC/DCP).

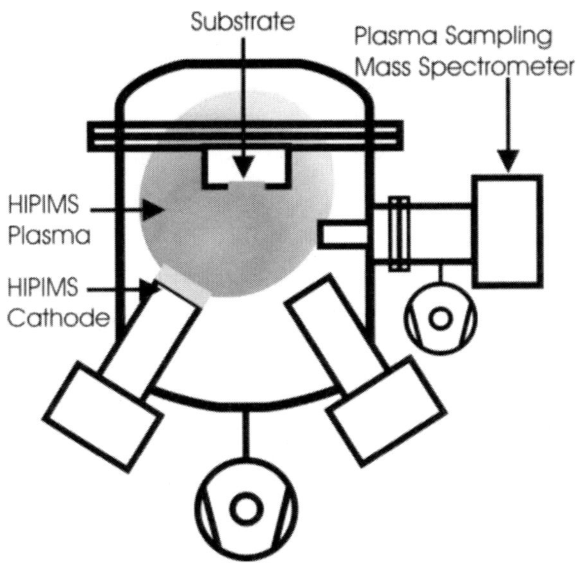

Figure 2. Schematic diagram of the high power impulse magnetron-sputtering unit (HIPIMS) used in the sputtering of metals on fabrics.

The targets were 5 cm in diameter, 99.99% pure from K. Lesker Ltd. UK. The HIPIMS was operated at 500 Hz with pulses of 100 microseconds separated by 1.9 ms. The HIPIMS short pulses avoid a glow-to-arc transition during plasma particle deposition.

The polyester used as the substrate for the sputtering corresponds to the EMPA test cloth sample No 407. It is a polyester Dacron, type 54 spun, plain weave ISO 105-F04 used for color fastness determinations. The calibration of the Ti, In, Cu and Ag film thickness was carried out on Si-wafers and determined with a profilometer (Alphastep500, TENCOR).

Bacterial Evaluation and Surface Characterization of the TiO_2-Metal Oxide Surfaces

Evaluation of the bacterial inactivation: The samples of *Escherichia coli* (*E. coli* K12) was obtained from the Deutsche Sammlung von Mikroorganismen und Zellkulturen GmbH (DSMZ) ATCC23716, Braunschweig, Germany, and were used to test the antibacterial activity. The polyester fabrics, both bare and sputtered ones, were sterilized by autoclaving at 121°C for 2h. 20 μL *E. coli* K12 culture aliquots, with an initial concentration of ~10^6 CFU mL^{-1} in NaCl/KCl (pH 7) were placed on coated as well as on bare (control) polyester fabric. The polyester is a micro-porous substrate and distributes the inoculum evenly on the film without needing an adsorption stage. A uniform contact was established between the sample and the well dispersed bacterial solution for all the investigated samples. The samples were then placed on Petri dishes covered with a lid to prevent evaporation. Afterwards, at scheduled time, each fabric sample was transferred into a sterile 2 mL Eppendorf tube containing 1 mL autoclaved NaCl/KCl saline solution. This solution was subsequently mixed thoroughly using a Vortex for 3 min. Serial dilutions were made with NaCl/KCl solution. An aliquot of 100 μL was pipetted onto a nutrient agar plate and then spread over the surface of the plate using standard plate method. Agar plates were incubated lid down, at 37 °C for 24h before colonies were counted. Three independent assays were performed for each sputtered textile sample.

The actinic irradiation lamps used as a source of white light were close to the indoor source of light in health facilities. The solar simulated light source was a solar simulator (Heraeus, Hannau, Germany) provided for a 100 W Xe-light resembling the solar spectrum with a light intensity of 50 mW/cm^2. This is equivalent to the average solar light dose reaching European countries.

X-Ray fluorescence analysis of samples (XRF): The TiO_2 and metal-content on the polyester was evaluated by X-ray fluorescence, by using an RFX, PANalytical PW2400 spectrometer.

Diffuse Reflectance Spectroscopy (DRS): Diffuse reflectance spectroscopy on polyester samples was carried out using a Perkin Elmer Lambda 900 UV–Vis NIR-spectrometer provided for with a PELA-1000 accessory within the wavelength range of 200–800 nm and a resolution of one nm. The absorption of the samples was plotted in Kubelka–Munk (KM) arbitrary units vs wavelength.

X-ray Photoelectron Spectroscopy (XPS): An AXIS NOVA photoelectron spectrometer (Kratos Analytical, Manchester, UK) equipped with monochromatic Al Ka (hν=1486.6 eV) anode was used. The electrostatic charge effects on the samples were compensated by means of the low-energy electron source working in combination with a magnetic immersion lens. The carbon C1s line with position at 284.6 eV was used as a reference to correct the charging effect. The quantitative surface atomic concentration of some elements was determined from peak areas using sensitivity factors [54]. Spectrum background was subtracted according to Shirley [55]. The XPS spectra were analysed and deconvoluted using CasaXPS-Vision 2 software, Kratos Analytical UK.

Inductively coupled plasma spectrometry (ICP-MS): the Cu released by the TiO_2/Cu(I)Cu(II)O during bacterial disinfection was determined since it is a sensitive analytical technique because due to the low background and high ion transmission. The FinniganTM ICPS used was equipped with a double focusing reverse geometry mass spectrometer able of a resolution of $1.2x10^5$ cps/ppb, detection limit of 0.2 ng/L. Samples 1 cm^2 were placed in Eppendorf tubes containing 1 ml (NaCl/KCl 0.08% solution). The solutions were then diluted 10 times to reach the volume necessary for the ICP-MS analyses.

Transmission electron microscopy (TEM): A Philips CM-12 (field emission gun, 300 kV, 0.17 nm resolution) microscope at 120 kV was used to measure grain size of the TiO_2/Cu films. The textiles were embedded in epoxy resin 45359 Fluka and the fabrics were cross-sectioned with an ultramicrotome (Ultracut E) and at a knife angle at 35°. High-Angle Annular Dark-Field (HAADF) imaging was used to map the scanning electron microscopy (STEM) using a FEI Tecnai TEM-STEM OSIRIS, 200kV. These images (Z-contrast images) are obtained by collecting the scattered electrons passing through the objective provided with an annular dark-field detector.

TiO_2/In_2O_3 Composite Films: Kinetic Evaluation and Surface Properties

The onjective of this study was to show the effect of In_2O_3 layers decorating TiO_2 layers on polyester acccelerating *E. coli* inactivation compared to TiO_2 alone irradiated under actinic light [56]. Figure 1 shows the bacterial inactivation under actinic light by TiO_2/In_2O_3 composite films sputtered for different times.

By inspection of Figure 3 it is readily seen that the sample sputtered for 10min from a Ti target in O_2 atmosphere and subsequently for 10s using an In target leads to the faster bacterial inactivation kinetics. The results presented in Figure 3, traces 3-5, suggest that longer sputtering times led to samples with larger In_2O_3 particle, thus decreasing the available contact surface with bacteria. In_2O_3 sputtering for 10s may results in optimal In_2O_3 cluster size, with the highest amount of In_2O_3-sites exposed on the previously sputtered TiO_2 layers on the polyester.

The accelerated bacterial inactivation demonstrated for the TiO_2-In_2O_3 compared to the bare TiO_2 samples may be related to the favourable electrostatic attraction existing between the positive charged Ti and interacting with the *E. coli* negative surface at pH 7. The *E. coli* presents a negative charge between pH 3-9 due to the large amount of negative carboxylic groups compared to the positively charged functional groups: amide I and amide II on the bilayer cell membrane.

Figure 4a shows the UV-Vis reflectance spectra of TiO_2 sputtered for increasing times on polyester. The Kubelka-Munk (KM/S) relations convert reflectance measurements into absorption spectra units. K and S are the absorption and scattering coefficients respectively of TiO_2 in Figure 4a.

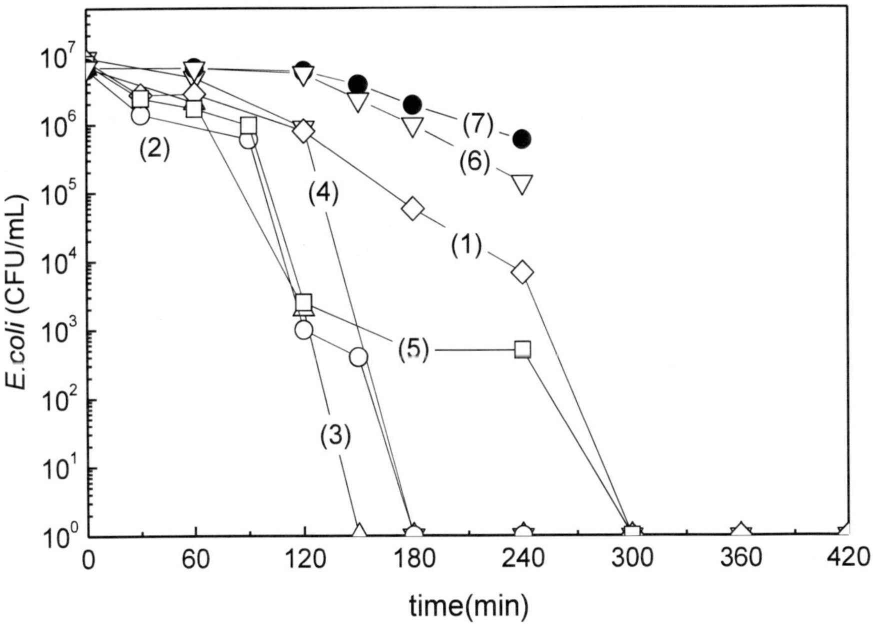

Figure 3. *E. coli* inactivation on samples under Osram Lumilux T8-L18W lamp (4.0 mW/cm^2) as noted in traces (1)-(6): 1) TiO_2 10 min, 2) TiO_2 10min /In_2O_3 5s, 3) TiO_2 10 min/In_2O_3 10s, 4) TiO_2 10 min/In_2O_3 20s, 5) TiO_2 10 min/ In_2O_3 40s, 6) In_2O_3 alone and 7) sample TiO_2 10 min/In_2O_3 10s in the dark.

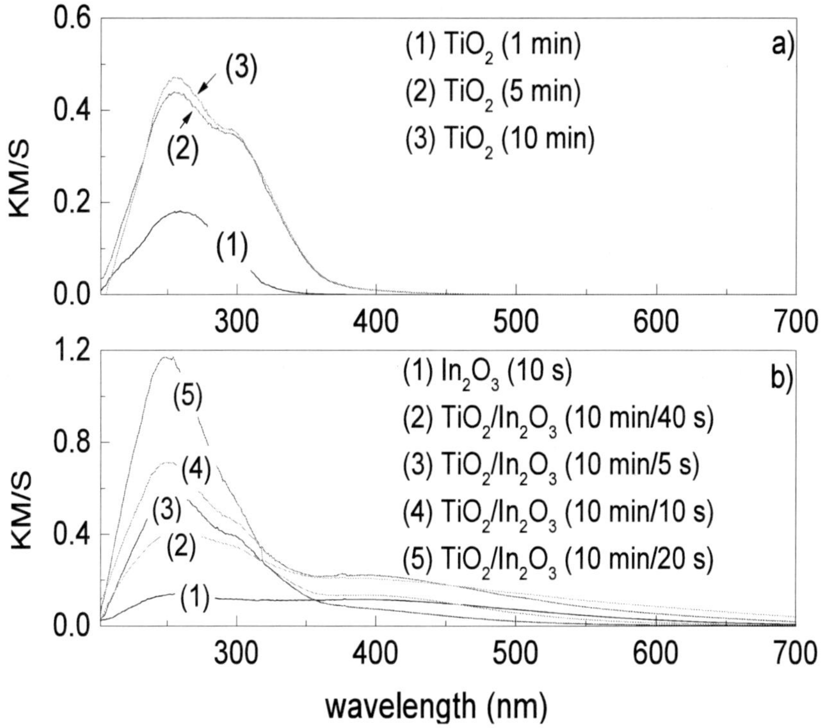

Figure 4. Diffuse reflectance spectra (DRS) of TiO_2 and TiO_2/In_2O_3 samples in Kubelka-Munk units: a) TiO_2 1 min, TiO_2 5 min and TiO_2 10 min. b) TiO_2 10 min/$In_2O_3$5s, TiO_2 10 min/$In_2O_3$10s, TiO_2 10 min/In_2O_3 20s, TiO_2 10 min/ In_2O_3 40s, $In_2O_3$10s.

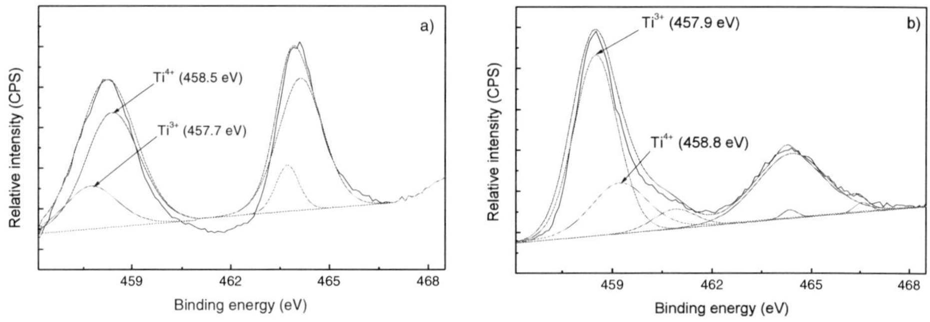

Figure 5. Deconvolution of the Ti2p XPS envelope for a TiO_2 10 min/$In_2O_3$10s sample: a) time zero and b) after 300 min bacterial inactivation under Osram Lumilux T8-L18W lamp irradiation.

The KM/S values for the samples were found proportional to the TiO_2 sputtering time. The faster bacterial inactivation induced by the TiO_2-In_2O_3 samples suggests electronic transfer from In_2O_3 to TiO_2 intervening in bacterial inactivation.

Figure 4b shows the increase in optical absorption (KM/S spectra) Samples sputtered with In_2O_3 for 5 to the 20s sputtered show low absorption intensities, compared to samples sputtered for 40s, forming larger clusters of In_2O_3. Figure 4b shows a weak absorption band from 400 to 500 nm, which can be attributed to the interfacial charge transfer (IFCT) from the TiO_2 to In_2O_3 layers. The weak absorption between 500 and 600 nm is due to the In_2O_3 interband indirect transitions at potential levels ≥ 2.09 eV [57]. These electronic transitions occur from the In_2O_3 valence band to two levels near the conduction band ≥ 2 eV. The TiO_2-In_2O_3 transition seems to originate from the In_2O_3 (In5s5p orbital) to the conduction band of TiO_2.

X-ray Photoelectron spectroscopy (XPS): Figure 5 shows the fitting of the Ti2p XPS envelopes for a TiO_2 10 min/$In_2O_3$10s sample at: a) time zero and b) after 300 min bacterial inactivation irradiated by an Osram Lumilux T8-L18W lamp. The Ti^{4+} doublet peak shifts from 458.5 eV at time zero to 458.8 eV after the 300 min during bacterial inactivation due to redox reactions. The Ti^{4+} shift is ≥ 0.2 eV [54-55] In Figure 5a and 5b, the Ti^{3+} peaks shifts from time zero to 300 min from 457.8 eV to 458.9 eV. Samples of TiO_2-In_2O_3 show two peaks at 444 eV and 451 eV, respectively, assigned to In3d5/2 and In3d3/2 that also shifted during the bacterial inactivation by 0.3 -0.4 eV (data not shown).

Interfacial charge transfer mechanism (IFCT) in TiO_2-In_2O_3 composite films: Coupling TiO_2-In_2O_3 induced a significant increase in the photocatalytic activity compared to TiO_2 alone and In_2O_3. This significant increase can be rationalized by the positions of the conduction band (cb) and valence band (vb) in In_2O_3 and TiO_2. The cb of In_2O_3 at -0.62 vs NHE [8-11] and is well positioned to transfer the cb electrons to TiO_2 with a cb at -0.2 eV vs NHE for anatase. The UV component of the actinic light source used generates charges in the TiO_2 transferring the holes to In_2O_3 vb as suggested below in Figure 6. The spatial separation of photogenerated charge carries should limit the detrimental e^-/h^+ recombination events in TiO_2. The scheme in Figure 6 suggests the IFCT in the TiO_2-In_2O_3 assuming that the generated charges have a sufficient lifetime to diffuse and inactivate bacteria.

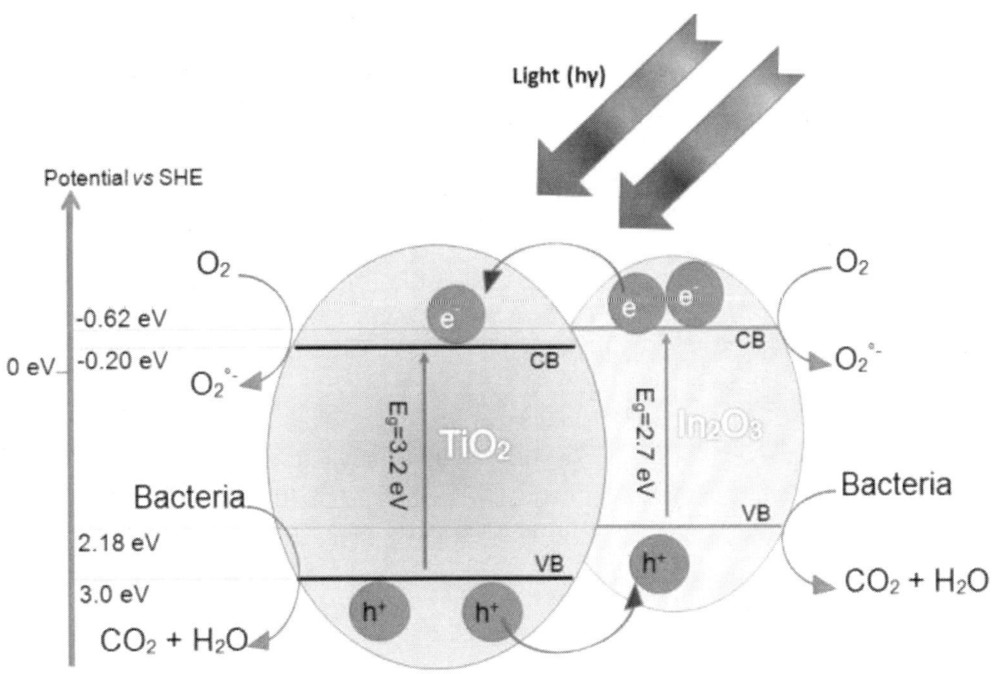

Figure 6. Charge injection during the interfacial charge transfer (IFCT) between In_2O_3 and TiO_2.

TiO_2/Cu(I)Cu(II)O Composite Films: Kinetic Evaluation and Surface Properties

Cu has been used in conjunction with TiO_2 for the purpose of preparing adhesive robust uniform films leading to an accelerated bacterial inactivation under simulated solar radiation as reported in some recent studies [27-28, 57-59]. The engineering of Cu-nanoparticulate films is an important area helping to reduce/eliminate the rate of HAI. We focus in this study on the deposition of Cu-films using high power impulse magnetron sputtering (HIPIMS) and compare the results obtained with recent work reported sputtering Cu on fabrics by DC/DCP [27-28] where a lower Cu-ions/CuO flux was available in the sputtering chamber.

In DCP-sputtering, the electronic density ~10^{16} e^-/m^3 and the ionization potential of the electrons is higher compared to DC. In the case of HIPIMS, pulses from one microsecond up to milliseconds generate current densities of ~10^{18} e^-/m^3 which is 10^4 times higher than that obtained with conventional DC-sputtering and the degree of ionization reaches about 50-80% [60]. Until now, HIPIMS sputtering to deposit bactericide metal nanoparticles films precluding bacterial inactivation has not been widely investigated. Low intensity actinic light led to complete bacterial inactivation (6log10) within the minute range. The small amount of Cu-ions determined by ICP-MS provide the evidence for an oligodynamic effect during Cu-mediated bacterial inactivation. The oligodynamic effect proceeds when materials induce effective bacterial inactivation at a very low level of reactant with concomitant low cytotoxicity [3].

HIPIMS sputtering generates high-density plasma ionizing the metal to a high degree at low pressures avoiding the formation of macro-particles that deteriorate the properties of the growing film [24-26]. The development of HIPIMS in the last decade is due to the growing demand for high quality anticorrosive films in aircraft, car industry and other industrial sectors. Recently, high power pulsed magnetron sputtering (HIPIMS) has gained increasing attention as a more advanced sputtering method to deposit uniform, robust films onto a variety of substrates. This is due to the high pulse power densities of several kW leading to elimination of growth defects and concomitant densification of the sputtered films. The enhanced densification attained by sputtering with HIPIMS is due to the high ion-to neutral-ratio found in the plasma during the deposition process compared to more traditional forms of sputtering. In conventional DC-sputtering, the degree of ionization is 1%-5% and the electronic density ~10^{14} e^-/m^3.

Figure 7a shows the bacterial inactivation kinetics by the HIPIMS TiO_2 sputtered polyester samples. As shown in Figure 7a no bacterial inactivation take place in the dark but the bacterial inactivation becomes faster for HIPIMS sputtering times between 1 min (trace 5) and 4 min (trace 2). Longer deposition times between 10 and 30 min did not accelerate the loss of viability kinetics due to the fact that an increased TiO_2 thickness > 12 nm sputtered within 4 min leads to: a) bulk inward diffusion of the charge carriers generated on TiO_2 under light leading to highly oxidative radicals [4-7] and b) longer sputtering times facilitate the TiO_2 inter-particle growth decreasing the TiO_2 contact surface with bacteria [60]. Figure 7b shows the *E. coli* inactivation within 60 min for HIPIMS Cu-sputtered samples within 15, 30, and 60s. The bacterial loss of viability in Figure 7b, trace 6 shows that no bacterial loss of viability occurs on polyester alone under light irradiation. Figure 7c shows that runs in the dark for samples sputtered for 150s induced a low loss of bacterial viability within 120 min, showing that the bacterial CFU reduction involves Cu-layers. Under actinic light radiation, traces 3 and 4 indicate that sputtering times of 30s and 60s induce faster bacterial loss of viability kinetics. A sputtering time of 150s induced the shortest inactivation time (trace 1). Sputtering for 300s induce bacterial inactivation taking longer times compared to samples sputtered for 150s. Therefore, the amount of CuO is not the main species leading to bacterial inactivation. Sample in trace 1) presents the highest amount of Cu-sites held in exposed positions interacting on the surface or close to the polyester surface with *E. coli* leading to bacterial loss of viability. The surface bactericide action seems to be due to a synergic effect introduced by the TiO_2/Cu layers since longer times was observed when sputtering TiO_2 as shown in Figure 7c.

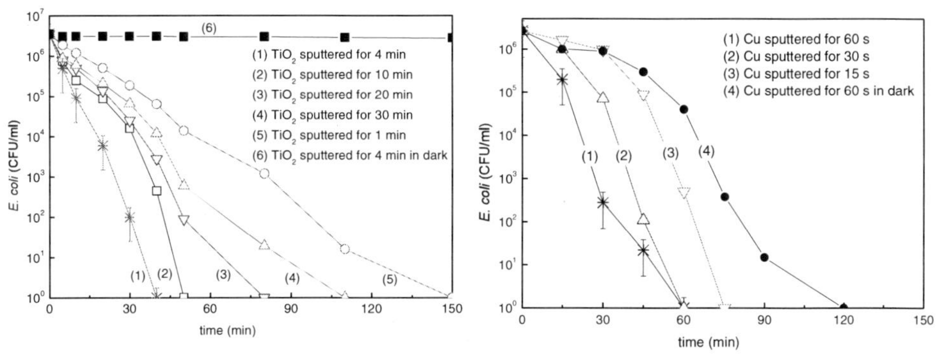

Figure 7a. *E. coli* survival on TiO_2 HIPIMS-sputtered (5 A) on polyester for different times in the dark and under solar simulated irradiation. b) *E. coli* survival on Cu HIPIMS-sputtered for different times in the dark and under Osram Lumilux 18W/827 actinic lamp (4 mW/cm^2).

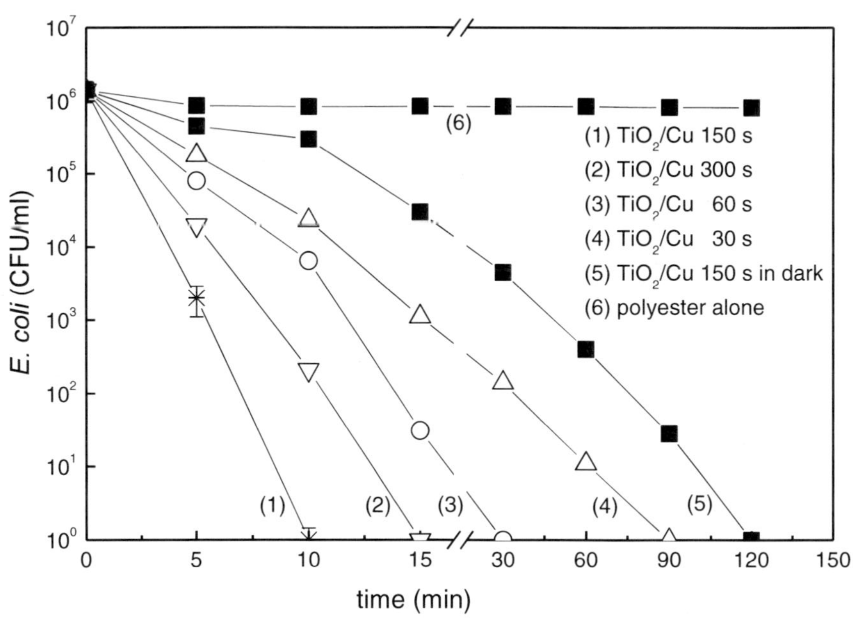

Figure 7c. *E. coli* survival on TiO_2/Cu HIPIMS-sputtered on polyester for different times in the dark and under Osram Lumilux 18W/827 actinic lamp (4 mW/cm^2).

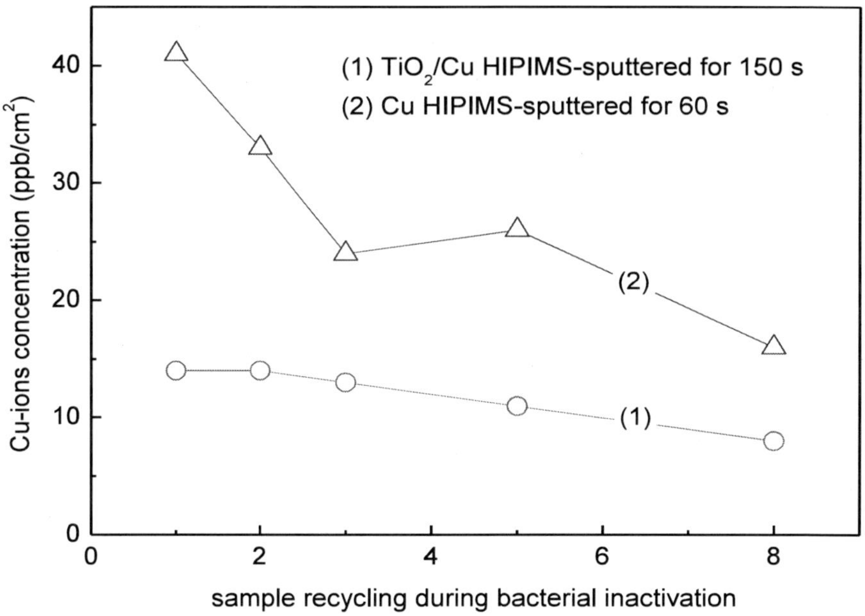

Figure 8. Concentration of ions eluted into solution determined by ICP-MS up to the 8th recycling of TiO_2/Cu samples inactivating *E. coli* K-12 under solar simulated light irradiation.

Figure 8 shows the release of Cu-ions as a function of catalyst reuse of TiO2/Cu sputtered for 150 s on polyester with respect to the Cu-sputtered on polyester for 60 s as measured by ICP-MS. The release of Cu- from the TiO_2/Cu samples was ~8 ppb/cm^2. This value is lower compared to the Cu-release from the Cu-sputtered samples reaching up to ~18 ppb Cu/cm^2 at the end of the 8th cycle. The small amounts of Cu released proceed through an oligodynamic effect to inactivate *E. coli* in the minute range. The Cu and TiO_2/Cu induced bacterial inactivation proceeds therefore in a way that it is not toxic to human health.

Electron Microscopy (TEM) and Immiscibility of Ti and Cu on TiO_2/Cu(I)Cu(II)O Samples

The particle size of the film nanoparticulate and the hydrophobic-hydrophilic balance determine to great extent the surface photocatalytic properties. Cu samples sputtered on polyester for 30s show Cu-nanoparticles between 8-15 nm. The TiO_2 samples sputtered for 150s present sizes between 8-12 nm, and the TiO_2/Cu samples sputtered for 150s show sizes of 5-10 nm. It seems that the role of TiO_2 is to bind, disperse and stabilize the Cu-clusters on the polyester surfaces due to its adhesive properties. The nanoparticles small size accounts partly for the favorable bacterial inactivation kinetics due to the large surface area per unit mass. The electronic transfer between the TiO_2/Cu sample and the *E. coli* depends on the charge diffusion length within the composite film. This in turn is a function of the TiO_2 and Cu particle size and shape [60].

Figure 9a (left hand side) presents TiO_2/Cu microscopy contrasted by high angular annular dark field (HAADF) showing the Cu-nanoparticles to be immiscible with TiO_2. Cu^{2+}

does not substitute Ti^{4+} in the TiO_2 lattice because of the significant difference in the radii of Ti^{4+} (0.53 Angström) and Cu^{2+} (1.28 Angström). Figure 9b shows the mapping of the Cu, Ti and O.

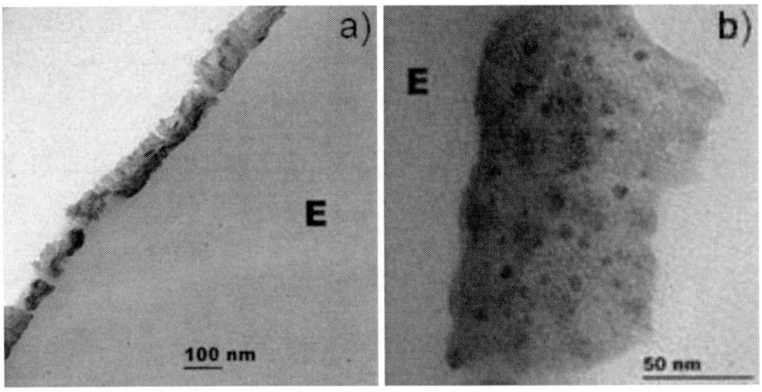

Figure 9a. Transmission electron microscopy (TEM) of: a) Cu sputtered for 150s by HIPIMS on polyester, b) TiO_2/Cu sputtered for 150s polyester with a scale of 50 nm using A Philips CM-12 (field emission gun, 300 kV, 0.17 nm resolution) microscope at 120 kV.

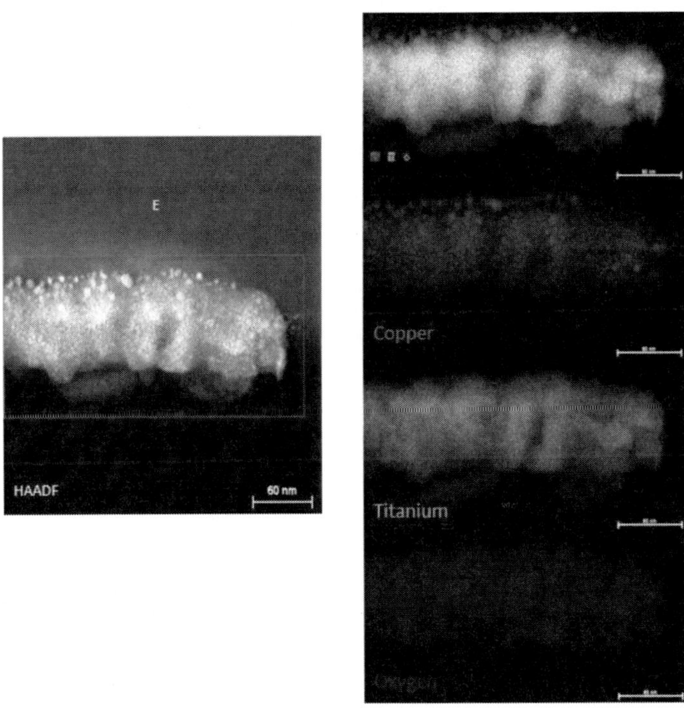

Figure 9b. High-Angle Annular Dark-Field imaging (HAADF) images of TiO_2/Cu HIPIMS sputtered for 150s showing the complete sample and the mapping of Cu, Ti and O by Z-contrast imaging in the TEM image using FEI Tecnai TEM-STEM OSIRIS, 200kV.

Due to its size, the CuO/Cu nanoparticles with particle size > 8 nm are not able to penetrate into the bacteria core through the cell wall porins with diameters of 1-1.3 nm [63]. Only Cu-ions are able to diffuse through bacterial porins leading to DNA damage responsible for the total loss of bacterial viability. The interfacial distances between TiO_2 and Cu/CuO on the polyester surface range from below 5 nm. This allows for interfacial charge transfer (IFCT) between both oxides to proceed with a high quantum efficiency [2, 7]. Quantum size effects have been shown to occur in particles with sizes 10 nm having about 10^4 atoms as presented by the TiO_2 particles with sizes ~10 nm. For CuO nanoparticles the (photo)-generated charge recombination takes place within shorter times due to the decrease in the available space for charge separation. Also, the decrease of the space charge layer decreases further the potential depth.

X-ray photoelectron spectroscopy (XPS) and Ar-etching of TiO_2/Cu films permit to show the atomic percentage concentration of Cu, Ti, O_2 and C of TiO_2/Cu samples sputtered for 150s as a function of penetration depth of the Ar-ions (Figure 10). It is readily seen that Cu, Ti and O decrease up to 240 Angströms due to the Ar-bombardment. The etching depth induced by the Ar-ions was referenced by the known etching value for Ta of 15 atomic layers per minute equivalent to ~30 Angstroms/min. The penetration of the Cu inside the sample protects the Cu-clusters inside the 130 microns thick polyester network during the *E. coli* inactivation process. The increase in the C-content in Figure 10 is due to the etching removing the TiO_2/Cu layers making available the C-content of the polyester. The inset in Figure 10 shows the significantly lower percentage of Cu and Ti for TiO_2/Cu sputtered by DC/DCP [27-28]. The concentration of Ti followed a different pattern compared to the one observed when sputtering by HIPIMS and increases beyond 100 Angströms because Ti deposition was hindered by the Cu-layers.

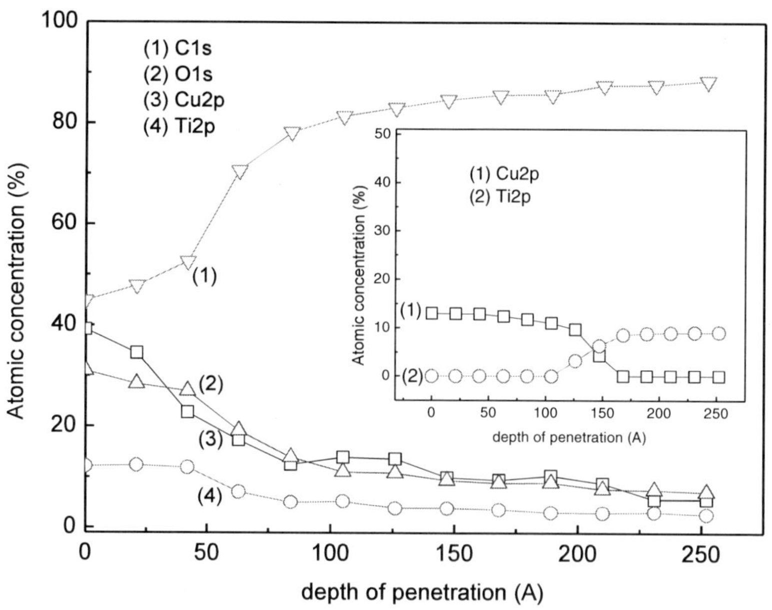

Figure 10. XPS etching showing the microstructure of TiO_2/Cu film sputtered by HIPIMS up to 240 Angstroms or ~120 atomic layers. Insert: depth profile of the TiO_2/Cu film sputtered by DC/DCP.

Figure 11 presents the loss of viability time (bacterial inactivation) vs thickness for DCP and HIPIMS TiO_2/Cu sputtered films. Figure 11 shows the much thinner TiO_2/Cu layer thickness necessary for complete bacterial inactivation sputtered by HIPIMS compared to

DC/DCP. A HIPIMS film with a thickness of 38 nm inactivated bacteria within ~10 min compared to a sputtered DC/DCP film 600 nm thick inducing inactivation bacterial inactivation within the same period of time.

Figure 12 shows the interfacial charge transfer between TiO_2 and Cu in the TiO_2/Cu photocatalyst TiO_2/Cu under simulated solar irradiation. In the TiO_2 semiconductor the solar irradiation induces both the e^- transfer and h^+ transfer from TiO_2 to CuO since the potential energy levels of the TiO_2cb and TiO_2vb lie above the CuOcb and CuOvb levels. The partial recombination of e^-/h^+ in the TiO_2 is hindered by the transfer of charges to the CuO facilitating the reactions occurring at the TiO_2cb and CuOcb as shown in Figure 12. Under simulated solar light, the CuO can be reduced to $Cu_2O$and the Cu_2O can reduce O_2 via a multi-electron process and re-oxidize to CuO. The charges generated by light in the TiO_2/Cu lead to the rapid loss of *E. coli* viability within 10 min (Figure 7c), along O_2 and CuO reduction at the CuOcb.

The interfacial charge transfer (IFCT) in the TiO_2/Cu sample seems to proceed with a high quantum efficiency since the bacterial inactivation proceeds within short times of about 10 min. But the magnitude of the increase in the IFCT absorption of the TiO_2/Cu seen by the DRS spectra is relatively small.

The conduction band of CuO at -0.30 V vs SCE (pH 7) is at a more negative potential than the potential required for the one electron oxygen reduction $O_2 + H^+ + e^- \rightarrow HO_2$ -0.22 V [7]. Furthermore, the Cu^{2+} can react with e^- (or O_2^-) $\rightarrow$ Cu^+ + (or O_2). The Cu^+ can reduce O_2 consuming electrons or be oxidized to Cu-ions by the photo-generated TiO_2 holes to Cu^{2+}.

The TiO_2vb holes react with the surface -OH of the TiO_2 releasing OH° radicals leading to inactivate/oxidize bacteria [1-3].

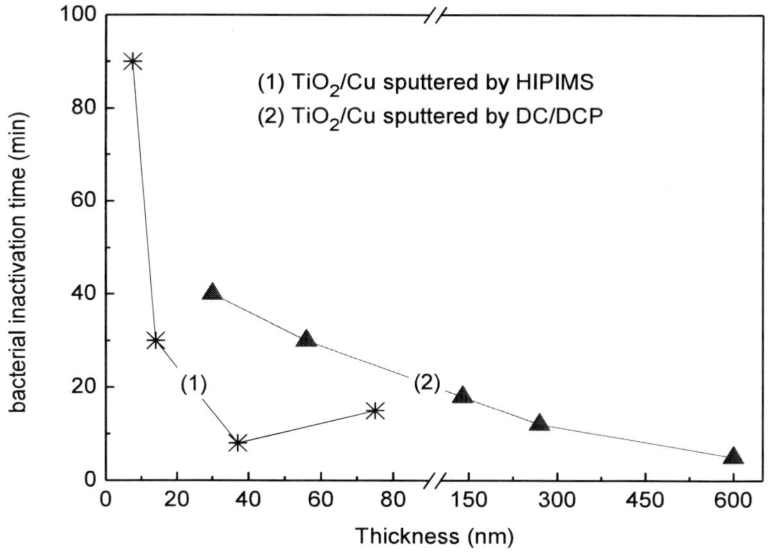

Figure 11. Bacterial inactivation time *vs* nominal thicknesses for HIPIMS sputtered TiO_2/Cu films and by DC/DCP sputtered layers under solar simulated irradiation.

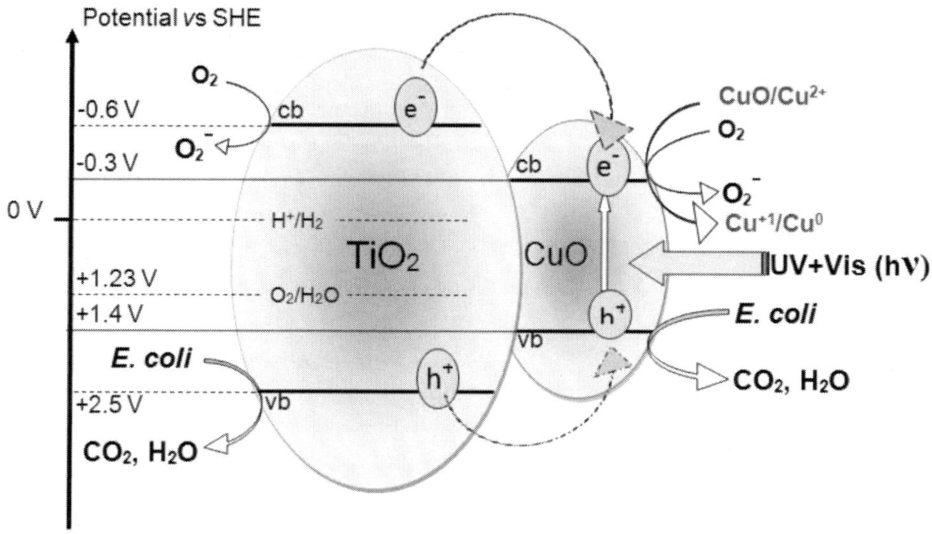

Figure 12. Diagram suggested for IFCT on TiO_2/Cu films on polyester under solar simulated light.

TiO_2/Ag_2O Composite Films: Kinetic Evaluation and Surface Properties of TiON and TiON-Ag Samples

Figure 13 shows the bacterial inactivation of *E. coli* by TiON/Ag films [62]. The TiON layers were sputtered for 4 min and the Ag-sputtering was applied up to 40s. A small bacterial inactivation was observed in the dark (Figure 13, trace 5). The bacterial inactivation kinetics increases as the Ag sputtering time was increased from 10 to 30 s and falls back at 40s possibly due to a) the formation of bigger Ag-clusters1b) to the surface Ag clusters hindering the diffusion of charges from the TiON underlayers and c) to a decreased formation of active catalytic sites on the Ag-clusters. The Ag-clusters are not necessarily crystalline and were shown to darken the sputtered samples as a function of time.

The *E. coli* survival on Ag-sputtered directly on the polyester fabric led to longer *E. coli* bacterial inactivation times for sputtering times of 10, 20 and 30s [46-48]. TiON-Ag (4min-30s) samples lead to 6log10 bacterial inactivation within 55 min. This is possible since the TiON thin layer on the polyester merges without forming boundaries with the polyester [61]. On the TiON coating, a topmost Ag-layer immiscible with TiON builds up as shown below in the EM section.

The TiON was DC-sputtered on polyester and Ag was subsequently sputtered in the case of the TiON-Ag samples by DCP sputtering. The polyester samples were 2x2 cm^2 in size. Before the deposition of the films, the residual pressure Pr in the sputtering chamber was Pr $\leq 10^{-4}$ Pa. The substrate to target distance was kept at 11 cm.

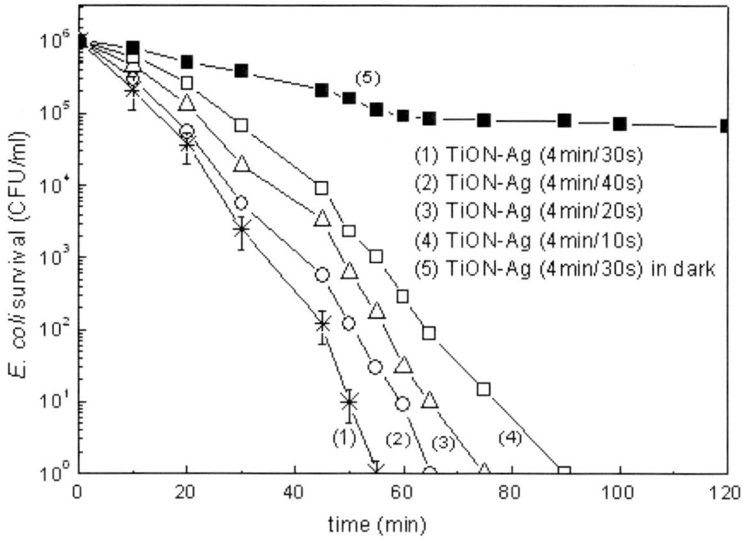

Figure 13. *E. coli* survival on TiON-Ag sputtered on polyester for different Ag deposition times on TiON layers. The light source is an Osram lamp L18W/827 (4 mW/cm^2, 400-700 nm). Reactive gas flow composition: Ar 90%: N_2 5%: O_2 5% and total P 0.5 Pa.

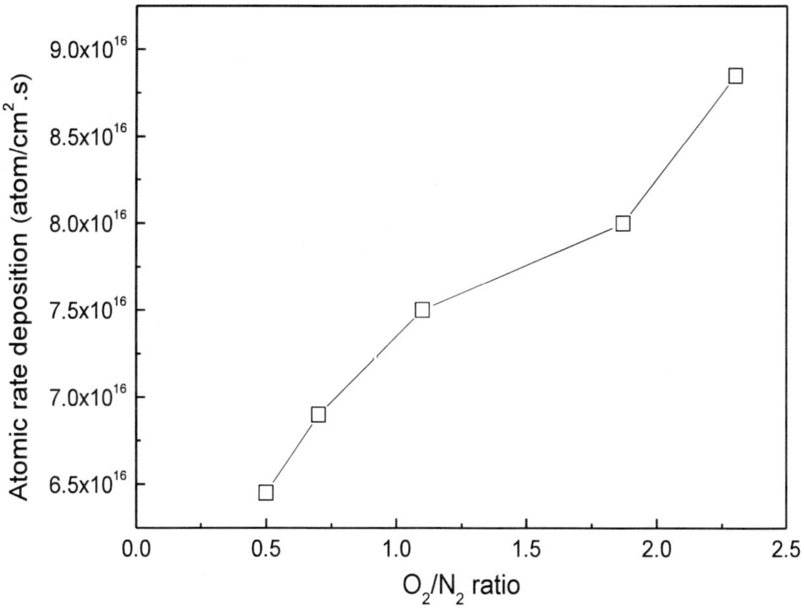

Figure 14. Atomic deposition rate of TiON films on polyester as a function of the O_2/N_2 ratio. Ar gas flow was 90% at a working pressure 0.5 Pa.

The TiON thin films were deposited by reactive DC-magnetron sputtering (DC) in an Ar + N_2 + O_2 gas flow from a 5 cm diameter Ti-target 99.99 % pure (Kurt J. Lesker, East Sussex, UK). The total pressure PT = (P_{Ar}+ P_{N2} + P_{O2}) was fixed at 0.5 Pa and after optimization the fastest bacterial inactivation kinetics was obtained with a gas flow Ar 90%: N_2 5%: O_2 5%.

The formation of TiO_2 when sputtering Ti in the presence of O_2 and N_2 can be understood in terms of: a) the partial oxidation of TiN takes place in the presence of an oxygen source due to the residual H_2O vapor in the sputtering chamber at the residual pressure Pr=10^{-4}Pa. This pressure is representative of about 10^{15} molecules/cm^2, there are sufficient O-radicals available to induce partial oxidation of TiN films and b) the films also oxidize after the deposition when exposed to air and during the sterilization process autoclaving at 121°C. Figure 14 shows the atomic deposition rate of TiON films as a function of the O_2/N_2 ratio. Sputtering for 4 min leads to layers ~70 nm thick. This is equivalent to 350 layers each containing 10^{15} atoms/cm^2 being deposited at a rate of ~$1.5x10^{15}$ atom/cm^{2}s. The TiON was sputtered first for 4 min followed by Ag-sputtering for 30s. The atomic deposition rate was observed to be a function of the O_2/N_2 ratio.

Figure 15 shows the TiON plasmon resonance in Kubelka-Munk (KM/S) units. The optical absorption of the TiON samples increases with longer sputtering times up to 4, 5 min. Increasing the sputtering times from 1 to 5 min in Figure 15 lead to TiON samples with a higher optical absorption due to the introduction of N-interstitial sites doping the TiO_2 with O-vacancies. This in turn leads to a larger amount charge transfer sites in the visible region. Samples sputtered for 4 ad 5 min are seen to present a similar optical absorption indicating saturation of the polyester surface by the sputtered TiON layers. This suggestion seems to be confirmed by the lack of further growth in the optical absorption by the samples sputtered for 10 minutes showing additionally a different shape for the TiON spectra.

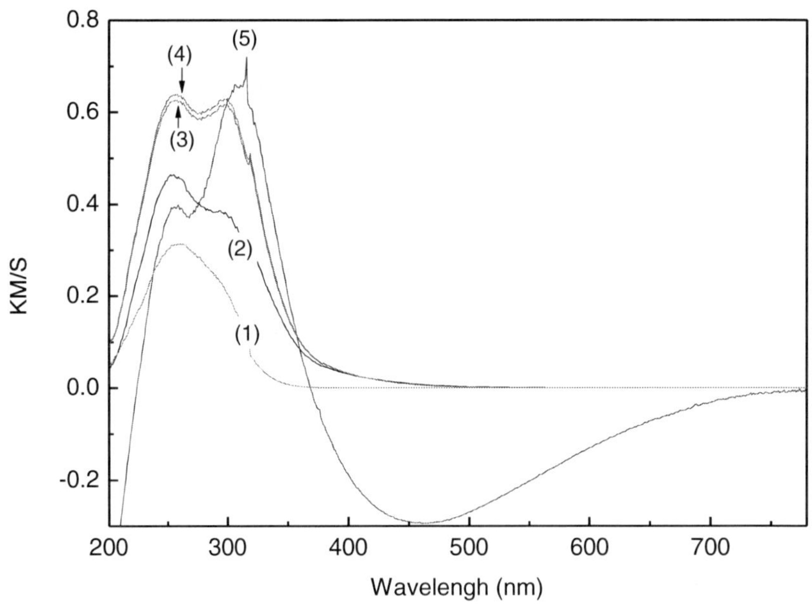

Figure 15. Diffuse reflectance spectra of TiON samples sputtered on polyester for different times: (1) 1 min, (2) 2 min, (3) 4 min, (4) 5 min and (5) 10 min.

This is due to the non-linear optics introduced by the high density of TiON layers on the polyester surface not following quantitatively the normal optical absorption increase beyond the saturation concentration.

Metal nanostructures from Ag, and Cu have generated significant attention because of their unique and tunable localized surface plasmon resonance (SPR). SPR is an optical phenomenon arising from the collective oscillation of conduction electrons, which is also responsible for the brilliant colors of metal colloids. At the plasmon resonant frequency, the absorber has a large absorption cross section leading to heating of metallic Ag, Cu NPs [50]. At the plasmon resonant frequency, the incident photons can either be scattered (at the same frequency but in all directions) or absorbed and subsequently converted into phonons or lattice vibrations. Therefore, the SPR metal peaks include both scattering and absorption components. Absorption component relates to the heating effect of metallic NP. Such phenomena are commonly referred to as photothermal heating or plasmonic heating and can be used to induce localized hyperthermia to kill cancer or bacterial cells.

Photothermal energy is able to heat nanoscale volumes without heating the bulk solution. Due to sharp heating of NPs a thin layer of supercritical water can be formed and this layer destroy bacteria and biofilms. The Ag, Cu surface photo-excited plasmon transfers the electron to the TiO_2cb by inelastic plasmon scattering in the metal/TiO_2 interface. Injection of electron to TiO_2cb leads to the metal-ions formation subsequently being released for bacterial inactivation. TiO_2 concomitantly also transfers the cb electrons to O_2 with O_2^- formation inducing damage to the bacterial cell.

Figure 16a presents in the left hand side the almost continuous dark TiON-Ag (4min-30s) deposit on the polyester fiber. The right hand side image with a higher magnification of 100 nm shows the immiscibility of the Ag-dark coating and the TiON coating grey layers. The Ag-particles present sizes between 20-40 nm within a TiON-Ag layer with a width of 70 ± 10 nm. Ag/Ag_2O particles of 20-40 nm will not pass through the bacterial cell protein wall with porin pores with a diameter of 1.1-1.5 nm [60]. This confirms once more that Ag bacterial inactivation is due the Ag-ions having sizes far below 1 nm, and not to the Ag-nanoparticule.

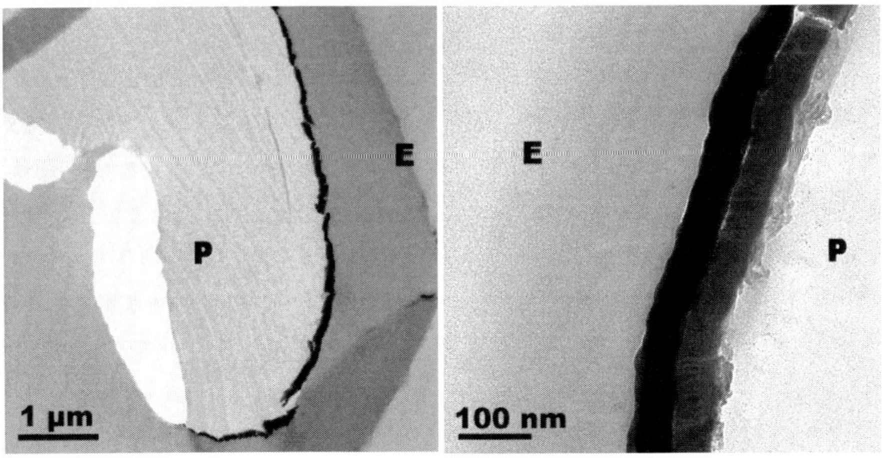

Figure 16a. Left hand side: Transmission electron microscopy of a TiON-Ag (4min-30s) DC-sputtered sample showing the continuity of the sputtered TiON-Ag layer around the polyester fiber. Right hand side: dense continuous Ag-layers being immiscible with the grey TiON layer. (P=polyester and E=epoxide).

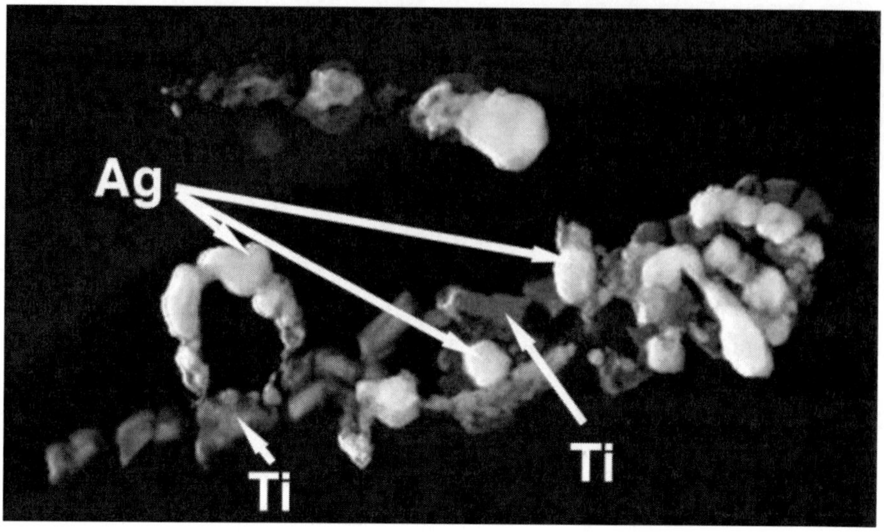

Figure 16b. TiON-Ag (4min-30s) determination by Bright Field (BF) showing Ti and Ag at the current beam position.

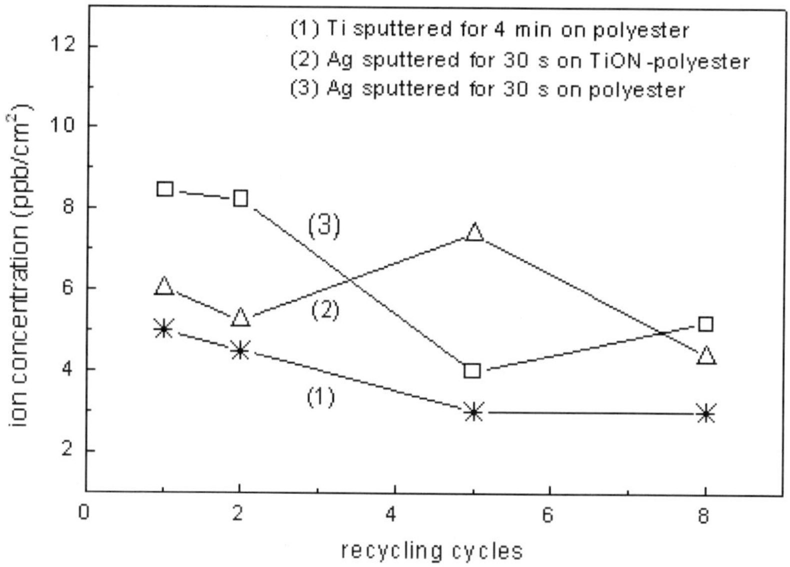

Figure 17. Ion-coupled plasma spectrometry (ICPS) determination of Ag ions and Ti-ions released during the recycling of (1) TiON sputtered 4 min on polyester, (2) a TiON-Ag (4 min-30s) polyester sample and (3) a sample of Ag-polyester sputtered 30s.

Figure 16b shows the image of a TiON-Ag (4min-30s) sample in Bright Field (BF). The immiscibility of the Ag and Ti on the sample surface is readily seen at the current beam position.

Ag-ions release during disinfection by TiON-Ag polyester was evaluated by ICP-MS. Figure 17 presents the release of Ti and Ag up to the 8th recycling during bacterial inactivation runs. The results for samples of TiON, TiON-Ag and Ag are shown in Figure 17. Trace 1 shows the slow release of inert Ti-ion up to the 8th recycling from the TiON 4 min sample. Trace 2 shows the Ag-ion release starting at a level of 6 ppb/cm^2 for TiON-Ag (4min-30s) sample and reaching ~5 ppb/cm^2 on the 8th recycling. Ag-ions are formed by oxidation of the Ag on the polyester in the reaction media. Ag-ions with a concentration > 0.1 ppb show a significant antimicrobial activity. Concentrations of Ag >35 ppb can be toxic to human cells [43]. Figure 17; trace 3 shows the release of Ag-ions sputtered for 30s on polyester. A high level of Ag-ions at time zero reaches on the 8th recycling about the same level as the Ag-released by the TiON-Ag (4min-30s) samples. The biocompatibility of TiON has been well documented in biomedical applications [63-64]. This is the reason for its interest and potential to be used in conjunction with Ag, Zn, Cu in disinfecting 2-D and RD surfaces The TiON inactivation kinetics with a concomitant low cytotoxicity are the two essential requirements desired for the application of antibacterial surfaces.

Mechanism of reaction: The Ag in the TiON-Ag (4min-30s) film forms AgOH on the film surface in contact with air. The AgOH decomposes spontaneously to Ag_2O (eq.1)[65]

$$2AgOH \rightarrow Ag_2O + H_2O \qquad (pK = 2.87) \tag{1}$$

This Ag_2O is stable in the thermodynamically stable region at pH 6-7 where the bacterial inactivation of *E. coli* proceeds. Visible / actinic light irradiation photo-activates Ag_2O with 1.46 < bg < 2.25 eV [66-68] as noted next in eq(2)

$$Ag_2O + light \rightarrow h^+ + e^- \tag{2}$$

For mechanistic considerations of the Ag_2O/TiO_2 transfer of charges we have to consider the position of the energy bands of Ag_2O and TiO_2. The TiO_2 presents a bg ~3.2 eV, a cb at -0.1 V NHE and a vb at +3.1 V NHE at pH 0. An interfacial charge transfer process (IFCT) proceeds upon visible light activation of the TiON-Ag samples. The Ag_2O seems to produce a microstructure: a) precluding to some extent of electron-hole recombination in the TiON layers and b) furthering the interfacial e^- transfer between the AgNP's and bacteria.

Under visible light irradiation, the transfer of charge from Ag_2O to TiO_2 is thermodynamically favorable since the Ag_2O cb -1.3 eV NHE at pH 0 and the vb of Ag_2O +0.2 V NHE at pH 0 lie above the TiO_2 levels: cb at -0.1 V and vb at +3.1 V. The Ag_2O cb electrons are transferred to the TiO_2 cb in a downward energetic process. The Ag_2O in eq(1) reacts with e^- induced by light irradiation eq(2)

$$Ag_2O + e^- \rightarrow Ag + O_2 \tag{3}$$

The O_2 shown in eq(3) promote reactions [5, 6] producing highly oxidative radicals, while the h^+ in eq(2) reacts with H_2O in eq(4).

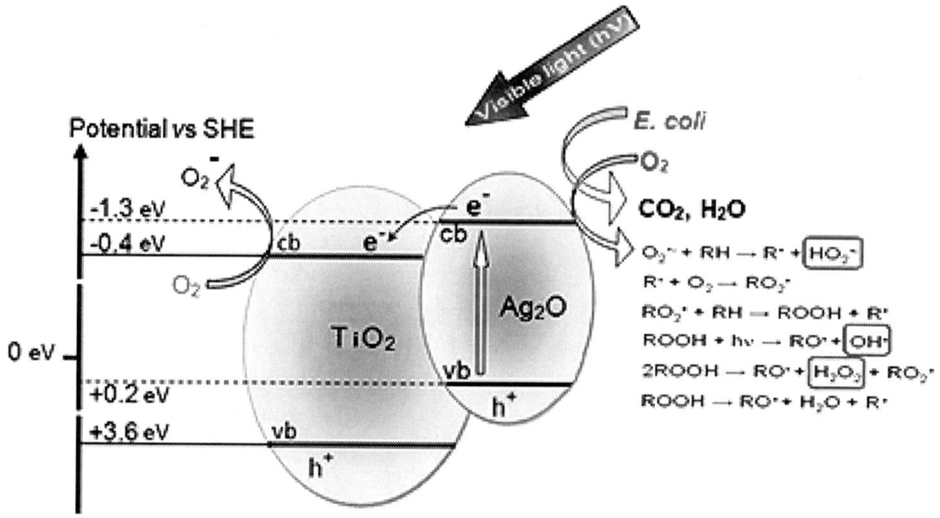

Figure 18. Scheme for the photoinduced interfacial charge transfer (IFCT) from Ag_2O to TiO_2 under visible light irradiation.

This reaction runs parallel with eq(5) generating OH^{o} radicals (or other highly reactive oxidative radicals) able to inactivate *E. coli*. The h^+ in eq(4) from eq (2) reacts

$$h^{+} + H_2O \rightarrow OH^{\circ} + H^{+} \tag{4}$$

$$e^{-} + H_2O + O_2 \rightarrow OH^{\circ} + OH^{-} \tag{5}$$

$$e^{-} + O_2 \rightarrow O_2^{\circ -} \tag{6}$$

Acknowledgment

We thank, the EPFL and the Swiss National Science Foundation Project (SNF 200021-143283) for financial support. Also, COST Actions MP1106 and MP1101 for interactive discussions during the course of this study.

References

[1] A. Kramer, I. Schwebke, G. Kampf, 2006, How long do nosocomial pathogens persist in on inanimate surfaces? *BMC Infect. Diseases*, 6, 137-146.

[2] K. Taylor, R. Roberts, J. Roberts, The challenge of hospital acquired infections (HAI). Nat. Audit Office, 2002.

[3] *Thüringer Surface and Biomaterials Kolloquium*, 13-15 September 2011, Zeulenroda, Germany.

[4] A. Foster, I. Ditta, S. Varghese, S. Steele, Photocatalytic disinfection using titanium dioxide: spectrum and mechanism of antimicrobial activity, *Appl. Microb. Biotech.., 90,* pp. 1847-1868, 2011.

[5] A. Fujishima, X. Zhang and D. Tryk, TiO_2 photocatalysis and related phenomena, *Surf. Sci. Reports,* 63, p. 515-582, 2008.

[6] S. Rtimi, C. Pulgarin, R. Sanjines and J. Kiwi, Innovative semi-transparent nanocomposite films presenting photo-switchable behavior and leading to a reduction of the risk of infection under sunlight, *RSC-Adv.* 3, pp. 16345-16348, 2013.

[7] S. Rtimi, R. Sanjines, M. Andrzejczuk, C. Pulgarin, A. Kulik, J. Kiwi, Innovative transparent non-scattering TiO_2 bactericide thin films inducing increased *E. coli* cell wall fluidity, Surf. and Coat. Technol., doi: 10.1016/j.surfcoat.2014.06.035.

[8] N. Novkovski, A. Tanusevski, Origin of the optical absorption of In_2O_3 thin films in the visible range, *Semicond. Sci. Technol.*, 23, 095012, 2008.

[9] J. B. Mathews, Epitaxial Growth Part B, IBM Thomas Watson Research Center, Academic Press, New York, 1975, pp. 382–436.

[10] A. Chakraborty, M. Kebede, Efficient decompositionof Organic Pollutant Over In_2O_3/TiO_2 Nanocomposite Photocatalyst Under Visible Light Irradiation, *J. Cluster Sci.*, 23, 247-257, 2012.

[11] W. Erbs, J. Kiwi, M. Gratzel, Light induced oxygen generation in aqueous dispersions of In_2O_3 particles and determination of their conduction band position, *Chem. Phys. Letts.,* 110, 648-650, 1984.

[12] J. Borkow, J. Gabbay, Biocidal textiles can help fight nosocomial infections, *Med. Hypothesis,* 70, 990-994, 2008.

[13] S. Malnick, L. Bardenstein, M. Huszar, J. Gabbay, J. Borkow, Pyjamas and sheets as potential source of nosocomial pathogens, *J. Hosp. Infect.*, 70, 89-92, 2008.

[14] J. Bandara, P. Bowen, L. Soare, W. Jardim, J. Kiwi, Photocatalytic Storing of O_2 as H_2O_2 Mediated by High Surface Area CuO, *Langmuir,* 21, 8554-8559, 2005.

[15] M. Paschoalino, N. Guedes, W. Jardim, E. Mielczarski, W. Mielczarsi, P. Bowen, J. Kiwi Inactivation of *E. coli* mediated by High surface Area CuO Accelerated by Light irradiation >360 nm, *J. Photochem. Photobiol. A.*, 199, 105-111, 2008.

[16] A. Torres, C. Ruales, C. Pulgarin, A. Aimable, P. Bowen, V. Sarria, J. Kiwi, Enhanced Inactivation of *E. coli* by RF-plasma Pretreated Cotton/CuO (65 m^2/g) under Visible Light, *Appl. Catal. B.*, 2, 2547-2552, 2010.

[17] D. Karin and Y. Gulneth in: *Progr. Inorg. Chem. S. Lippard.*, Ed. 35, 220-237, 1987.

[18] W. Tolman, Making and Breaking the O-O Bond: New Insights in Copper Complexes. *Acc. Chem. Res.,* 30, 227-240, 1997.

[19] J. Park, Y. Jung, J. Cho, W., Choi, *Appl. Surf. Sci.,* 252, 5877-5891, 2006.

[20] J. Kiwi, S. Rtimi, C. Pulgarin, v1, p.74-82, 2013, Chapter: Cu, Cu/TiO_2 in films sputtered by up to date methods on non-thermal thin resistant substrates leading to bacterial inactivation, "Microbial pathogens and strategies for combating them: science, technology and education" Editor: A. Méndez-Vilas Publisher: Formatex Research Center ISBN.

[21] K. Page, M. Wilson, P. I. Parkin, Antimicrobial surfaces and their potential in reducing the role of the inanimate environment in the incidence, *J. Mater. Chem.,* 19, 3819-3831, 2009.

[22] S. Dancer, The role of the environmental cleaning in the control of hospital acquired infections, *J. Hosp. Infect.*, 73, 378-386, 2009.

[23] L Zhang, R. Dillert, D. Bahnemann, M. Vormoor, photoinduced hydrophilicity and self-cleaning: models and reality, *En. and Environ. Sci.*, 5, 7491-7507, 2012.

[24] P. A. Ehiasarian, A. Vetushka, A. Hecimovic, S. Konstantinidis, *J. Appl. Physics*, 104, 083305, 2008.

[25] P. J. Kelly, H. Li, K. A. Whitehead, J. Verran, R. D. Arnell, I. iordanova, A study of the antimicrobial and tribological properties of TiN/Ag nanocomposite coatings, *Surf and Coat. Technol.*, 204, 1137-1141, 2009.

[26] K. Sarakinos, J. Alami, D. Konstantinidis, High power pulsed magnetron sputtering: A review on scientific and engineering state of the art, *Surf and Coat. Technol.*, 204, 1661-1684, 2010.

[27] C. Castro, R. Sanjines, C. Pulgarin, P. Osorio, A. Giraldo, J. Kiwi, Structure-reactivity relations of the Cu-cotton sputtered layers during *E. coli* inactivation in the dark and under light, *J. Photochem. Photobiol. A.*, 216, 295-302, 2010.

[28] P. Osorio, R. Sanjines, C. Ruales, C. Castro, C. Pulgarin, J.-C. Lavanchy, J. Kiwi, Antimicrobial Cu-functionalized surfaces prepared by bipolar asymmetric DC-pulsed magnetron sputtering (PMS), *J. Photochem. Photobiol. A.*, 220, 70-76, 2011.

[29] E. Kusiak-Nejman, A. Morawski, A. P. Ehiasarian, C. Pulgarin, E. Mielczarski, J. Mielczarski, B. Baghriche, A. Kulik, J. Kiwi, *E. Coli* Inactivation by High Power Impulse Magnetron Sputtered (HIPIMS) Cu-Surfaces. *J. Phys. Chem. C,* 115, 21113-21119, 2011.

[30] K. Sunada, T. Watanabe, K. Hashimoto, Bactericidal Activity of Copper-Deposited TiO_2 Film under UV Light, *Environ. Sci. Technol.*, 37, 4785-4789, 2003.

[31] K. Hashimoto, H. Irie, A. Fujishima, TiO_2 Photocatalysis: A Historical Overview and Future Prospects, *Japanese J. Appl. Phys.,* 44 8269-8285, 2005.

[32] K. Hashimoto, Highly efficient light photocatalysis: direct excitation of TiO_2 valence band electron to surface anchored Cu(II) or Fe(II). *SP3 Third International Conference on Semiconductor Photochemistry,* Univ. of Stratchlyde, Glasgow, Scotland, 12-16 April, 2010, p. 26.

[33] M. Haenle, M. Fritshe, C. Zietz, R. Bader, F. Heidenau, An extended spectrum bactericidal titanium dioxide coating for metallic implants: in vitro effectiveness against MRSA J. Mater. Sci: *Mater. Med. JMSMM.*, 22, 381-387, 2011.

[34] H. A. Foster, P. Sheel, D. W. Sheel, P. Evans S. Varghese, N. Rutschke, J. H. M. Yates, Antimicrobial activity of titania/silver and titania/copper films prepared by CVD, *J. Photochem. Photobiol. A.,* 216, 283-289, 2010.

[35] H. M. Yates, L. A. Brook, I. B. Ditta, P. Evans, H. A. Foster, D. W. Sheel, A. Steele, Photo-induced self-cleaning and biocidal behaviour of titania and copper oxide multilayers, *J. Photochem. Photobiol. A.,* 197, 197-2005, 2008.

[36] P. S. M. Dunlop, P. C. Sheeran, M. J. A. Byrne, S. A. M. McMahon, M. A. Boyle, J. K. G. McGuigan, Inactivation of clinically relevant pathogens by photocatalytic coatings, *J. Photochem. Photobiol. A.,* 216, 303- 312, 2010.

[37] L. A. Brook, P. Evans, H. A. Foster, M. E. Pemble, A. Steele, D. W. Sheel, H. M. Yates, Highly bioactive silver and silver titania composite films grown by chemical vapour deposition, *J. Photochem. Photobiol. A.,* 187, 53-63, 2007.

[38] V. Stranak, H. Wulff, H. Rebl, C. Tietz, K. Arndt, R. Bogdanowizs, N. Bebe, R. Bader, Z. Zubicka R. Hippler, Deposition of thin titanium-copper films with antimicrobial effect by advanced magnetron sputtering methods, *Mat. Sci. and Eng. C.,* 32, 1512-1519, 2011.

[39] Y. Dong, X. Li, T. Bell, R. Sammons, H. Dong, Surface microstructure and antibacterial property of an active screen plasma alloyed stainless steel surface with Cu and N, *Biomedical Mats.,* 5, 054105-054112, 2010.

[40] W. Zhang J. Ji, Y. Zhang, Q. Yan, E. Kurmaev, A. Moewes, J. Zhao, P. Chu, Effects of NH_3, O ,N co-implantation on Cu diffusion and antimicrobial properties of Cu-polyethylene, *Appl. Surf. Sci.,* 253, 8981-8985, 2007.

[41] T. Yuranova, G. Rincon, A. Bozzi, S. Parra, C. Pulgarin, P. Albers and J. Kiwi, Antibacterial textiles prepared by RF-plasma and vacuum-UV mediated deposition of silver. *J. Photochem. Photobiol. A. Chem.*, 161, 27-34, 2003.

[42] T. Yuranova, A. G. Rincon, C. Pulgarin, D. Laub, N. Xanthopoulos, H.-J. Mathieu and J. Kiwi, Bactericide cotton textiles active in *E. coli* abatement prepared under mild preparation conditions. *J. Photochem. Photobiol. A.*, 181, 363-369, 2006.

[43] M. I. Mejía, G. Restrepo, J. M. Marín, R. Sanjines, C. Pulgarín, E. Mielczarski, J. Mielczarski, J. Kiwi, Magnetron-Sputtered Ag-Modified Cotton Textiles Active in the Inactivation of Airborne Bacteria, *ACS Appl. Mater. and Interf.,* 2, 230-235, 2010.

[44] A. Fujishima, K. Hashimoto, T. Watanabe, TiO_2 photocatalysis Fundamentals and applications, Bkc Inc, 2000, Tokyo.

[45] S. Rtimi, C. Pulgarin, M. Bensimon, J. Kiwi, Evidence for TiON sputtered surfaces showing accelerated antibacterial activity under simulated solar irradiation, *Solar Energy,* 93, 55–62, 2013.

[46] O. Baghriche, A. P. Ehiasarian, E. Kusiak-Nejman, A. W. Morawski, C. Pulgarin, R. Sanjines, J. Kiwi, High power impulse magnetron sputtering (HIPIMS) and traditional pulsed sputtering (DCMSP) Ag-surfaces leading to *E. coli* inactivation *J. Photochem. Photobiol. A.,* 227, 11-17, 2012.

[47] Oualid Baghriche, Arutiun P. Ehiasarian, Ewelina Kusiak-Nejman, Antoni W. Morawski, Cesar Pulgarina, Rosendo Sanjines, John Kiwi, Advantages of high power impulse magnetron sputtering (HIPIMS) of silver for improved *E. coli* inactivation, *Thin Solid Films,* 520, 3567-3573, 2012.

[48] O. Baghriche, R. Sanjines, C. Ruales, C. Pulgarin, I. Stolitchnov, A. Zertal, J. Kiwi, Ag-surfaces sputtered by DC and pulsed DC-magnetron sputtering effective in bacterial inactivation: Testing and characterization *Surf. and Coat. Technol.*, 206, 2410-2416, 2012.

[49] Y. Ida, S. Watase, T., Shinagawa, M. Watanabe, M. Chigane, M. Inaba, A. Tasaka, M. Izaki, *Mater. Chem.*, 20, 1254-1256, 2008.

[50] R. A. Smith. Semiconductors, Cambridge Iniversity press, Uk, 1976.

[51] A. Bozzi, T. Yuranova, J. Kiwi, Self-cleaning of wool-polyamide and Polyester Textiles Due to Surface-TiO_2 rutile Modification under Daylight Irradiation, *J. Photochem. Photobiol. A.,* 172, 27-34, 2005.

[52] P. Osorio, R. Sanjines, R. Ruales, C. Castro, C. Pulgarin, A.-J. Rengifo, J.-C. Lavanchy J. Kiwi, Antimicrobial Cu-functionalized surfaces prepared by bipolar asymmetric DC-pulsed magnetron sputtering (PMS), *J. Photochem. Photobiol. A.,* 220, 70-76, 2011.

[53] S. Rtimi, O. Baghriche, C. Pulgarin, J.-C. Lavanchy, J. Kiwi, Growth of TiO_2/Cu by HIPIMS for accelerated bacterial loss of viability, *Surf. and Coat. Technol.*, 232, 804-813 2013.

[54] C. D. Wagner, M. W. Riggs, E. L. Davis, G. E. Müllenberg (Eds), Handbook of X-ray Photoelectron Spectroscopy, Perkin-Elmer Corporation Physical Electronics Division, Minnesota, 1979.

[55] A. D. Shirley, 1972, Corrections of electrostatic charged species in XSP-spectroscopy, *Phys. Rev.,* B5, 4709-4716.

[56] F. Petronella, S. Rtimi, O. Baghriche, C. Pulgarin, R. Sanjines, J.-C. Lavanchy, J. Kiwi, Uniform In_2O_3/TiO_2 Films Effective in Bacterial Inactivation Under Low Intensity Solar Irradiation, *Photochem. Photobiol. A.,* 279, 1-7, 2014.

[57] O. Baghriche, R. Sanjines, C. Pulgarin, S. Rtimi, J. Kiwi, Effect of the spectral properties of TiO_2, Cu, TiO_2/Cu sputtered films on the bacterial inactivation under low intensity actinic light, *J. Photochem. Photobiol. A.,* 251, 50-56, 2013.

[58] S. Rtimi, R. Sanjines, M. Bensimon, C. Pulgarin, .J. Kiwi, Accelerated *Escherichia coli* inactivation in the dark on uniform copper flexible surfaces, *Biointerphases,* 9, 029012, 2014.

[59] S. Rtimi, O. Baghriche, C. Pulgarin, R. Sanjines, J. Kiwi, Innovative TiO_2/Cu surfaces inactivating bacteria < 5 min under low intensity visible/actinic light, *ACS Appl. Mater. and Interf.*, 4, 5234-5240, 2012.

[60] J. H. Nikaido, Prevention of Drug Access to Bacterial Targets, Permeability Barriers and Active Flux, *Biol. Chem.*, 269, 3905-3909, 1994.

[61] J. Mathews, Epitaxial Growth Part B, IBM Thomas Watson Research Center, Academic ress, New York, 1975, p. 382-436.

[62] S. Rtimi, O, Baghriche, R. Sanjines, C. Pulgarin, M. Ben-Simon, J.-C. Lavanchy, J. Kiwi, TiON and TiON-Ag sputtered textile showing antibacterial activity induced by simulated-solar-light, *J. Photochem. Photobiol. A.,* 256, 52-63, 2013.

[63] J. Probst, U. Gbureck, R. Thull, Binary nitride and oxynitride PVD coatings on titanium for biomedical applications, *Surf. and Coat. Technol.*, 148, 226-233, 2001.

[64] B. Subramanian, C. Muraleedharan, R. Annanthakumar, M. Jayachandran, A comparative study of TiN, TiON and TiAlN as surface coatings for bioimplants. *Surf. and Coat. Tecnol.*, 205, 5014-5020, 2011.

[65] G. Biedermann, G. Sillèn, Studies of the hydrolysisof Metal-ions.Part 30. A critical survey of solubility equilibria of Ag_2O. *Acta Chem. Scand.,* 14, 14-0717, 1969.

[66] Y. Ida, T. Watase, M. Shinagawa, M. Watabanbe, M. Chigane, M. Inaba, A. Tasaka, M. Izaki, Direct Electrodeposition of 1.46 eV Band-gap Silver(I) Oxide Semiconductor Films by Electrogenerated Acid, *Chem. Mater.,* 20, 1254-1256, 2008.

[67] A. Varkey, Some optical properties of silver peroxide (AgO) and silver oxide (Ag_2O) films produced by chemical-bath deposition Solar, *En. Mater, and Solar Cells,* 29, 253-259, 1993.

[68] J. Yuand, J. Ran, Some optical properties of silver peroxide (AgO) and silver oxide (Ag_2O) films produced by chemical-bath deposition Energy and Env. *Environ. Sci.,* 4, 1364-1371, 2011.

In: Advances in Semiconductor Research
Editors: D. Persano Adorno and S. Pokutnyi
ISBN: 978-1-63321-755-3

Chapter 11

PLASMA WAVE EXCITATION BY TERAHERTZ ELECTRIC SIGNAL IN HIGH ELECTRON MOBILITY TRANSISTORS

***A. Mahi*[1,2,3], *C. Palermo*[1,2,*], *H. Marinchio*[1,2], *G. Sabatini*[1,2], *A. Belghachi*[3] and *L. Varani*[1,2]**
[1]Univ. Montpellier, IES, Montpellier, France
[2]CNRS, IES, Montpellier, France
[3]Laboratory of Physics of Semiconductor Devices, University of Bechar, Algeria

Abstract

One of the most promising strategies to exploit FETs as THz sources and detectors lies in the plasmonic approach where the physical mechanism at the basis of the THz working operation is an electron-gas plasma-wave excitation. However the most used theoretical approach to study the plasma modes in a FET channel is based on the analytical solution of carrier concentration and mean velocity conservation equations in the framework of the gradual channel approximation (GCA). To be far off this critical limitation approximation and to describe the plasma wave oscillation in complex structures, we propose a numerical hydrodynamic model, coupled to a pseudo-2D Poisson equation, able to take into account the influence of gate voltage, channel-to-gate distance and ionized donor concentration in the channel. We study the drain current response of a HEMT to a THz signal applied to its gate and/or to its drain contacts in order to obtain the optimal configuration in terms of detection. We show that an incident THz radiation producing an oscillating voltage on the gate and/or the drain contacts of a HEMT leads to the excitation of plasma modes in the THz range and the transition from a 2D to a 3D plasma modes behavior. We conclude that the contribution of the 3D plasma resonance due to the presence of the homojunctions and also the increase of the drain-source voltage decreases the amplitude of the resonance frequency peaks. We demonstrate that the amplitudes of the harmonic and average drain-current responses associated with the presence of hybrid plasma modes in the channel, strongly depend on which transistor terminal collects the incident THz radiation and that a maximum DC response can be obtained by appropriately dephasing the

[*]E-mail address: cpalermo@um2.fr

two electrode signals. These results open up new possibilities for the improvement of the performances of plasmonic THz detectors by appropriately dephasing the electrode signals.

PACS: 73.50.Mx, 73.50.Fq, 73.61.Ey

Keywords: Terahertz, plasma wave, simulation, modeling, hydrodynamic, field-effect transistor

1. Introduction

Generation and detection of electromagnetic radiation in the terahertz (THz) domain is a subject matter of great interest because of its potential applications in different areas, such as security, medicine, broadband communications, etc. [1]. In this context, the field effect transistor (FET) is among the most promising devices for THz radiation generation and detection [2, 3]. An intriguing strategy to exploit FETs as THz sources and detectors lies in the plasmonic approach where the physical mechanism at the basis of the THz working operation is an electron-gas plasma-wave excitation. The investigation of plasma oscillations in two-dimensional (2D) electron-gas channels was initiated at the beginning of 1990s by Dyakonov and Shur [2, 4]. The resonant detection of THz radiation by 2D plasma waves was experimentally demonstrated using InAlAs/InGaAs High Electron Mobility Transistors [5, 6, 7]. Recently, the same kind of detection was also experimentally obtained using a MOSFET detector with integrated on-chip antennas [8] and transistors where the gate and/or drain electrodes contain an integrated antenna [9, 10].

It is clear that the improvement of the transistor sensitivity strongly depends on the optimization of the electromagnetic coupling between the FET channel, where plasma waves are excited, and the THz incident radiation. However, in most of the cases presented in the literature the FET detector is not connected to an external antenna and the THz radiation is assumed to be simply coupled to the channel through the metallic electrodes (source, drain and gate) without any deep investigation of their effects on the plasma wave excitations in the channel. A complete understanding of these physical processes should require a self-consistent solution of Maxwell equations describing the propagating THz radiation incident on the FET (with or without an antenna) and of the Boltzmann equation describing the transport processes in the transistor. At the present stage, such kind of numerical simulations represent a formidable numerical task that is out of the scope of our investigation. Nevertheless it is extremely important to understand how the oscillating voltage produced by the incident THz radiation on the FETs electrodes affects the DC signal, typically measured at the drain contact, and if some strategy to enhance such signals can be envisaged. In this chapter, we consider a plasmonic THz detector based on a High Electron Mobility Transistor (HEMT) and, by means of a numerical hydrodynamic (HD) model, we study the drain current response to a THz signal applied on its gate and/or on its drain contacts in order to obtain the optimal configuration in terms of detection.

Section 2 is devoted to the presentation of the numerical model and the considered structure. In particular, a pseudo-bidimensional expression of the Poisson equation is derived, which takes into account the planar structure within a one-dimensional description. The HD equations used to calculate the quantities of interest are also presented. In section

3, the modeling of the channel excitation by the THz wave is described. Three ways of wave detection are considered, which are a captation (i) only by the drain, (ii) only by the gate and (iii) by both gate and drain, involving or not a phase-shift between the two voltage signals. Before concluding, section 4 exposes the obtained results and the more efficient way to excite the electron channel is discussed. Moreover, the influence of the gate on the detected modes behavior is investigated.

2. Numerical Model

2.1. THz Wave-HEMT Channel Interaction

The aim of our numerical model is to provide the simulation of an antenna response to a THz wave applied on the gate, on the drain and on both terminals. The wave-HEMT interaction is introduced through the oscillation of one or both applied drain and gate voltages, at the incident radiation frequency. Then, the drain current response to different kinds of excitation is calculated.

Technically speaking, the HEMT channel is submitted to symmetric boundary conditions: the average potentials of both drain and source are fixed or controlled, so that the simulation of wave captation by the drain is made possible. The device is then working under voltage-driven operation-mode and the time-dependent drain current can be expressed by neglecting the harmonics of order higher than one as [11]

$$I_d(t) = I_{d_0} + \overline{\Delta I_d}(f) + \Delta I_d(f)\cos(2\pi f + \varphi_I) \tag{1}$$

where I_{d_0} is the drain dark current, that is in the absence of THz excitation, $\Delta I_d(f)\cos(2\pi f + \varphi_I)$ the harmonic response at excitation frequency, and $\overline{\Delta I_d}(f)$ the additional dc current due to the rectification of the induced current oscillation. This last quantity is of particular interest as it quantifies the detection ability (that is the THz-to-dc conversion) of the device and can be measured with an adapted experimental protocol.

2.2. Hydrodynamic Equations

As discussed in Refs. [6] and [12], the precise calculation of the non-linear response in such a device requires to take into account many non-linearities sources (effects of non-uniform quantities along the channel, thermal effects, depleted zone at the junction between ungated and gated regions etc.). To this aim, we consider the 1D hydrodynamic equations in the transport direction, here taken as the $x-$axis, relative to the electron density $n(x,t)$, the mean velocity $v(x,t)$ and the mean energy $\epsilon(x,t)$

$$\begin{cases} \dfrac{\partial n}{\partial t} = \dfrac{\partial (nv)}{\partial x} \\ \dfrac{\partial v}{\partial t} = -v\dfrac{\partial v}{\partial x} - \dfrac{e}{m^*(\epsilon)}E - \dfrac{1}{n}\dfrac{\partial(\overline{\delta v^2}(\epsilon)n)}{\partial x} - v\nu_v(\epsilon) \\ \dfrac{\partial \epsilon}{\partial t} = -v\dfrac{\partial \epsilon}{\partial x} - eEv - \dfrac{1}{n}\dfrac{\partial\left(\overline{\delta v \delta \epsilon}(\epsilon)n\right)}{\partial x} - (\epsilon - \epsilon_0)\nu_\epsilon(\epsilon) \end{cases} \tag{2}$$

where e is the elementary charge, E the x electric field component and ϵ_0 is the equilibrium mean energy. The velocity and energy relaxation rates ν_v, ν_ϵ, the effective mass

m^*, the velocity variance $\overline{\delta v^2}$ and the energy-velocity covariance $\overline{\delta v \delta \epsilon}$ depend on the local mean energy as reported on Fig. 1 and are calculated by a Monte Carlo (MC) simulator of bulk $In_{0.53}Ga_{0.47}As$ at 300 K, in application of the hierarchical approach for electron device modeling. Ionized impurities, absorption and emission of polar and non polar op-

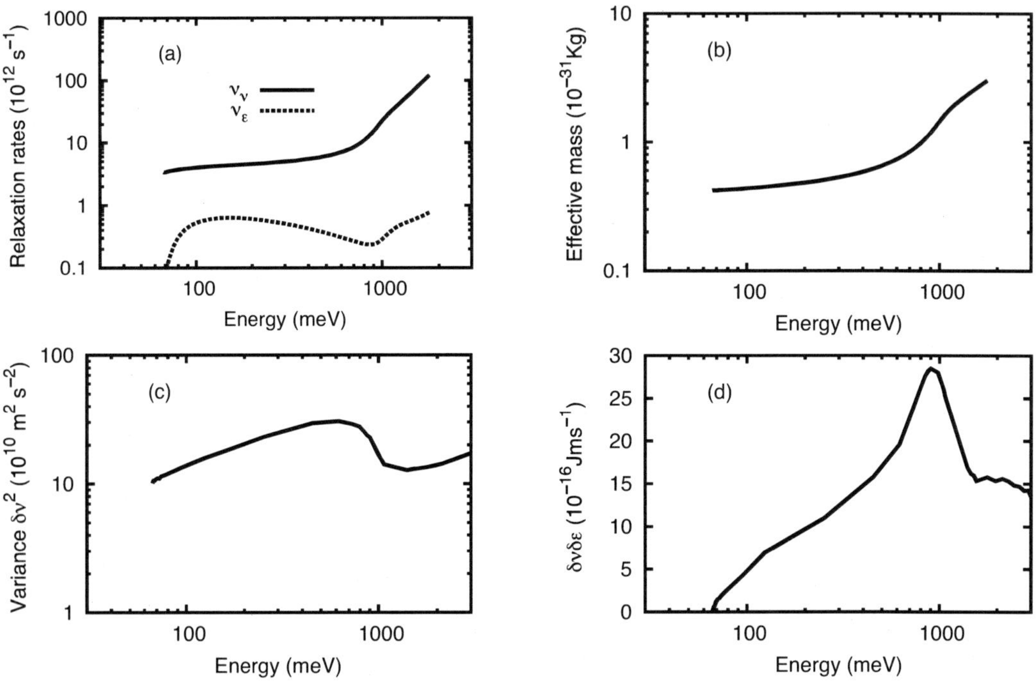

Figure 1. (a) Velocity and energy relaxation rates, (b) mean effective mass, (c) velocity variance and (d) velocity-energy covariance as functions of the electron mean energy, calculated with a Monte Carlo simulation of bulk $In_{0.53}Ga_{0.47}As$ at temperature 300 K.

tical phonons, acoustic elastic phonons, intervalley and alloy scatterings [13] were taken into account to provide these results. Note that the physical parameters describing bulk $In_{0.53}Ga_{0.47}As$ were interpolated from data available for the corresponding binary materials [14, 15, 16, 17, 18].

2.3. Poisson Equation

2.3.1. Pseudo Bidimensional (P2D) Approach

Even if the transport description is performed through a one-dimensional approach, the resolution of the Poisson equation which describes the self-consistent electric field inside the whole three-terminals device remains *a priori* 2D [19]. Indeed, the widely used gradual channel approximation (GCA) only considers the transverse effects of the electric field [20, 21, 22] which are induced by the gate voltage, and leads then to neglect the longitudinal drain-source effects. Moreover, depending on the complexity of the device, different conditions influencing the electric field must be considered, such as ungated and gated zones, cap layers, doping plans, etc. Nevertheless, since the distributions of applied potentials and dopings governing electron transport are usually placed in layers planar with respect to the channel, it is reasonable to formulate a 1D approximation of the 2D Poisson equation which allows to take into account their influence. As a matter of fact, approaches of this kind have been proposed to model the channel region of both Metal-Semiconductor Transistors [23] and HEMT [24]. We generalize then such works to provide a more complete formulation able to simulate more complex structures by involving only a simple set of parameters (that is gate-to-channel distance, doping concentration, channel width,etc.).

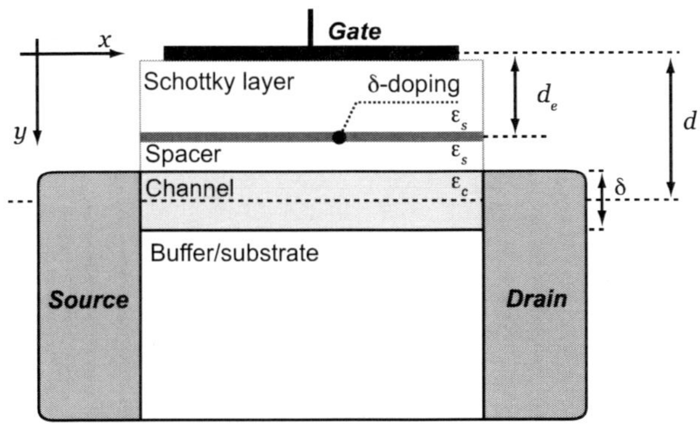

Figure 2. Schematic representation of a HEMT with, from top to bottom: the gate, the Schottky layer (of width d_e and dielectric coefficient ϵ_s), the δ-doping plan (of donor surface density N_+^{2D}), the spacer (of dielectric coefficient ϵ_s), the channel (of width δ, dielectric coefficient ϵ_c, doping concentration n_+^{3D} and free electron density n^{3D}) and the substrate. The distance between the gate and the center of the channel (gate-to-channel distance) is d and the width of the transistor (dimension not represented in the figure) is W.

To this aim, we consider the planar device shown in Fig. 2. A doped layer with surface concentration of ionized donors N_+^{2D} and a thin channel of thickness δ with intrinsic donor concentration n_+^{3D} are placed at distances d_e and d from the gate, respectively. A sufficiently thick substrate is placed under the channel. It is supposed that the free carriers characterized by volume concentration n^{3D} are inside the channel only. The spatial distribution of the potential $\psi(x,y)$ of such a structure is described by the 2D Poisson equation:

$$\frac{\partial}{\partial x}(\varepsilon\frac{\partial}{\partial x}\psi) + \frac{\partial}{\partial y}(\varepsilon\frac{\partial}{\partial y}\psi) = e\left[n^{3D}(x) - n_+^{3D} - N_+^{2D}\delta(y-d_e)\right] \tag{3}$$

where ε is the dielectric constant of the medium. By integrating Eq. (3) inside the channel in the transverse direction one obtains:

$$\varepsilon_c\frac{\partial^2}{\partial x^2}V(x) = e[n(x) - n_+^{3D} - \overline{n}_+^{3D}(x)] \tag{4}$$

where ε_c is the dielectric constant in the channel,

$$V(x) = \frac{1}{\delta}\int_{d-\delta/2}^{d+\delta/2}\psi(x,y)dy \tag{5}$$

and

$$n(x) = \frac{1}{\delta}\int_{d-\delta/2}^{d+\delta/2} n^{3D}(x,y)dy \qquad (6)$$

are, respectively, the potential and free-carrier concentration averaged over the channel thickness. The quantity

$$\overline{n}_{+}^{3D}(x) = \frac{\varepsilon_s}{e\delta}\left[\frac{\partial}{\partial y}\psi(x, d+\delta/2) - \frac{\partial}{\partial y}\psi(x, d-\delta/2)\right] \qquad (7)$$

is the density of induced electrical charge in the channel which determines the jump of the transverse component of the electric field between vertical boundaries of the channel. Equation (4) coincides in form with the 1D Poisson equation describing the self-consistent distributions of the potential $V(x)$ and free-carrier concentration $n(x)$ along the channel which are formed at the background of an effective donors density distribution

$$N'_D(x) = n_{+}^{3D} + \overline{n}_{+}^{3D}(x) \qquad (8)$$

so that a general pseudo-2D Poisson equation can be expressed as

$$\varepsilon_c \frac{\partial^2}{\partial x^2}V(x) = e[n(x) - N'_D(x)] \qquad (9)$$

When the longitudinal electric field, i.e. $\partial V(x)/\partial x$, has practically no influence on the transverse field distribution outside the channel, i.e. $\partial\psi(x,y)/\partial y$, Eqs. (4) and (7) can be considered as mutually independent, and Eq. (9) provides a 1D approximation of Eq. (3).

2.3.2. Expression of the Effective Donor Density N'_D

Let us now make $N'_D(x)$ explicit as a function of the different parameters of the device structure. For the considered planar structure, it is reasonable to assume that outside the channel the longitudinal component of the electric field is negligible with respect to the transverse one, i.e., $\partial\psi/\partial x \ll \partial\psi/\partial y$. In this case, the transverse dependence of the potential with respect to the channel can be approximated by a continuous linear function of y that changes its slope on surfaces of a given potential (the gate) and doped layers as reported on Fig. 3. By integrating Eq. (3) from the gate to the channel one finds the following relation between potentials at the gate, $V_g = \psi(x, 0)$, the doped layer, $\psi(x, d_e)$ and the channel, $V(x)$:

$$\frac{V(x) - \psi(x, d_e)}{d - d_e} - \frac{\psi(x, d_e) - V_g}{d_e} = -\frac{e}{\varepsilon_s}N_{+}^{2D} \qquad (10)$$

where ε_s is the dielectric constant outside the channel. By assuming that potential variations inside a deep substrate are small enough, one can use the following expressions for the transverse electric field at the channel boundaries:

$$\frac{\partial}{\partial y}\psi\left(x, d+\frac{\delta}{2}\right) = 0, \quad \frac{\partial}{\partial y}\psi\left(x, d-\frac{\delta}{2}\right) = \frac{V(x) - \psi(x, d_e)}{d - d_e} \qquad (11)$$

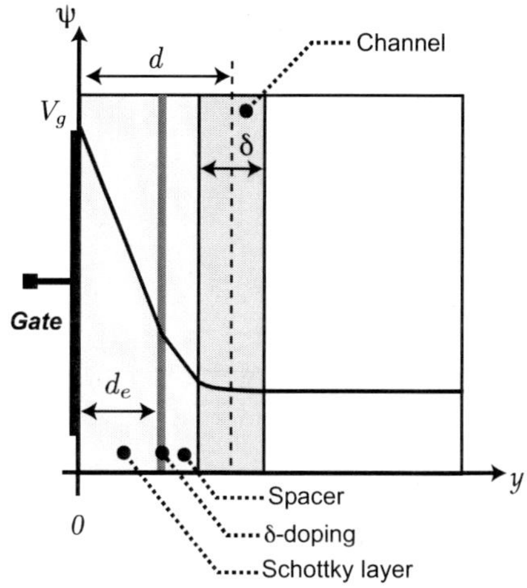

Figure 3. Transverse profile (y-axis) of the potential for a section of HEMT (fixed value of x).

By substituting Eq. (11) into Eq. (7) and then by using Eqs. (10) and (11) to eliminate from consideration the influence of the potential $\psi(x, d_e)$ on the doped layer in explicit form, one obtains an expression for the carrier density in the channel induced by external conditions:

$$N'(x) = n_+^{3D} + \frac{d_e}{\delta d} N_+^{2D} + \frac{\varepsilon_s}{e} \frac{V_g - V(x)}{\delta d} \tag{12}$$

which creates the background on which the free-carrier distribution in the channel is formed. Thus, by combining Eqs. (9) and (12), one obtain in the framework of the above used assumptions, a closed 1D approximation of the 2D Poisson equation

$$\varepsilon_c \frac{\partial^2}{\partial x^2} V(x) + \varepsilon_s \frac{V_g - V(x)}{\delta d} = e \left[n(x) - n_+^{3D} - \frac{d_e}{\delta d} N_+^{2D} \right] \tag{13}$$

This last equation can be specified for each particular case.

- For a HEMT without doping plan, the consideration of $N_+^{2D} = 0$ leads to the relation:

$$\varepsilon_c \frac{\partial^2}{\partial x^2} V(x) + \varepsilon_s \frac{V_g - V(x)}{\delta d} = e \left[n(x) - n_+^{3D} \right] \tag{14}$$

- Within the ungated zones, or for an ungated device, $d_e \to \infty$ and $d \to \infty$ which leads to

$$\varepsilon_c \frac{\partial^2}{\partial x^2} V(x) = e \left[n(x) - n_+^{3D} - \frac{N_+^{2D}}{\delta} \right] \tag{15}$$

If we consider moreover the absence of a doping plan, we retrieve a 1D structure and the basic 1D Poisson equation

$$\varepsilon_c \frac{\partial^2}{\partial x^2} V(x) = e \left[n(x) - n_+^{3D} \right] \tag{16}$$

- If one multiply Eq. (13) by δ and makes $\delta \to 0$, the appearing quantity $n(x)\delta = n^{2D}(x)$ is a surface density which keeps then a finite value, so that the Poisson equation is reduced to the so called gradual channel approximation widely used in analytical models of FETs [19]:

$$\varepsilon_s \frac{V_g - V(x)}{d} = e\left[n^{2D}(x) - \frac{d_e}{d} N_+^{2D}\right] \tag{17}$$

2.4. Stationary Results

We consider a HEMT whose channel parameters are reported on Fig. 4. The calculated

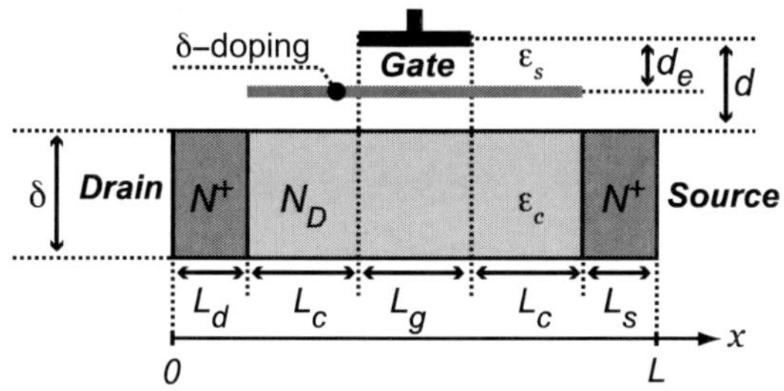

Figure 4. Design and parameters of the simulated HEMTs. Doping densities are $N^+ = 5 \times 10^{18}$ cm^{-3}, $N_D = 10^{18}$ cm^{-3} and $N_+^{2D} = 10^{12}$ cm^{-2}. The considered distances are $d_e = 10$ nm, $\delta = 10$ nm and $d = 22$ nm. Dielectric permittivities are $\varepsilon_c = 13.4\varepsilon_0$ and $\varepsilon_s = 11.8\varepsilon_0$, ε_0 being the vacuum permittivity.

static profiles for $L_g = L_c = 300$ nm and $L_s = L_d = 200$ nm of the main transport parameters, which are the electron density, electric field and mean velocity, are reported in Fig. 5 for different applied drain-source voltages V_{ds_0} and for $V_g = -50$ mV. These results confirm that the 1D-HD model associated to our P2D Poisson equation is able to reproduce the standard electrical behavior of the HEMT.

3. Numerical Protocol

Let now simulate the electrical response to an incoming THz wave by calculating in different conditions the drain current $I_d(x,t)$ expressed by Eq. (1). To this aim, we consider an InAlAs/InGaAs/InP HEMT whose structure is identical to that described in Ref. [12] and reported in Fig. 4 and with $L_g = 100$ nm, $L_c = 50$ nm and $L_s = L_d = 50$ nm submitted to symmetrical boundary conditions (that is operating in voltage-driven mode). We take into consideration three kinds of excitation of the channel electron gas: (i) the wave is detected only by the gate contact ; (ii) the wave is detected only by the drain contact and (iii) the wave is detected by both gate and drain contacts with different values of the phase-shift φ.

3.1. Captation by the Gate

The captation of the THz incoming wave only by the gate is introduced in our model as initially described by Dyakonov and Shur [25] and Veksler [21]. Assuming that the gate

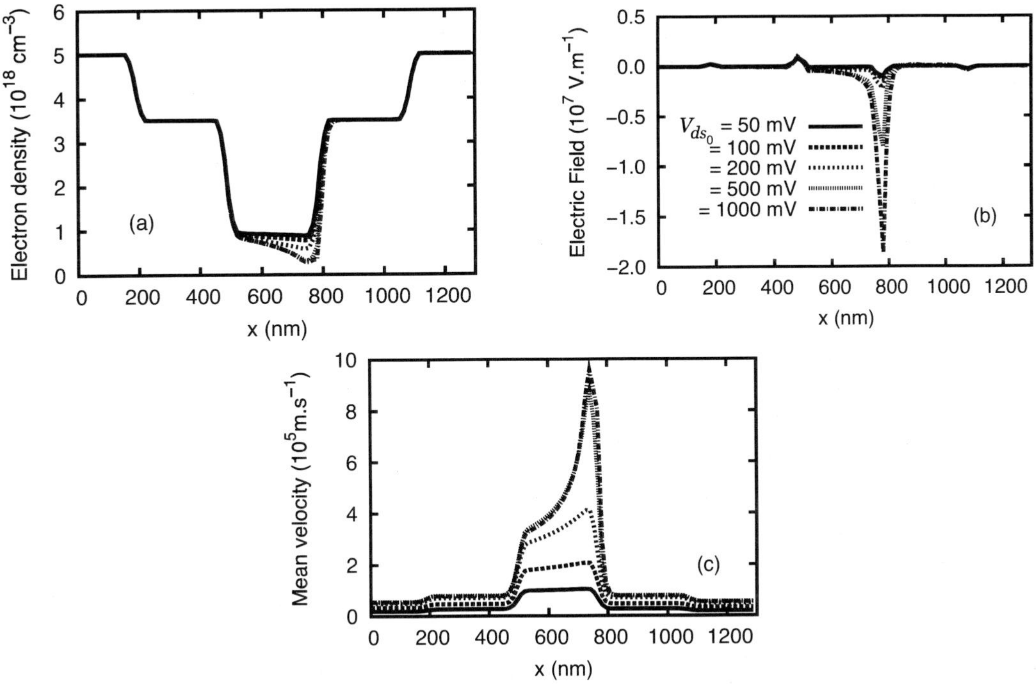

Figure 5. Profiles of electron density (a), electric field (b), and mean velocity (c) along the HEMT for different values of the applied drain-source voltages V_{ds_0}.

is initially biased at voltage $V_g = V_{g_0}$, the THz radiation received by the gate antenna (that is an antenna welded onto the gate of the HEMT) is described by the superposition of an harmonic component $\Delta V_g(t)$, so that the gate-to-source potential is at every time t equal to

$$V_g(t) = V_{g_0} + \delta V_g(t) \tag{18}$$

with

$$\delta V_g(t) = \delta V_{g_0} \cos(2\pi f t) \tag{19}$$

where f is the frequency of the incident THz signal and δV_{g_0} its amplitude. In this approach, both drain and sources contacts are constantly biased, so that $V_{ds}(t) = V_{ds_0}$, and the expression of the gate voltage (18) is accounted for each time the Poisson equation (13) is solved.

3.2. Captation by the Drain

In order to simulate the detection of the incoming wave only by the drain, the gate is considered as constantly biased and then $V_g(t) = V_{g_0}$. The channel is excited at its drain side, so that

$$V_{ds}(t) = V(x = L, t) = V_{ds_0} + \delta V_{DS}(t) \tag{20}$$

where V_{ds_0} is the average drain voltage and $\delta V_{ds}(t)$ is the harmonic component of THz excitation

$$\delta V_{ds}(t) = \delta V_{ds_0} \cos(2\pi f t + \varphi) \tag{21}$$

where f is the frequency of the incident THz wave and δV_{ds_0} its amplitude. Technically, Eq. (20) is used as a boundary condition for the Poisson equation solution.

3.3. Simultaneous Excitation

The simulation of the simultaneous excitation of both gate and drain is made by considering both Eqs (18) and (20) within the Poisson equation resolution process. The influence of the phase-shift φ on the drain current is investigated.

3.4. Theoretical Guideline

Considering symmetrical boundary conditions and P2D Poisson equation, the pth plasma mode frequency can be analytically linked to the fundamental 2D mode f_{2D} *via* the expression [26]:

$$f_p = f_{2D}\frac{p}{\sqrt{1+\left(\frac{p\gamma\pi}{2}\right)^2}} \tag{22}$$

where the capacitance ratio $\gamma = \sqrt{C_{ds}/C_g}$ describes both transverse and longitudinal effects of the field. We give for the drain-source capacitance per unit of gate width $C_{ds} = \varepsilon_c\delta/L_g$ and for the gate-channel one $C_g = \varepsilon_s L_g/d$. For the considered structure, $\gamma = 0.158$. The fundamental frequency f_{2D} is calculated using the symmetrical boundary conditions within the GCA and reads

$$f_{2D} = \frac{1}{2L_g}\sqrt{\frac{eU_0}{m^*}} \tag{23}$$

where $U_0 = V_{g_0} - V_{ds_0} - V_{th}$ is the gate voltage swing, V_{th} being the threshold voltage with $V_{th} = -298$ mV for the considered structure.

4. Results

4.1. Gate Excitation

We consider firstly the case for which the incoming wave is detected only by the gate. For this sake, the drain-source voltage is fixed to $V_{ds_0} = 40$ mV with $\delta V_{ds_0} = 0$ and the incoming wave is supposed to induce a harmonic bias of amplitude $\delta V_{g_0} = 1$ mV superposed to the initial gate bias V_{g_0}. We have reported in Fig. 6 both average value $\overline{\Delta I_d}$ and amplitude of the harmonic part ΔI_d of the drain current response for different values of V_{g_0}. The spectrum of the response to an excitation of the gate voltage shows three resonance peaks. The two first ones match well with the two first 2D plasma resonances ($f_1^{2D} = 4.8$ THz and $f_2^{2D} = 7.6$ THz) calculated with Eq. (22). Interestingly, the third one corresponds to the 3D plasma resonance $f_{3D} = 12$ THz obtained by considering $p \to \infty$ in Eq. (22). This emphasizes the transition form a 2D to a 3D behavior of the electron gas, revealed when the wavelength associated with the excitation becomes negligible as compared to the transverse dimension [27, 26].

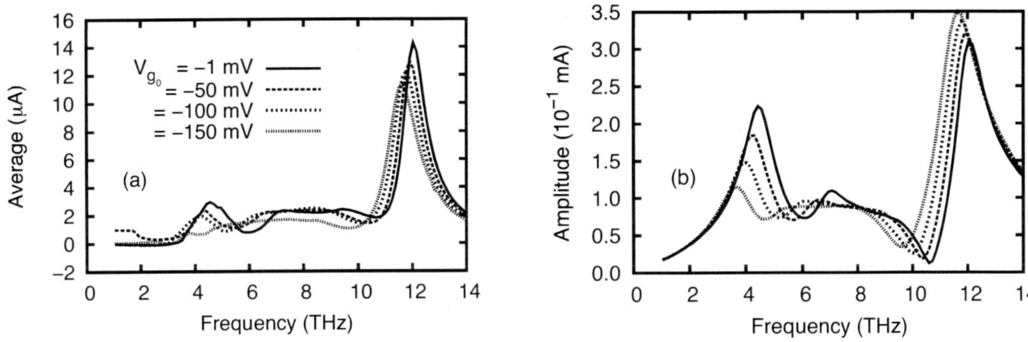

Figure 6. (a) Average response and (b) amplitude of the harmonic response of the drain-current as functions of the radiation frequency for an excitation applied on the gate of the HEMT, for a drain-source voltage V_{ds_0} = 40 mV and different gate mean biases V_{g0}.

We observe that the resonance frequencies decrease when the gate voltage increases, since the electron concentration in the channel becomes closer to that of the access regions. Moreover, the amplitude of the 2D plasma peaks decreases. This last effect is not reported on the 3D plasma peak.

4.2. Drain Excitation

We consider now the case for which the excitation is applied only to the drain. The gate is then kept at a constant bias $V_{g0} = -1$ mV with $\delta V_{g0} = 0$. As concerns the drain, a harmonic signal of amplitude $\delta V_{ds0} = 1$ mV is superposed to the constant bias V_{ds0}. In these conditions, Fig. 7 shows the average value $\overline{\Delta I_d}$ and the amplitude of the harmonic part ΔI_d of the drain current response for different values of V_{ds0} at room temperature and as functions of the THz excitation frequency.

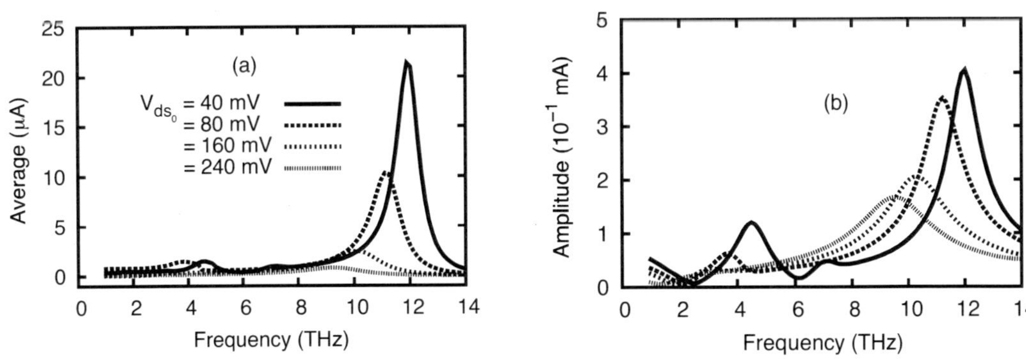

Figure 7. (a) Average response and (b) amplitude of the harmonic response of the drain-current as functions of the radiation frequency for an excitation applied on the drain of the HEMT, for gate bias $V_{g0} = -1$ mV and different values of the drain-source biases V_{ds0}.

An observation of the spectra for a drain-source voltage fixed at $V_{ds_0} = 40$ mV allows to identify the same three resonance peaks observed in the case of the gate excitation. For both considered quantities, an increase of the drain bias leads to a downshift of the plasma frequencies, related to an increase of the plasma transit time within the channel. When the drain-source voltage V_{ds_0} exceeds 240 mV, the plasma oscillations are suppressed. This damping effect is also observed for the 3D mode, the increase of the drain-source voltage leading to the suppression of both 2D and 3D resonances.

4.3. Combined Excitation

We finally consider the case of a simultaneous excitation of the gate and the drain. For this sake, we consider at the same time the voltage perturbations described by both Eqs. (18) and (20). We take for the bias values $V_{ds_0} = 40$ mV and $V_{g_0} = -1$ mV. The considered THz harmonic part amplitudes are $\delta V_{g_0} = -1$ mV and $\delta V_{ds_0} = 1$ mV.

4.3.1. Comparison of the Excitation Modes

As a first step, we consider a null phase-shift $\varphi = 0$ between the drain and gate excitations and the obtained results are reported in Fig. 8 (curve 3). A comparison is made with the previous obtained results, that is for gate and drain stimulations (reported as curves 1 and 2, respectively). We remark that the domain of 2D plasma resonances (that is from 3 THz to 10 THz) of the average response (Fig. 8. (a)) is more sensitive to the gate excitation mode while the region of the 3D resonance peak is more sensitive to the drain excitation mode. In both cases, the values of the resonance frequencies are similar at the same bias because the excited plasma mode properties depend only on static parameters (that is gate length, carrier density, gate and drain biases, etc.) and are independent on the physical mechanism of plasma wave excitation. In term of detection, it appears then that an excitation on the gate will be more efficient at lower frequencies while an excitation on the drain will be preferable for a detection exploiting the 3D resonance. Moreover, contrary to the experiments performed on non-resonant devices reported in Ref. [28], the poorest results in term of detection is obtained when both excitations are applied in phase because of a weak excitation of the plasma resonances.

In order to understand why the harmonic response (and therefore the average current) is so weak in the case of a simultaneous excitation, we calculate and show on Fig. 9 the phase angle $\varphi_I(f)$ between the harmonic current and each voltage excitation as defined in Eq. (1). It is evident that the current responses corresponding to these two ways of excitation are almost in opposite phase when all three resonances occur (for $f \approx f_1$, f_2 or f_{3D}): the contribution of both excitations are mutually destructive leading to weak harmonic and average currents i.e., to poor detection performances. This explains why the plasma wave is only lightly excited by the combined excitation mode, leading to the results reported in Fig. 8.

4.3.2. Influence of the Excitation Phase Difference φ

Starting from the previous results and conclusions, it is obvious that the phase-shift between the drain and gate excitations must be adjusted to achieve a better control on the

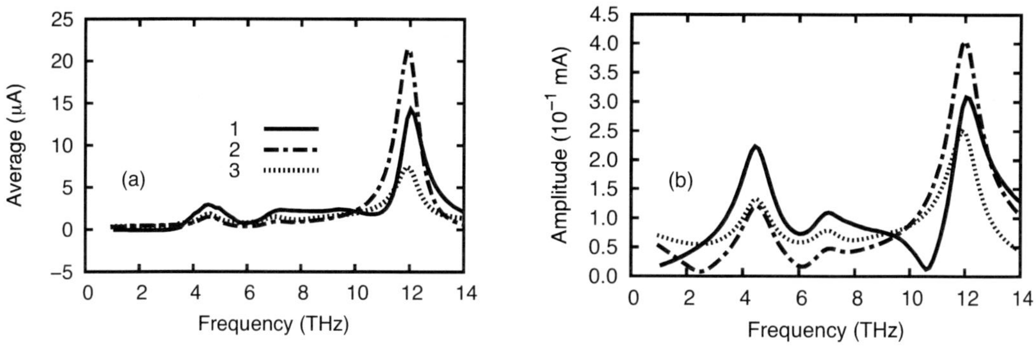

Figure 8. (a) Average response and (b) amplitude of the harmonic response of the drain current as functions of the radiation frequency and for an excitation applied: (1) to the gate, (2) to the drain and (3) to both drain and gate with a null phase-shift ($\varphi = 0$). Bias values are $V_{ds_0} = 40$ mV and $V_{g_0} = -1$ mV.

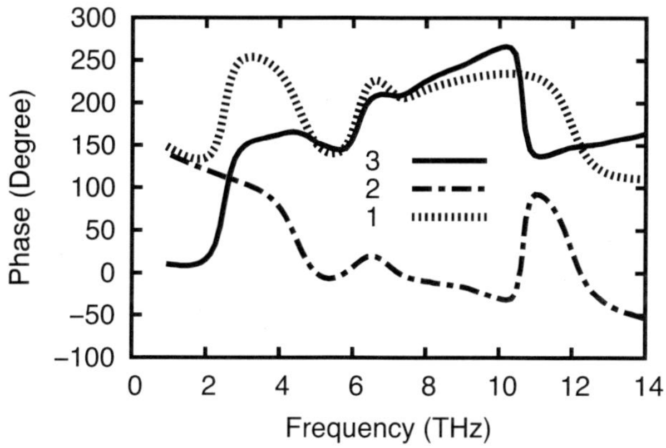

Figure 9. Phase angle φ_I between the harmonic drain current and the excitation as a function of the radiation frequency for (1) an excitation applied to the drain and (2) an excitation applied to the gate. The curve (3) represents the difference between these two quantities.

stimulation of the plasma wave. To lead a quantitative study, we apply in this section a non-zero phase-shift φ between the two excitations and report in Fig. 10 the average response and the amplitude of the harmonic response of the drain current as functions of the detected frequency and for increasing values of the phase-shift from zero to the opposite phase. We observe that the amplitudes of the current responses strongly depend on the phase angle and tends to increase with it. This way, the gate and drain contributions to the current response, and then to the channel excitation, work in a constructive way: by comparing to the spectra presented on Figs. 8 and 10, it is obvious that the dephased combined excitations give significantly better results compared to a single excitation. In particular, the current response takes its maximum values by imposing a phase-shift around $130 - 180°$: for the frequency region within the range $6 - 10$ THz, the current response is maximum for a phase

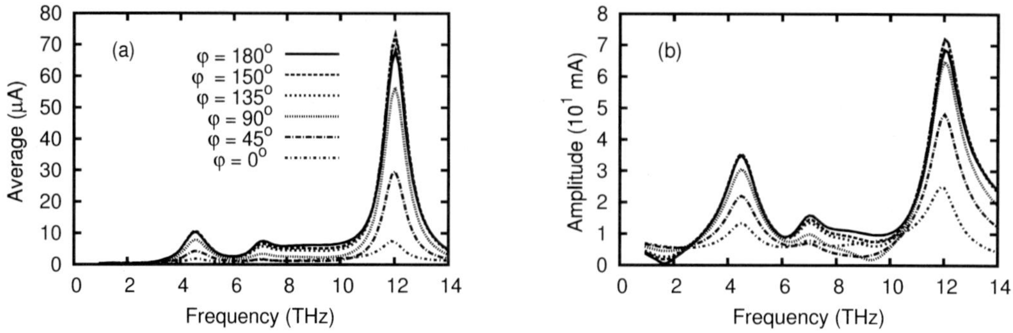

Figure 10. (a) Average response and (b) amplitude of the harmonic response of the drain current as functions of the excitation frequency for different values of the phase-shift φ between drain and gate excitations.

angle equal to $180°$ and, for the rest of the spectrum, it is maximum for a phase angle equal to $150°$. As exhibited by Fig. 9, these conditions tend to those in which the intrinsic phase-shifts between the current response and the THz excitation are compensated.

Technically, by using such a phase-shift between the gate and drain excitations, it is possible to achieve a DC response that is approximatively 11 times higher than that obtained for two excitations in phase and 3.5 higher than that obtained for a single drain excitation. This result clearly indicates the critical importance of a carefully designed and connected external antenna and the possibility to include appropriate phase shifters to enhance the THz detection [28, 29, 30].

4.3.3. Influence of the Channel-to-Gate Distance

Considering the simultaneous excitations, we investigate the influence of the channel-to-gate distance on the current quantities. To this purpose, we report in Fig. 11 the average response and the amplitude of the harmonic response of the drain current as functions of the excitation frequency and for different values of the effective distance (gate-to-channel distance d). We consider the best detection conditions, that is, as exposed previously, the case of a phase-shift equal to $\varphi = 150°$. As concerns the biases and perturbations, the considered values are $V_{ds_0} = 40$ mV, $V_{g_0} = -20$ mV, $\delta V_{g_0} = -1$ mV and $\delta V_{ds_0} = 1$mV. As the gate is more distant, the observed effects are different depending on the considered frequency range. On one hand, for the lower part of the frequency range $2 - 10$ THz, the peak frequencies increase before fading out, since the gate looses its influence, leading to an increase of the electron density and then a disappearance of the 2D modes. On the other hand, we observe a clear increase of the amplitude, frequency and quality factor of the 3D plasma mode. As a matter of fact, as the gate influence is decreasing, the HEMT structure tends to that of a $n^+ - n - n^+$ diode. The 3D resonances are exhausted while the influence of the access zones grows [31].

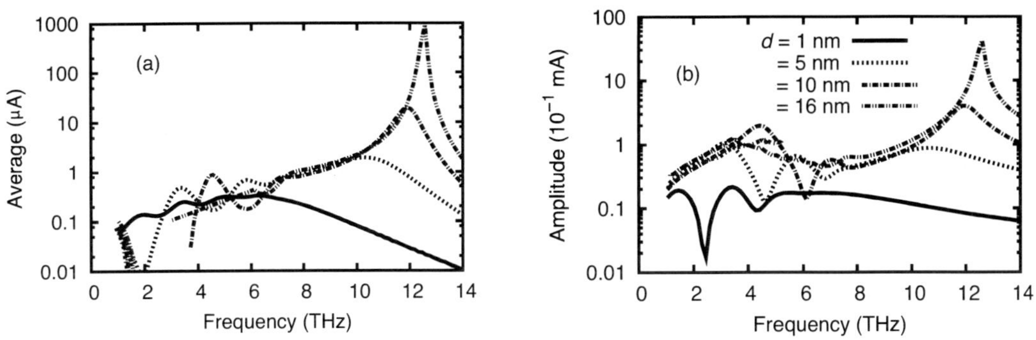

Figure 11. (a) Average response and (b) amplitude of the harmonic response of the drain current as functions of the excitation frequency and for different values of the effective gate-channel distance (d). Combined excitation with voltage $V_{ds_0} = 40$ mV, $V_{g_0} = -20$ mV, $\delta V_{g_0} = -1$ mV and $\delta V_{d_0} = 1$ mV.

4.3.4. n^+nn^+ Region Influence

In order to interpret the increase of the plasma mode frequencies when the gate-channel increases and to evidence the influence of the access zones, we modify the doping concentration N^+ of Fig. 4 and let constant the channel doping concentration to $N_D = 10^{18}$ cm^{-3}. The gate-to-channel distance is $d = 10$ nm. We report on Fig. 12 the average response and the amplitude of the harmonic response of the drain current as functions of the excitation frequency and for different values of the access region concentration ($N^+ = 1 - 9 \times 10$ ^{18} cm^{-3}). As previously, we consider the best detection conditions, that is a combined excitation with a phase-shift $\varphi = 150°$.

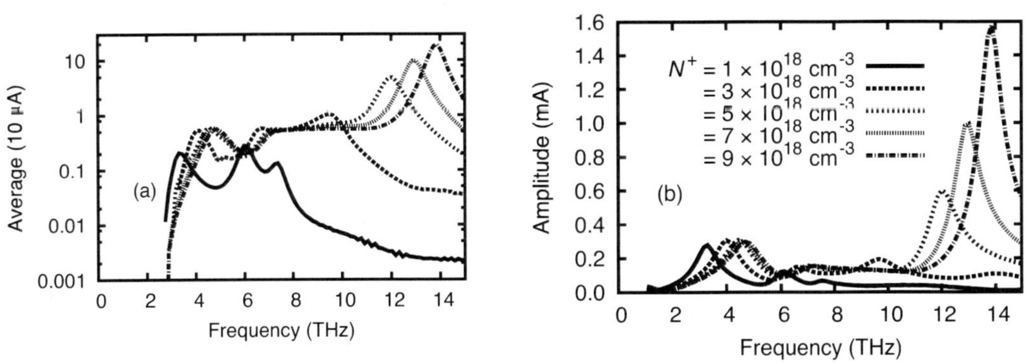

Figure 12. (a) Average response and (b) amplitude of the harmonic response of the drain current as functions of the excitation frequency and for different values of the access region concentration N^+. Combined excitation with bias values $V_{ds_0} = 40$ mV, $V_{g_0} = -20$ mV and excitation amplitudes $\delta V_{g_0} = -1$ mV and $\delta V_{ds_0} = 1$ mV.

On one hand, we observe that an higher access region electron density increases the 3D plasma mode frequency as well as its amplitude and its quality factor. This confirms the coupling effect between the two zones: the long-range Coulomb interaction leads to mutual oscillations of the blocks of electrons with different carrier concentrations N^+ and N. The gate tends to suppress this coupling between the zones, which becomes then more important when d is higher (see Fig. 11). Moreover, for a carrier concentration in the access region almost equal to that of the channel ($N^+ \approx N_D$), the 3D plasma oscillation is damped which agrees with the previous results reported in Ref. [32].

On the other hand, we observe that an increase of the carrier density of the access regions rises the 2D plasma frequency resonance even if, in this case, the peak amplitudes remain practically constant. This frequency increase can be related to a higher insertion in the channel of the access regions carriers. Indeed, this leads to a higher velocity of the channel plasma wave since, in a gated 2D medium, this quantity is proportional to $\sqrt{n}$.

5. Conclusion

We set up a numerical hydrodynamic model able to simulate the response of the drain current of a HEMT to an incoming THz wave. Different methods of coupling between the channel electron gas and the electromagnetic wave were investigated through a simple voltage perturbation approach. The complex planar structure of the transistor was modeled within a one-dimensional framework through a pseudo-bidimensional expression of the Poisson equation.

We theoretically showed that an incident THz wave producing an oscillating voltage on the gate and/or the drain contacts is able to excite the 2D and 3D plasma modes in the THz range, the former being more sensitive to the gate excitation, and the latter to the drain perturbation. Our study of the best conditions of detection allowed to conclude that a simultaneous detection by both drain and gate, that is with a zero phase-shift, leads to a counterproductive excitation of the electrons and then to a deterioration of the sensitivity of the device. However, the insertion of a delay between both signals, that is a phase-shift of about $180°$, enhances the detection ability of the device within the whole considered THz range.

Finally, the study of both gate and doping concentration influences on the detected spectrum showed that the 2D modes are essentially controlled by the gate and are finally weakly depending on the access zones. As concerns the 3D plasma peak, however, we observed that its amplitude, frequency and quality factor are strongly related to the access zones.

References

[1] D. L. Woolard, E. R. Brown, M. Pepper, and M. Kemp, *Proc. IEEE* 93, 1722 (2005).

[2] M. Dyakonov and M. S. Shur, *Phys. Rev. Lett.* 71, 2465 (1993).

[3] J. Lusakowski et al., *J. Appl. Phys.* 97, 064307 (2005).

[4] M. Dyakonov and M. S. Shur, *IEEE Trans. Electron Dev.* 43, 380 (1996).

[5] F. Teppe et al., *Appl. Phys. Lett.* 89, 222109 (2006).

[6] P. Nouvel et al., *J. Appl. Phys.* 106, 013717 (2009).

[7] J. Mateos and T. Gonzalez, *Terahertz Science and Technology, IEEE Transactions on* 2, 562 (2012).

[8] F. Schuster et al., *Opt. Express.* 19, 7827 (2011).

[9] A. Lisauskas et al., *J. Appl. Phys.* 105, 114511 (2009).

[10] G. C. Dyer et al., *Appl. Phys. Lett.* 100, 083506 (2012).

[11] H. Marinchio et al., *Appl. Phys. Lett.* 96 (2010).

[12] H. Marinchio et al., *Appl. Phys. Lett.* 94, 192109 (2009).

[13] C. Jacoboni and P. Lugli, The Monte Carlo Method for Semiconductor Device Simulation, *Computational Microelectronics*, Springer, 1989.

[14] J. Mateos, T. Gonzalez, D. Pardo, V. Hoel, and A. Cappy, *IEEE Trans. Electron Devices* 47, 250 (2000).

[15] K. F. Brennan and D. H. Park, *J. Appl. Phys.* 65, 1156 (1989).

[16] O. Madelung, *Semiconductors Data Handbook*, Springer Berlin, 2003.

[17] S. Adachi, *Physical Properties of III-V Semiconductors InGaAsP*, Wiley New York, 1992.

[18] M. V. Fischetti, *IEEE Transactions on Electron Devices* 38, 634 (1991).

[19] S. Selberherr, *Analysis and Simulation of Semiconductor Devices*, Springer-Verlag, 1984.

[20] M. Dyakonov and M. Shur, *Phys. Rev. Lett.* 71, 2465 (1993).

[21] D. Veksler et al., *Phys. Rev. B* 73, 125328 (2006).

[22] V. Ryzhii et al., *J. Appl. Phys.* 92, 5756 (2002).

[23] P. Sandborn, J. East, and G. Haddad, *IEEE Trans. Electron Devices* 34, 985 (1987).

[24] R. Drury and C. Snowden, *IEEE Trans. Electron Devices* 42, 1026 (1995).

[25] M. Dyakonov and M. Shur, *Electron Devices, IEEE Transactions on* 43, 380 (1996).

[26] H. Marinchio et al., *Applied Physics Letters* 98 (2011).

[27] A. Mahi, H. Marinchio, C. Palermo, A. Belghachi, and L. Varani, *Electron Device Letters, IEEE* 34, 795 (2013).

[28] S. Preu et al., *IEEE Trans. Terahertz Sci. Technol.* 2, 278 (2012).

[29] H.-T. Chen et al., *Nat. Photon.* 03, 148 (2009).

[30] C.-F. Hsieh, H.-L. Chen, C.-Y. Chen, R.-P. Pan, and C.-L. Pan, Voltage controlled liquid crystal terahertz quarter wave plate, in *Lasers and Electro-Optics Society, 2005. LEOS 2005. The 18th Annual Meeting of the IEEE*, pages 985–986, 2005.

[31] P. Ziadé et al., *Semiconductor Science and Technology* 25, 075012 (2010).

[32] A. Mahi et al., Terahertz detection and electronic noise in field-effect transistors, in *Noise and Fluctuations (ICNF), 2013 22nd International Conference on*, pages 1–4, 2013.

In: Advances in Semiconductor Research
Editors: D. Persano Adorno and S. Pokutnyi

ISBN: 978-1-63321-755-3

Chapter 12

THE P-TYPE DOPING OF ZnO: MIRAGE OR REALITY?

Mauro Mosca*[*], *Roberto Macaluso, Fulvio Caruso, Valerio Lo Muzzo and Claudio Calì
Department of Energy, Information engineering, and Mathematical models (DEIM), Thin Films Laboratory, University of Palermo, Palermo, Italy

Abstract

This chapter deals with a critical review on *p*-type doping of ZnO. In the past 15 years, ZnO has attracted considerable attention due to its unique properties, which make it a promising material for optoelectronic devices applications. However, a reliable *p*-type ZnO doping remains a major challenge because of self-compensation effects; thus, despite the advantages of these devices, the fabrication of ZnO-based devices is hampered by the lack of a stable *p*-type doping. A careful and critical analysis of the results reported in literature raises many doubts about the correctness of the doping-type assignments and, in general, the values of the electrical parameters reported. A historical survey of researchers' attempts of doping ZnO *p*-type is given here, together with several possible explanations about the causes of the failures.

Keywords: ZnO films, *p*-type doping, thin film electrical characterization, ZnO-based LED

Introduction

Since the middle of the last century, ZnO began to attract interest for its outstanding properties, such as transparency, piezoelectricity, conductivity, ferromagnetism, and a wide energy band gap.

[*] E-mail address: mauro.mosca@unipa.it. Corresponding author: Mauro Mosca. Department of Energy, Information engineering, and Mathematical models (DEIM), Thin Films Laboratory, University of Palermo, viale delle Scienze, Building 9 – I90128 Palermo (Italy).

The first investigations on ZnO lattice parameters were conducted by *Bunn* on 1935 [1]. Twenty years later, *Mollwo* [2] carried out many important studies on optical properties, eventually completed by *Damen* on 1966 [3], who resumed ZnO vibrational properties studied by Raman scattering.

In the age of the electronic expansion, the semiconductive nature of ZnO stimulated new perspectives in research on electronic devices fabrication. In 1965 *Mead* [4] made-up the first Schottky junction, where the semiconductor consisted of ZnO. After three years *Drapak* [5] fabricated an LED based on a *n*-ZnO/*p*-Cu_2O heterostructure.

Since then, the interest of ZnO in the field of electronic (and optoelectronic) devices has been confirmed by important technological researches like the fabrication of the first metal-insulator-semiconductor structures [6], the studies on ohmic contacts (Al/Au) [7] and on electroluminescent diodes based on *n*-ZnO/*p*-ZnTe [8].

In the last decade a renewed interest on ZnO has come back to the fore, since it has been seen as a possible competitor of GaN material system in the fabrication of blue and UV optoelectronic devices. ZnO is not only confined to the field of optoelectronics applications, but it is also considered attractive for making varistors, acoustic and piezoelectric devices, gas sensors, displays, solar cells, power electronic devices and for spintronic applications (see for example [9, 10]). However, there is no doubt that the most promising applications of ZnO are the same of those of GaN: The goal is to develop a material system which mainly allows developing light sources like the ones based on GaN, in particular the blue laser [11].

Because of its large exciton binding energy (60 meV) at room temperature, in contrast with other compounds (26 meV for GaN and 20 meV for ZnSe) a very strong intensity of photoluminescence (PL) is expected. This means that it could be possible to obtain lasing at room temperature, at least from a theoretical point of view.

There are other advantages over GaN: (i) availability of large-area ZnO substrates, (ii) amenability to wet chemical etching, (iii) high radiation resistance, (iv) relatively low cost of the material, (v) simpler crystal growth, and (vi) ease to grow nanostructured films.

Unfortunately, despite all of the advantages mentioned above, ZnO suffers of poor *p*-type doping, in terms of reliability, stability, and reproducibility: This represents a major disadvantage for a material which is intended to be used for the realization of optoelectronic devices. On the other hand, as-grown non-intentionally doped (n.i.d.) ZnO films show native *n*-type doping. For many years, the native *n*-type doping of ZnO films was attributed to oxygen vacancies (V_O) which are created during the growth process. However, recent studies show that the phenomenon is much more complex and that V_O act as deep donors. As a result, the *p*-type doping is difficult to obtain because these vacancies compensate the acceptors which are introduced into the material for obtaining a *p*-type doping.

Over the years, different ways for realizing *p*-type ZnO have been undertaken, often with arguable results and poor reproducibility. Some of these are still less convincing concerning the high hole concentration and mobility, not in line both with standard electron transport phenomena in ZnO and with the majority of experimental works published, which report values for hole concentration and Hall mobility at room temperature ranging from 10^{16} to 10^{17} cm^{-3} and from few to some 10 cm^2/V s, respectively. In addition, no highly performing ZnO-based light sources have being fabricated to date.

In the following, a review of all these results will be carried out on the ground of new theoretical and experimental studies. A critical analysis of these results should suggest adopting a bit of prudence for carrier type identification.

The Native N-Type Doping

When non-intentionally doped, ZnO films show an *n*-type behavior. For a long time this phenomenon has been attributed to the non-stoichiometric growth of ZnO films which causes oxygen vacancies (V_O) and zinc interstitials (Zn_i). In 1996 Vanheusden et al. [12] proved that oxygen vacancies in ZnO films behave as deep donor, consequently they cannot act as the cause of the *n*-type doping for the n.i.d. films. Hence, nothing was left to assume but to credit the native *n*-type doping with the presence of interstitial zinc atoms.

However, even this hypothesis has been rapidly put aside, due to the high formation energy of Zn_i atoms required at room temperature [13].

A new model, where hydrogen atoms were responsible of the non-intentional *n*-type doping, was proposed in 2000 by Van de Walle [14, 15]: Hydrogen atoms are always present during a growth process and they can diffuse into the lattice of the growing layer, due to their high mobility at temperatures less than 600°C.

Hydrogen atoms form strong O – H bonds with oxygen atoms, and this couple has been considered as a "new" *n*-type doping element for ZnO [16]. This theoretical model has been hereafter confirmed by several experimental studies [17, 18].

A complex consisting of an oxygen vacancy and a hydrogen atom, where the latter is located close to the center of the vacancy (within a distance of 0.05 Å), also behaves as a shallow donor. While these vacancies would be neutral and electrically inactive, the addition of hydrogen turns them into shallow donors (0.8 eV) [14]. This explains why ZnO growth of films in an oxygen-deficient environment leads to *n*-type conductive films.

The nature of the native doping of ZnO films should be taken into account with due consideration during *p*-type doping process, since *n*-type atoms can operate as compensation centers neutralizing completely – or in large part – the *p*-type doping process.

Unfortunately, the native doping process is not entirely controllable since it also depends on the type of substrate, the growth technology, and the deposition parameters (reactor pressure, substrate temperature, etc.). These parameters often have a strong influence on *n*-type carrier concentration and, consequently, on film conductivity.

In addition to stoichiometric defects, the native *n*-type doping can be incremented due the presence of some impurities in the source material, or in the reaction gas supplied in the deposition chamber; sometimes, impurities could be originated from an imperfect cleaning of the chamber, or the substrate itself. Typical chemical elements that run as *n*-type doping centers belong to the following groups:

- IIIA group: Ga, In, B, Al;
- IVA group: Si, Ge, Sn;
- VIIA group: Cl, F, I;
- transition metals: Sc, Y.

A good review of the different kind of defects in ZnO was published by McCluskey and Jokela in 2009 [19].

Main Causes of Failure of P-Type Doping

A list of the main causes that interfere with *p*-type doping of the ZnO is the following:

- presence of lattice defects that run as some trap sites and pin the Fermi Energy level;
- limited solubility of doping elements and formation of precipitates of the dopants;
- deep acceptor levels introduced by the *p*-type dopants (that cannot be ionized at room temperature).

Besides the above-mentioned manifest difficulties in *p*-type doping of ZnO, one should also consider how to maintain the *p*-type doping of the film stable for an unlimited time. Several authors, in fact, demonstrated that a *p*-type film of ZnO can spontaneously turn in *n*-type after a short lapse of time [20, 21].

Probably, the most challenging problem in ZnO (and, in general, in wide band-gap semiconductors) is the doping compensation. The graph in Figure 1 explains schematically how the formation energy of charged compensating defects depends on the Fermi energy position. The incorporation of acceptors (donors) into a semiconductor makes the Fermi level E_F move towards the valence band maximum E_{VBM} (conduction band minimum E_{CBM}); this shift leads to the spontaneous formation of compensating charged defects, whose formation energy depends on the Fermi energy position. As an example, if a ZnO film is doped with an acceptor impurity, the Fermi level shifts towards the valence band maximum; as shown in Figure 1, the formation energy of the charged *n*-type defects decreases. Consequently, the formation of *n*-type defects is favored, resulting in compensating the acceptor impurity [22], [23].

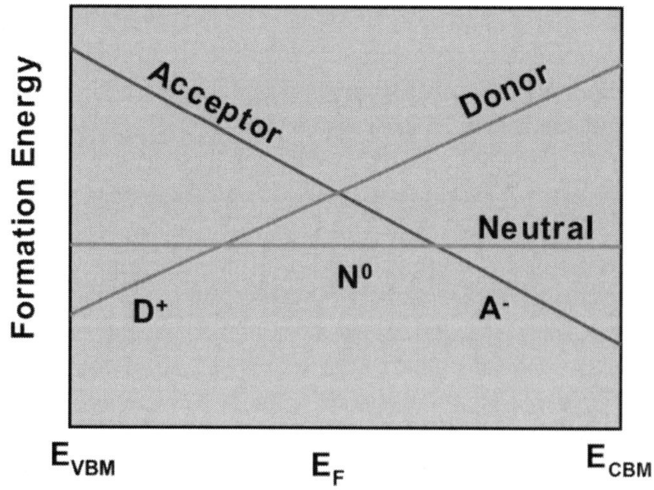

Courtesy of Y. Yan and S.-H. Wei [22].

Figure 1. Schematic dependence of the formation energy of charged defects on the Fermi energy position.

In order to understand how to minimize the compensating effects of these lattice defects, it is useful to make a distinction: They can be divided in intrinsic defects, only due to growth imperfections, and extrinsic defects, due to external causes, such as impurities present in the reactor during the deposition process, external diffusion from the substrate, etc.

A list of intrinsic and extrinsic defects for ZnO is reported below.

Intrinsic Defects

- Oxygen vacancies: As already said, these defects behave as deep donors and exhibit a low formation energy in *p*-type ZnO films; they can strongly compensate the effects of the acceptor dopants.
- Zinc interstitials: They are shallow donors, exhibiting a high formation energy in *n*-type ZnO films; therefore, their effects is negligible in *n*-type films. On the contrary, this energy largely decreases when the Fermi level shifts towards the valence band maximum; consequently, the defects compensate the intended acceptor levels.
- Zinc antisites: These defects are deep donors but unlike the oxygen vacancies, they exhibit a very large formation energy that scarcely compensate the intended acceptor levels. Indicated as A^{V}_{Zn}, they are formed when a doping element of the VA group (A = P, N, As, Sb) takes the place of the Zn atom instead of the oxygen.
- *n*-type islands: Depending on the growth conditions and the type of substrate, a 3D-growth of ZnO films can occur (Volmer-Weber or Stranski-Krastanov modes). However, a 2-D growth of the films prevents the formation of *n*-type islands that work as compensation centers, while it is in general favored both by the lattice mismatch between the film and the substrate, and the high growth temperature. For instance, it has been observed the formation of ZnO nanorods in films grown at 10^{-2} mbar of oxygen pressure by pulsed laser deposition (PLD) with the substrate maintained at 650°C (Figure 2.*a*), while at 600°C no nanostructures appear on the film surface (Figure 2.*b*); at 150°C the surface appears to be smooth (Figure 2.*c*). Krtschil et al. [24] proved that 2-D-grown ZnO films, *p*-doped with As and N atoms, result in a less *n*-type concentration of defects. *n*-type defects, in fact, are associated with 3-D islands, as evidenced by atomic force microscope (AFM) and scanning capacitance microscope (SCM) analysis (Figure 3.*a* -*d*).

Growth defects (where impurities are preferentially incorporated) limit the use for devices and have to be further decreased in density [25].

In order to limit the concentration of *n*-type defects a possible solution is to grow ZnO films on ZnO substrates.

While these substrate are moderately expensive, this expedient may prevent the formation of defects and islands that compensate the acceptor elements.

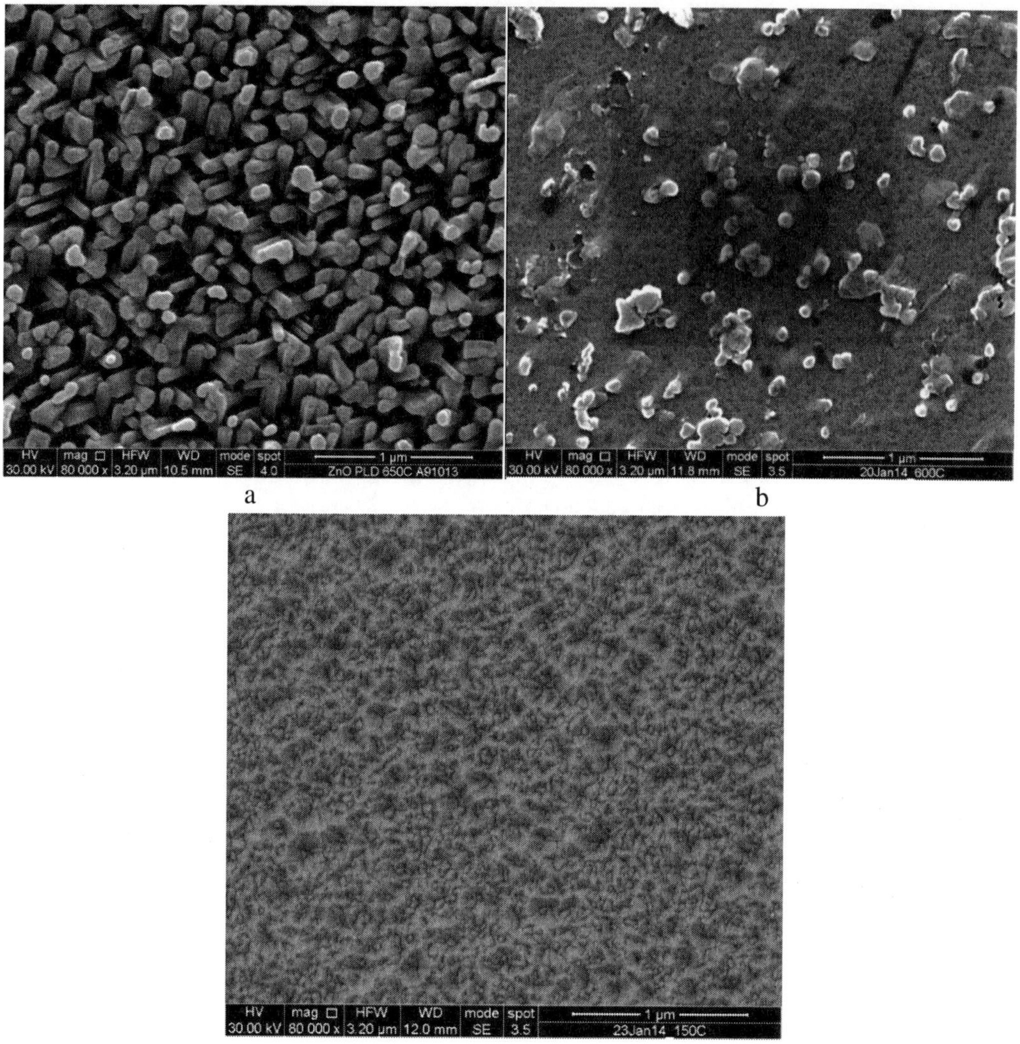

c

Figure 2. Scanning Electron Microscope (SEM) images of ZnO films grown by PLD at 10^{-2} mbar of oxygen pressure and substrate temperature equals to (*a*) 650°C, (*b*) 600°C, (*c*) 150°C.

Extrinsic Defects

- Hydrogen atoms: As previously discussed, hydrogen always behaves as a donor and it is ubiquitous during any growth process. This non-intentional doping compensates the intended acceptor levels [14, 19].
- Diffusion from the substrate: The adoption of some substrates such as GaN, GaAs, or Al_2O_3 can give rise to an *n*-type non-intentional doping, which compensates the effects of the intended acceptor dopants. In fact, some elements such as aluminum, nitrogen, or arsenic can diffuse from the substrate to the film during growth process [26].

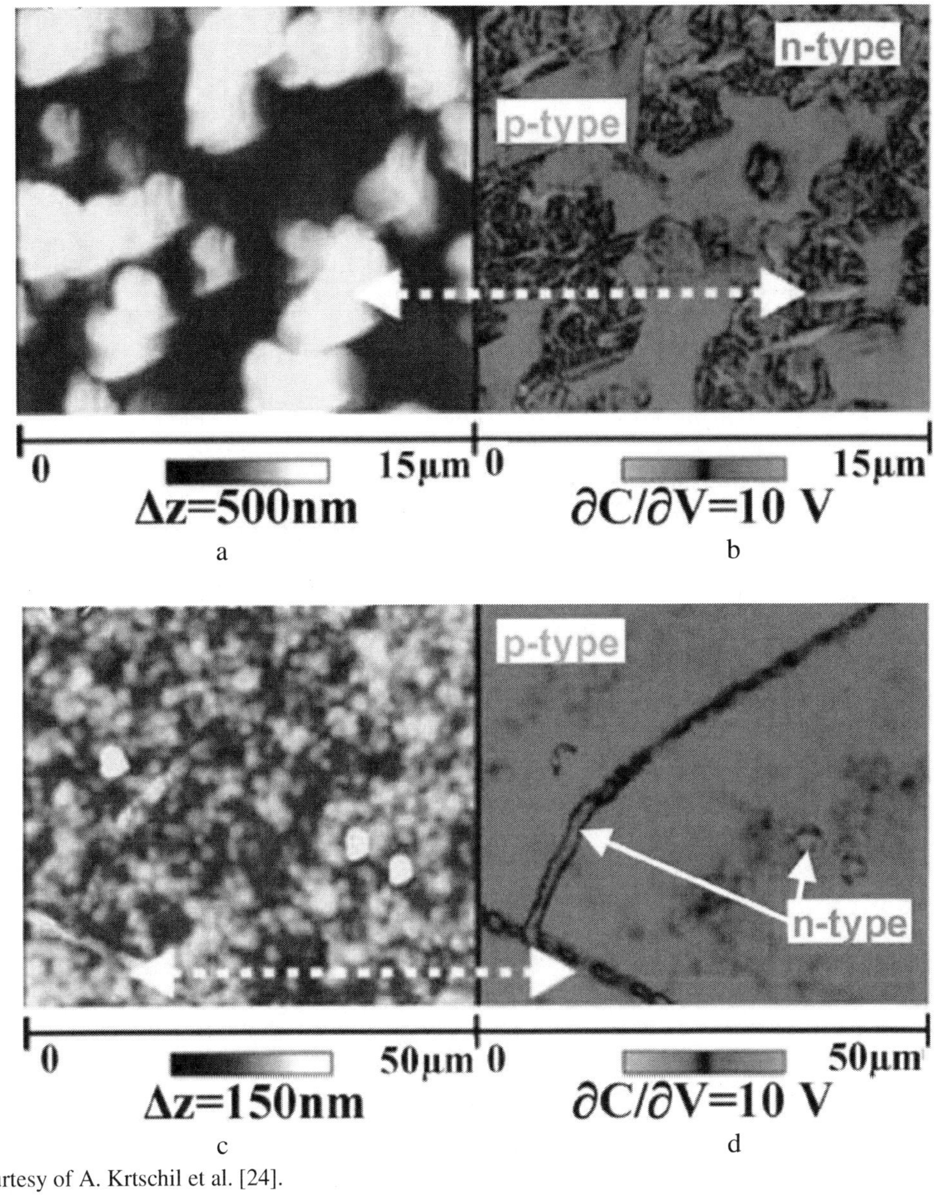

Courtesy of A. Krtschil et al. [24].

Figure 3. Typical AFM (*a* and *c*) and SCM (*b* and *d*) images of Metal-Organic Chemical Vapor Deposition (MOCVD) ZnO layers co-doped with both Zn and As but with different surface morphologies. Image (*a* and *b*) shows a typical sample with mixed conductivity, i.e.,with dominant 3D islands which are of *n*-type in SCM and with smoother, *p*-type regions. In contrast, image (*c* and *d*) shows a sample grown under conditions yielding very flat surface reveals dominant smooth *p*-type domains which are only partly disturbed by very few microcracks and 3D islands which are all *n*-type (depicted by arrows).

Principal P-Type Dopant Elements

Most of the studies on *p*-type doping of ZnO were originated by the paper of *H. Kawazoe* et al. of Tokyo Institute of Technology, Japan, published in 1997 on *Nature* [27].

He hypothesized that several *p*-type wide bandgap semiconductor oxides, such as $CuAlO_2$ (and hereafter $SrCu_2O_2$ [28]), could be effectively used with intrinsic *n*-type ZnO to form violet LEDs. At the end of the last century, several papers were dedicated on new kind of *p*-type oxides [29, 32], *n*-ZnO/*p*-type oxide heterostructures [32, 33], and their possible applications as light-emitting devices [34]. However, due to the low quality of these *p*-type oxides and the remarkable lattice mismatch with the underlying ZnO, these diodes showed very low efficiency. As a consequence, research strongly moved towards *p*-type doping of ZnO itself, since it was believed that a homostructure would improve the growth quality of the *p*-type layer. However, the challenge was more arduous than expected!

To date, the elements known as acceptors for ZnO are the following:

- Zn vacancies;
- IA group: Li, Na, K;
- IB group: Cu, Ag, Au;
- VA group: N, P, As, Sb.

Zn Vacancies

Although Zn vacancies work as acceptors, it is difficult to grow ZnO with such vacancies, since it would be necessary a careful control of film stoichiometry during the growth process; furthermore, these defects would be unstable. Nevertheless, some results are reported in literature [35, 36].

IA Group Elements

The stability of the doping process improves if one substitutes a Zn vacancy with an external element, such as those of IA group.

According to several theoretical calculations, the elements Li, Na, and K introduce some acceptor levels near to the maximum of the valence band, being respectively 0.09 eV, 0.17 eV, and 0.32 eV above it. Unfortunately, due to the high diffusivity these elements do not seem to replace zinc atoms (Li_{Zn}, Na_{Zn}, K_{Zn}) as wished, but turn out to be interstitial impurities (Li_i, Na_i, K_i): As Zn interstitial, these impurities behave as donors. [37]. In addition, in order to obtain a *p*-type doping with IA group elements, during the growth process a very high oxygen pressure (not achievable using the common reactors) should be required, as reported in [38]. One of the reasons why the *p*-doping with IA group elements is so hard to attain is mainly due to the high lattice mismatch that IA group elements (especially Li and Na) introduce when substituting Zn atoms in the crystal structure. Given their lower ionic radius compared with that of Zn, these atoms induce lattice stress points that trap electrons and compensate the acceptors. Moreover, due to their very high diffusivities, the IA group atoms may occupy interstitial positions where they act as donors, inducing a self-compensation effect [39]. Finally, it should be mentioned that the strong reactivity of IA group elements (especially Li) makes them difficult to work with. Table 1 shows a survey of properties of *p*-type ZnO doped with IA group elements as reported in literature.

Table 1. Survey of properties of *p*-type ZnO doped with IA group elements

Year	Dopant	Substrate	Growth method	p (cm^{-3})	μ_h (cm^2 V^{-1} s^{-1})	References
2005	Li	glass	DC-sputtering	1.01×10^{17}	3.47	[40]
2006	Li	quartz	PLD	1.8×10^{16} - 1.4×10^{18}	0.134-0.965	[41]
2006	Li	glass	DC-sputtering	$1.23\text{-}1.68 \times 10^{15}$	0.66-1.08	[42]
2006	Li	quartz	PLD	6.0×10^{17}	1.75	[43]
2009	Li	*n*-Si (100)	sol-gel	3.98×10^{16} - 5.32×10^{18}	2.87-155	[44]
2010	Li	glass	rf-sputtering	6.67×10^{17}	0.77	[45]
2011	Li	glass	rf-sputtering	2.47×10^{18}	0.22	[46]
2007	Na	Si (111)	DC-sputtering	2.5×10^{17}	0.41	[47]
2008	Na	glass quartz	PLD	$0.35\text{-}7.99 \times 10^{17}$ $0.37\text{-}1.67 \times 10^{18}$	0.11-2.86 0.31-1.00	[48]
2012	Na	*c*-plane sapphire	PA-MBE/PLD[a]	1.94×10^{15}	7.54	[49]
2012	Na	*r*-plane sapphire	PLD	5.19×10^{16}	1.41	[50]
2014	Na	*r*-plane sapphire	PLD	3.31×10^{17}	0.28	[51]
2008	K	*c*-plane sapphire	rf-sputtering	$0.32\text{-}8.9 \times 10^{17}$	0.75-3.89	[52]
2014	K	*c*-plane sapphire	rf-sputtering	8.86×10^{15} (6% K) 1.06×10^{18} (8% K) 4.66×10^{16} (11% K)	50.52 (6% K) 3.27 (8% K) 8.02 (11% K)	[53]

[a]ZnO were grown by Plasma-Assisted Molecular Beam Epitaxy (PA-MBE). An Na layer was deposited on the surface of ZnO film by PLD After growth, diffusion was carried out by annealing at temperatures ranging from 650 to 950 °C.

IB Group Elements

For the above-mentioned reasons many researchers focus on IB group elements: in fact, compared with Li and Na, they introduce a lower mismatch when substituting Zn atoms in the lattice (given their higher ionic radius), reducing the effects of self-compensation.

Besides, the formation energy of IB group elements when substituting Zn atoms is low, whereas the interstitial formation energy is rather high.

IB group elements used as *p*-type dopants are Cu, Ag, and Au; in order to introduce acceptor levels in the forbidden bandgap they should substitute Zn atoms in the lattice, forming Cu_{Zn}, Ag_{Zn}, and Au_{Zn} impurities. On the other hand, the acceptor levels introduced by IB group elements are deeper than the levels introduced by IA group elements.

Table 2 shows a survey of properties of *p*-type ZnO doped with IB group elements as reported in literature.

Table 2. Survey of properties of *p*-type ZnO doped with IB group elements

Year	Dopant	Substrate	Growth method	p (cm^{-3})	μ_h (cm^2 V^{-1} s^{-1})	References
2007	Cu	ITO on glass	rf-sputtering	$0.18\text{-}2.8 \times 10^{20}$	Not reported	[54]
2010	Cu	quartz	rf-sputtering	$4.31\text{–}5.78 \times 10^{19}$	0.32-0.49	[55]
2012	Cu	ITO or FTO on glass	electrochemical synthesis	not reported	Not reported	[56]
2006	Ag	*c*-plane sapphire	PLD	$0.49\text{-}6.0 \times 10^{17}$	0.29-2.32	[57]
2008	Ag	glass	rf-sputtering	$0.55\text{-}2.2\times 10^{16}$	1.65-1.83	[58]
2009	Ag	glass	e-beam	$3.98\text{-}5.09 \times 10^{19}$	0.45-1.69	[59]
2011	Ag	*n*-Si (100)	rf-sputtering	$1.1\text{-}1.6 \times 10^{17}$	1.9-5.1	[60]
2012	Ag	*c*-plane sapphire	PLD	2.29×10^{18}	3.03	[61]
2014	Au	quartz	rf-sputtering	2.38×10^{15}	0.16	[62]

VA Group Elements

One of the first elements studied as possible *p*-type dopant for ZnO was nitrogen, which had been used in the past as a dopant element for other wide-band gap semiconductors, such as ZnSe [63]. Already in 1983, Kobayashi et al. [64] proposed N as a possible *p*-type dopant for ZnO; ionic implantation should have been used given the low solubility of the nitrogen in ZnO.

Unlike IA and IB group elements, VA group elements should substitute O atoms in ZnO lattice. Among VA group elements, nitrogen exhibits a ionic radius not too different from the one of oxygen, minimizing lattice mismatch and lattice self-compensation stress centers.

The ionization energy of N levels was at first supposed to be 0.4 eV [37]: Although this value was higher than that of IA group elements, it is the lowest of the other VA group elements, this fact explaining why many researchers focused on nitrogen rather than other VA group elements. The evaluation of the exact value of ionization energy of levels introduced by N-doping was controversial and, as shown in the following section, has led to a higher energy value than the one previously reported. However, many works report on weak (and sometimes unstable) *p*-type doping obtained by using N as a dopant.

Table 3 shows a survey of properties of *p*-type ZnO doped with nitrogen as reported in literature.

Because of the high ionization energy value of N levels, much attention was paid to the other elements of VA group, especially phosphorus. According to Park et al. [37], P levels would be too deep to be ionized at room temperature.

Further, P atoms would tend to substitute Zn atoms rather than O atoms, due to the large difference between their ionic radii; consequently, these P_{Zn} antisites behave as some donors, compensating any intended acceptor level. In particular, antisites formation is favored in oxygen-rich growth process.

Despite these drawbacks, many journals have been reporting on ZnO *p*-type doping with phosphorus since 2000. Zn_3P_2, P_2O_5, or other P-based salts were used for the doping process.

Table 3. Survey of properties of *p*-type ZnO doped with nitrogen

Year	N Source	Substrate	Growth method	p (cm^{-3})	μ_h (cm^2 V^{-1} s^{-1})	References
2001	N_2O	fused silica	PLD	$3\text{-}6 \times 10^{18}$	0.1-0.4	[65]
2002	N_2	*c*-plane Li:ZnO	MBE[a]	9×10^{16}	2	[66]
2003	ammonium acetate	α-Al_2O_3	CVD[b]	9.8×10^{17}	0.97	[67]
2004	N^+ ions	Si_3N_4 on Si	N^+ implantation	$0.50\text{-}7.3 \times 10^{17}$	2.51-6.02	[68]
2004	NO	glass	MOCVD[c]	4.6×10^{15} - 1.7×10^{17}	0.203-2.32	[69]
2005	NO	sapphire	MBE	1.26×10^{18}	0.53	[70]
2005	Zn_3N_2	Zn_3N_2 on quartz, sapphire, GaN, ZnO	Zn_3N_2 oxidation	~ 10^{17}	~ 10	[71]
2005	zinc acetate and ammonium acetate	Si (100)	ultrasonic spray pyrolysis	~ 10^{18}	~ 100	[72]
2005	NH_3	fused silica	PE-CVD[d] and annealing	3.5×10^{13} (500°C-annealed) 2.7×10^{16} (600°C-annealed)	0.8 (500°C-annealed) 0.3 (600°C-annealed)	[73]
2006	N_2	glass	DC-sputtering	5.3×10^{14} - 2.4×10^{17}	0.052-0.505	[74]
2006	N_2	corning glass	rf-sputtering	1.9×10^{15} - 2.1×10^{19}	0.09-20.8	[75]
2006	NO	corning glass, Si	MOCVD	2.9×10^{14} - 8.4×10^{17}	4.55-579	[76]
2006	NO	*c*-plane sapphire	MBE	4.45×10^{17}	0.18	[77]
2007	NH_3	sapphire	ALD[e]	3.41×10^{16}	0.81	[78]
2007	CH_3COONH_4	glass	sol-gel	5.14×10^{17}	1.70	[79]
2007	N_2	Si, quartz, glass	PLD	$0.18\text{-}1.6 \times 10^{17}$	12.3-16.2	[80]
2007	N_2	*a*-, *r*-, *c*-plane sapphire	PLD	6×10^{15}	7	[81]
2007	NO	α-Al_2O_3	rf-sputtering	2.4×10^{18}	0.74	[82]
2007	NO, N_2O	glass	MOCVD	3.9×10^{14} - 6.1×10^{16}	0.31-0.96	[83]
2007	DMHy(Ga)	*c*-plane sapphire	MOCVD	2.41×10^{18}	4.29	[84]
2007	NO	*c*-plane sapphire	MBE	6.9×10^{16} - 1.2×10^{18}	0.22-0.96	[85]
2007	N_2	GaN, GaAs, ZnTe	MBE	6.7×10^{17}	95.2	[86]

Table 3. Continued

Year	N Source	Substrate	Growth method	p (cm^{-3})	μ_h (cm^2 V^{-1} s^{-1})	References
2008	N^+ ions	glass	N^+ implantation	$6.84\text{-}8.44 \times 10^{19}$	0.87-1.11	[87]
2008	NH_3	fused silica	ALD	3×10^{11} - 1×10^{15}	0.2-6	[88]
2008	N_2O	*c*-plane sapphire	MOCVD	~ 4×10^{19}	5	[89]
2008	NH_3	*c*-plane sapphire	MBE	2.2×10^{16}	0.82	[90]
2009	NO	semi-insulating ZnSe	radical beam gettering epitaxy	$1.4\text{-}7.2 \times 10^{17}$	0.9-1.2	[91]
2009	CH_3COONH_4	quartz	sol-gel	$1.9\text{-}3.2 \times 10^{18}$	26.4-29.2	[92]
2009	N_2	Si (001)	DC-sputtering	2.5×10^{17}	~ 60	[93]
2009	N_2	glass	rf-sputtering	3.32×10^{18}	1.31	[94]
2009	N_2	*c*-plane sapphire	MBE	$0.44\text{-}1.8 \times 10^{17}$	0.18-4.12	[95]
2010	N_2	glass, Si	rf-sputtering	8.2×10^{15}	2.3	[96]
2010	urea	glass	PLD	$0.14\text{-}1.3 \times 10^{17}$	1.34-9.24	[97]
2011	$(CH_2)_6N_4$	*c*-plane sapphire	hydrothermal	$0.22\text{-}4.4 \times 10^{16}$	0.8-8.6	[98]
2011	NH_3	quartz	rf-sputtering	$8.3\text{-}9.2 \times 10^{18}$	0.011-0.015	[99]
2011	NO	sapphire	MBE	2.21×10^{16}	1.33	[100]
2011	NO	sapphire	MBE	1.49×10^{16}	1.72	[101]
2012	N_2	alumina Si (100) quartz	rf-sputtering	$1.45\text{-}1.95 \times 10^{17}$	0.2-58	[102]
2014	N_2	quartz	rf-sputtering	1.3×10^{17} (650°C-annealed) 1.5×10^{16} (700°C-annealed)	18 (650°C-annealed) 31 (700°C-annealed)	[103]

[a]Molecular-Beam Epitaxy (MBE).
[b]Chemical Vapor Deposition (CVD).
[c]Metal-Organic Chemical Vapor Deposition (MOCVD).
[d]Plasma-Enhanced Chemical Vapor Deposition (PE-CVD).
[e]Atomic Layer Deposition (ALD).

Table 4 shows a survey of properties of *p*-type ZnO doped with phosphorus as reported in literature.

As and Sb show ionization energies still higher than P; nevertheless, *p*-type doping with As and Sb was reported. Limpijumnong et al. [126] theorized that VA group elements could form a complex, indicated as A_{Zn}–$2V_{Zn}$ (A = P, As, or Sb), formed by the dopant element A (which would substitute a Zn atom) and two adjacent Zn vacancies. According to this model, the coupling of Zn vacancies with the A_{Zn} antisites induce a lowering of the acceptor levels.

Table 4. Survey of properties of *p*-type ZnO doped with phosphorus

Year	P Source	Substrate	Growth method	p (cm^{-3})	μ_h ($cm^2 V^{-1} s^{-1}$)	References
2003	P_2O_5	sapphire	rf-sputtering	1.0×10^{17} - 1.7×10^{19}	0.53-3.51	[104]
2005	Zn_3P_2	Al_2O_3	PLD	5.1×10^{14} - 1.5×10^{17}	2.38-39.3	[105]
2005	P_2O_5	glass	MOCVD	$0.2\text{-}1.6 \times 10^{18}$	0.189-0.838	[106]
2006	P_2O_5	n-Si	rf-sputtering	$0.27\text{-}2.2 \times 10^{17}$	4-13	[107]
2006	Zn_3P_2	glass n-Si	rf-sputtering	1.9×10^{16} - 3.8×10^{19}	6.9-40.1	[108]
2006	GaP	*r*-plane sapphire	MBE	6.0×10^{18}	1.5	[109]
2007	P_2O_5	sapphire	rf-sputtering	3.3×10^{17} - 1.75×10^{19}	2.3-3.25	[110]
2007	P_2O_5	*c*-plane sapphire	PLD	1.3×10^{17}	1	[111]
2008	P_2O_5	*c*-plane sapphire	rf-sputtering	4.71×10^{18}	1.15	[112]
2008	P	n^+-Si(100)	rf-sputtering	1.28×10^{19}	6.10	[113]
2008	P_2O_5	*c*-plane sapphire	rf-sputtering	2.36×10^{17}	2.14	[114]
2008	P	n^+-Si(100)	rf-sputtering	$0.18\text{-}1.3 \times 10^{19}$	6.08-13.1	[115]
2008	P_2O_5	glass	MOCVD	2.2×10^{16} - 2.6×10^{18}	0.14-1.56	[116]
2008	P_2O_5	n^+-Si(100), quartz	MOCVD	1.84×10^{18}	0.277	[117]
2009	Zn_3P_2	*n*-Si, quartz	rf-sputtering	9×10^{15} - 4.2×10^{16}	2.91-3.39	[118]
2009	P_2O_5	sapphire	PLD	$5.1\text{-}6.7 \times 10^{17}$	1.52-6.63	[119]
2010	P_2O_5	sapphire	rf-sputtering	$0.39\text{-}1.5 \times 10^{17}$	2.5-18.2	[120]
2010	P_2O_5	-	PLD	1.7×10^{17}	13	[121]
2011	P_2O_5	Si (111)	thermal evaporation	~ 10^{18}	not reported	[122]
2011	P_2O_5	α-Al_2O_3 (0001)	PLD	5×10^{17}	0.3	[123]
2011	P^+ ions	fused silica	P^+ implantation	5.0×10^{15} (no annealing) 2.3×10^{18} (500°C-annealed)	6.4 (no annealing) 4.5 (500°C-annealed)	[124]
2012	P	P-heavily doped Si	rf-sputtering and diffusion	$0.64\text{-}9.0 \times 10^{17}$	9.6-32.4	[125]

Table 5 shows a survey of properties of *p*-type ZnO doped with phosphorus as reported in literature.

Table 5. Survey of properties of *p*-type ZnO doped with arsenic or antimonium

Year	Dopant	Substrate	Growth method	p (cm^{-3})	μ_h (cm^2 V^{-1} s^{-1})	References
2000	As	GaAs	PLD and diffusion	10^{18}-10^{21}	0.1-50	[127]
2004	As	quartz	rf-sputtering	4×10^{18}	4	[128]
2006	As	sapphire (0001)	PLD	10^{18}-10^{19}	10-100	[129]
2006	As	GaAs	rf-sputtering and diffusion	0.36-1.3×10^{20}	5.5-29	[130]
2007	As	plastic	filtered cathodic vacuum arc technique	4×10^{19}	2	[131]
2008	As	glass	rf-sputtering	1.64-7.40×10^{16}	0.605-1.53	[132]
2008	As	SiO_2/Si	rf-sputtering	0.212-3.03×10^{17}	1.32-6.08	[133]
2008	As	glass	e-beam	3.63×10^{17}	4.21	[134]
2009	As	sapphire	rf-sputtering	5×10^{12} - 1×10^{16}	Not reported	[135]
2010	As	GaAs	PLD and diffusion	1.43-4.475×10^{20}	17.17-39.73	[136]
2010	As	sapphire	rf-sputtering	$1.6 \times 10^{15} – 2.16 \times 10^{17}$	1.30-5.85	[137]
2011	As	GaAs	ALD	3.0-3.4×10^{20}	24.2-27.2	[138]
2011	As	n-Si (100)	As^+ implantation	0.608-1.98×10^{19}	19-32.9	[139]
2002	Sb	n-Si (111)	excimer laser irradiation	5×10^{20}	1.5	[140]
2005	Sb	n-Si (100)	MBE	0.6-1.7×10^{18}	20-25.9	[141]
2006	Sb	p-Si (111)	MBE	1.0×10^{16}	10	[142]
2006	Sb	n-Si (100)	MBE	1.0×10^{18}	22.4	[143]
2007	Sb	quartz	PLD	5×10^{16} - 2.9×10^{18}	0.012-1.75	[144]
2007	Sb	sapphire	PLD	1.9-2.2×10^{17}	0.5-7.7	[145]
2008	Sb	n-Si (100)	MBE	9.6×10^{18}	16	[146]
2009	Sb	n-Si (001)	PLD	3.78×10^{16}	3.74	[147]
2009	Sb	sapphire	rf-sputtering	4.412×10^{17}	Not reported	[148]
2010	Sb	Al_2O_3 (0001)	PLD	4.88-6.25×10^{18}	20.6-57.44	[149]
2011	Sb	Si (100)	rf-sputtering	1.12×10^{18}	2.02	[150]

Other P-Type Doping Methods

Diffusion from the Substrate

A number of papers report on ZnO *p*-type doping after thermal treatment. Some authors [127, 151-153] observed that the conductivity of a ZnO film deposited on a GaAs substrate in oxygen ambient converted from *n*- to *p*-type after annealing at a temperature ranging from 500 to 600°C. The *p*-type conductivity of ZnO films was attributed to arsenic atoms diffusion into ZnO films as shallow acceptors. The process is sketched in Figure 4.

The films obtained with this simple technique appeared stable with doping type considerably depending on the post-annealing temperature. Typical acceptor concentration and mobility values of 10^{20} cm^{-3} and 25-30 $cm^2/(V\ s)$ respectively were obtained (with some record value of 10^{21} cm^{-3} as hole concentration and 50 $cm^2/(V\ s)$ as hole mobility [127]).

The As diffusion in ZnO films could be obtained using simple quartz substrates, considerably less expensive than GaAs ones.

Look et al. [128] proposed to deposit a film of Zn_3As_2 at 350°C, then to sputter ZnO on top of the Zn_3As_2 at 450 °C. After the annealing, the Zn_3As_2 film resulted semi-insulating while the ZnO film turned out to be *p*-type, suggesting that the ZnO layer was doped with As by a gas-phase reaction.

Adopting the same technique, ZnO *p*-type films were obtained by phosphorus diffusion from InP substrates to deposited ZnO layers [154-156].

An annealing treatment at temperatures ranging from 500 to 600°C made the conductivity of ZnO films convert from *n*- to *p*-type. Since P-outdiffusion from InP occurs after 450°C, in order to avoid P-impoverishment in the InP substrate prior to ZnO deposition, growth temperature was fixed at 400°C. Typical acceptor concentration and mobility values of more than 10^{19} cm^{-3} and ~10 $cm^2/(V\ s)$ respectively were obtained.

Co-Doping

A promising approach to ZnO *p*-type doping is the co-doping technique. Co-doping means that the film is doped not only with acceptors but also with impurities that behave as donors. Although this seems a contradiction, the presence of the donors – together with the acceptors – makes the Fermi level move from the proximity of the valence band maximum (when only acceptors are present) to the center of the bandgap.

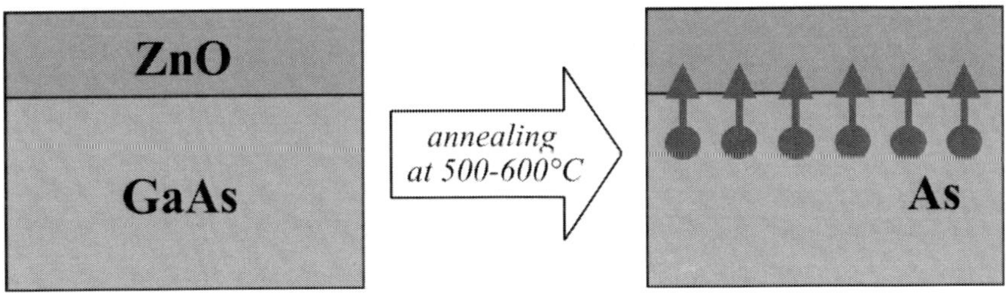

Figure 4. Outdiffusion of As atoms from a GaAs substrate to the above ZnO layer following an annealing process in oxygen ambient.

As shown previously in Figure 1, this leads to an increase of the donors formation energy and, at the same time, a decrease of the acceptors formation energy; as a consequence, the solubility of the acceptor dopants increases [23]. However, due to the compensation effects of the donors, the grown film cannot behave as a *p*-type unless the donors are removed after the growth process. Often, it is convenient to work with donors having high values of diffusivity, in order to be easily removed by a post-growth annealing.

In the co-doping process hydrogen could be a good candidate as a donor, given its extremely high diffusivity. Hydrogen co-doping has been shown to be effective to neutralize interstitial compensating centers. In ZnO co-doped with H impurities, the formation of compensating interstitials was severely suppressed, and the acceptor solubility was greatly enhanced by forming H-acceptor complexes, as reported by Lee and Chang [157].

Using NH_3 as a dopant gas, many authors reported on the growth of *p*-type ZnO films by different deposition methods. The first attempt to dope ZnO with acceptors (nitrogen) performed in 1997 by Minegishi et al. [158] was probably a co-doping with N-H. Other ZnO depositions, carried out in NH_3 ambient, revealed a change in doping type, from *n*- to *p*-type, after a post-growth annealing.

Another approach for co-doping envisages the use of acceptors and donors as "reactive" co-dopants, in a ratio of 2:1 respectively [159]. The goal is to enhance the incorporation of the acceptors reducing the mutual repulsive interactions. In fact, one should consider that the incorporation of donors (together with acceptors) causes strong attractive interactions between dopants that dominate over the repulsive interactions between the acceptors. A secondary beneficial effect, leading to an enhanced acceptors ionization, is that the acceptors energy level lowers towards the maximum of the valence band (while the donors level rises towards the minimum of the conduction band), as depicted in Figure 5. It is worth to point out that in this case the donors are not *p*-type killers (and they are not to be removed) but play an important role to activate the acceptors.

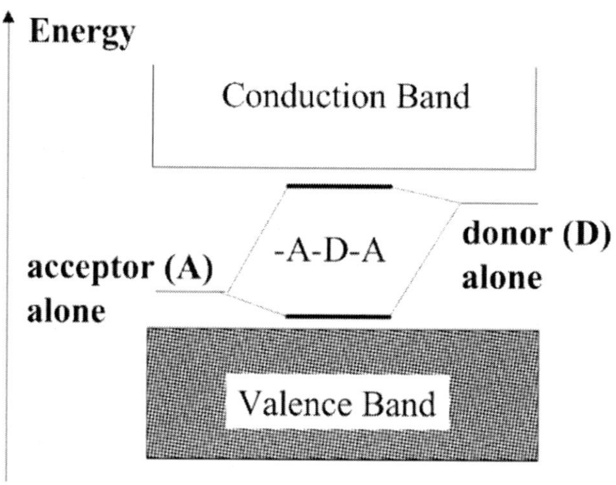

Courtesy of T. Yamamoto [159].

Figure 5. Energy level diagram for a *p*-type co-doped semiconductors. The acceptor level (A) is lowered and the donor level is raised (D) with the formation of acceptor–donor–acceptor complexes upon co-doping.

Co-doping processes were carried out using different acceptors-donors pairs, e.g. H-N [160-162], Al-N [163-169], Ga-N [170-172], In-N [173], O-N [174], Mg-N [175], Sb-N [176], F-Li [177, 178], In-C [179], Ag-S [180], Al-As [181], Ga-P [182], etc. Recently, several theoretical calculations suggested that the co-doping with Al-N should be the best candidate among various co-doping techniques [183]. Theoretical studies have shown that ZnO films co-doped with Al-N form Al–2N complexes, which enhance N solubility and *p*-type conductivity [184]. Other experimental studies [167] confirmed that doping with only N atoms causes a strong repulsive potential, thereby creating a large distance (6.14 Å) between two N atoms, which causes the low solubility and instability of N atoms in ZnO films. In addition, without co-doping there is also a possibility of non-uniform distribution of N atoms in ZnO films. Al-N co-doping in ZnO reduces the repulsive potential between N atoms, causing the distance to reduce at 4.57 Å, thus increasing the capability of incorporating N acceptors. Moreover, co-doping with Al-N, not only enhances the incorporation of N atoms but also should give rise to a shallower N-acceptor level in the bandgap of ZnO films [184].

ZnO Homojunction-Based Devices

After the first evidences of ZnO *p*-type doping, many researchers started to work on feasibility of ZnO *p-n* junction and, in particular, on ZnO homojuction-based light-emitting devices.

The first ZnO *p-n* junction was fabricated in 2000 by PLD by Ryu et al. [185]. As-doped ZnO were used for the *p*-type sides and Al-doped ZnO for the *n*-type sides of the *p-n* junction. As-doping of ZnO film was obtained by diffusion of As atoms from an underlying Zn:GaAs substrate (annealed at 450°C). A gold electrode was deposited on Al:ZnO film while the *p*-contact was obtained by depositing an indium film onto the Zn:GaAs substrate backside. Although the rectifying properties were poor (one could measure a current of 1 mA at –1 V), the characteristic was approximately exponential with a turn-on voltage unexpectedly low (< 1 V). Another "pioneering" ZnO *p-n* junction experiment was performed in 2001 by Tüzemen et al. [186] who changed ZnO conductivity from *n*- to *p*-type by adjusting the oxygen/argon ratio in the sputtering plasma. Ohmic contacts were fabricated in Al/Au; however, also in this case, the turn-on voltage was low (about 1 V). Better results were obtained by Ryu et al. [187] who deposited As-doped ZnO on n.i.d. ZnO film, which in turn had been deposited on *n*-type SiC. They proved that Ni/Au contacts on ZnO:As behave as ohmic, while In/Au and Ti/Au on ZnO:As are Schottky-type. The turn-on voltage raised to values as high as 6-8 V.

Other ohmic contacts were fabricated in Pt/Au on $Zn_{0.9}Mg_{0.1}O$ (with turn-on voltage of 4 V) [188], TiN on N:ZnO (turn-on voltage: 3 V) [189], Ni/Au on $Zn_{0.9}Mg_{0.1}O$ (turn-on voltage: 6.5 V) [190]. The first work reporting on electroluminescence from ZnO-based *p-n* homojunctions was published in 2001 by Guo et al. [191]. A structure of Au/*p*-ZnO (film)/*n*-ZnO (single crystal)/In was fabricated by plasma-enhanced pulsed laser reactive deposition (PE-PLRD) using a pure metal Zn target in N_2O plasma.

While PL spectra (at 77 K) of *p*-ZnO film and *n*-ZnO single crystal displayed a line at around 380 nm and a wide band in visible range, a broad emission in the region of 600 to 800 nm was mainly excited in the electroluminescence (EL) spectrum. This broad emission most likely corresponded to defect states in the LED structure.

Only a weak emission in the region of 370 to 380 nm, corresponding to the free-exciton band, was also slightly excited in the EL spectrum, this fact demonstrating the poor quality of the *p*-type ZnO film. Better results were obtained years later by Tsukazaki et al. [192]. A ZnO *p-i-n* structure was grown on lattice-matched (0001) $ScAlMgO_4$ substrates by laser molecular-beam epitaxy; then, an 100-nm-thick ZnO buffer layer was deposited at 650°C on the substrate and annealed in-situ at 1000°C. All of the successive layers were grown in a layer-by-layer stack. For the *n*-side of the junction, a 400-nm-thick ZnO doped with Ga was deposited, followed by a 10 nm undoped ZnO layer, and finally a *p*-type ZnO doped with N was grown with the repeated temperature modulation (RTM) technique [193]. A good rectification was seen with a threshold voltage of 5.4 V. The EL spectrum showed three different peaks at 395, 420 and 500 nm and its shape resembled that of the PL band observed for the *p*-type ZnO:N, indicating that most of the radiative recombination occurred in the *p*-type layer. Jiao et al. fabricated ZnO homojunctions LEDs on sapphire by plasma-assisted molecular beam epitaxy (PA-MBE) [194]. A nitrogen monoxide plasma was used to grow the *p*-type ZnO layer of the LED, while the *n*-type layer was grown by using Zn and O_2 as precursors and controlling the parameters without donor doping. A threshold voltage of about 4.0 V under forward bias was measured. EL spectrum of the LED was located at the blue-violet region and exhibited the same shape at temperature below 80 K, proving that also in this case the emission mainly occurred in the *p*-side of the junction.

Table 6. Survey of ZnO-based homojunction LEDs

Year	Structure	Growth Method	Emission Color (nm)	Turn-on Voltage (V)	Emitted Power (mW)	References
2001	N:ZnO (film)/ *n*-ZnO (single crystal)	PE-PLRD[a]	broad bluish-white band (600-800) UV (370-380 - weak)	0.3	Not reported	[191]
2004	N_2O implanted ZnO/n-ZnO/Al_2O_3	CVD	violet (388)	3	not reported	[200]
2005	N:ZnO/n.i.d. ZnO/Ga:ZnO/ $ScAlMgO_4$	RTM[b]	violet (395) blue (420) greenish (500)	5.4	Not reported	[192]
2006	N:ZnO/n.i.d. ZnO/sapphire	PA-MBE	no emission at 300 K – blue (410) at 11 K	ohmic at 300 K 4 at 11 K	Not reported	[194]
2006	P:ZnO/Ga:ZnO/ sapphire	rf-sputtering	broad red band (640) UV (380)	3.2	Not reported	[201]
2006	N-In:ZnO/n.i.d. ZnO/GaAs	ultrasonic spray pyrolysis	broad green band (508) blue (428 - weak)	4	Not reported	[195]
2006	N:ZnO/n.i.d. ZnO (single crystal)	MOCVD	broad yellow-blue band (430-600) UV (375 - weak)	2.3	Not reported	[202]
2006	N:ZnO/n.i.d. ZnO/ZnO (single crystal)	MOCVD	green (520) blue (435) UV (372 nm - weak)	80 Schottky contacts	Not reported	[196]
2006	n-ZnO/n-$Be_{0.3}Zn_{0.7}O$/BeZnO-ZnO MQW/p-$Be_{0.3}Zn_{0.7}O$/p-ZnO	Hybrid beam deposition	green (550) violet (388)	10	not reported	[203]
2006	ZnO/ N:ZnO/Sapphire	MOCVD	violet (384)	unannealed: 3 (almost ohmic), annealed: 3.3 (rectifying)	not reported	[204]
2007	N-Al:ZnO/ Al:ZnO/ n.i.d. ZnO/*n*-Si	rf-sputtering	no emission at 300 K – blue-green (481) violet (390) at 110 K	3.3	Not reported	[205]

Year	Structure	Growth Method	Emission Color (nm)	Turn-on Voltage (V)	Emitted Power (mW)	References
2007	n.i.d. ZnO/ N:ZnO/ sapphire	PA-MBE	blue (443)	20	Not reported	[206]
2007	Sb:ZnO/ Ga:ZnO/ *n*-Si(100)	MBE	broad yellow band (605) violet (383) at 300 K UV (373.5) at 9 K	6	Not reported	[197]
2007	*n*-MgZnO/ *p*-MgZnO	MOCVD	blue-green (440-540)	3.5	not reported	[207]
2007	*n*-ZnO/N:ZnO/ c-plane sapphire	MBE	visible (423-523)	3.7	not reported	[174]
2007	n.i.d. ZnO/ As:ZnO/ GaAs	MOCVD	greenish (496) violet (387)	4	not reported	[208]
2007	Al:ZnO/As:ZnO/ *n*+-GaAs	PLD	greenish (496)	4.5	not reported	[209]
2008	N:ZnO/Ga:ZnO/ n.i.d. ZnO/ *c*-plane sapphire	MOCVD	green (517) violet (387.5)	3.4	Not reported	[210]
2008	Sb:ZnO/Ga:ZnO/ MgO/*n*-Si	MBE	violet (388)	2.5	not reported	[211]
2008	Sb:ZnO/ n.i.d. ZnO/ *c*-plane sapphire	MOCVD	visible (536, 710) blue (410)	3.3	not reported	[212]
2008	P:ZnO/MgZnO/ ZnO/MgZnO/ *n*-ZnO	PLD	visible (500-1000) violet (385)	2.2	not reported	[213]
2008	N:ZnO/ bulk *n*-ZnO	MOCVD	UV (379)	2	not reported	[214]
2011	n.i.d. ZnO/ As:ZnO/GaAs/ ITO on glass	photo-assisted MOCVD	white (410, 450, 510)	2.5	Not reported	[199]
2011	n.i.d. ZnO/ As:ZnO/GaAs/ p^+-Si	photo-assisted MOCVD	white (385-580)	4	Not reported	[215]

[a]Plasma_Enhanced Pulsed Laser Reactive Deposition (PE-PLRD).
[b]Repeated Temperature Modulation (RTM).

A peculiarity of this device is that the EL spectrum weakened significantly with the increase of temperature and subsequent quenching at room temperature. This behavior was probably connected to the instability of the *p*-type ZnO layer, in fact Hall-effect measurements at 200 K showed that N-doped ZnO films were *p*-type, whereas with the increase of temperature *p*-type conduction became gradually unstable.

Other LED homostructures were fabricated by Du et al. [195] (using N-In co-doped ZnO as *p*-type layer) where the EL spectrum consisted of a wide band with a maximum at 2.44 eV (λ = 508 nm) and a weak shoulder centered at 2.9 eV (λ = 428 nm), proving that the emission was dominated by deep-level transitions; Liu et al. [196] (using N-doped ZnO as *p*-type layer) where the EL spectrum (measured at 80 V of forward voltage) exhibited an intense band at 520 nm, a weaker one at 435 nm, and a very low peak at 372 nm, very similar to the PL emission of the *p*-type layer; Chu et al. [197] (using Sb-doped ZnO as *p*-type layer) where the

EL spectrum at 9 K was dominated by a peak at 3.32 eV (λ = 373.5 nm) (assigned to the transition between free electrons and a neutral acceptor level [198]), whereas at room temperature the EL spectrum mainly consisted of a wide yellow band centered at 605 nm (attributed to oxygen interstitials) and a narrower peak at 383.3 nm corresponding to band-to-band transition; Zhao et al. [199] (using As-doped ZnO as *p*-type layer) where EL spectrum was more similar to that of a white LED rather than a blue one (three bands, each at 410 nm, 450 nm, and 510 nm).

Table 6 shows a survey of ZnO-based homojunction LEDs as reported in literature.

Enthusiasm and Failures

Given these huge expectations one could wonder why no emitting ZnO-based devices have been commercially available so far.

Nowadays, ZnO *p*-type doping seems to be a consolidate technique, and the first ZnO-based LEDs have been obtained in research laboratory. This temporal gap between research and commercialization is probably due to several difficulties in material development and technology optimization; otherwise, there could be another reason, simpler but more serious in terms of the results obtained so far.

As in the Andersen fable in which "the king is naked", so no one has ever admitted that probably there was no evident *p*-doping of ZnO. Taking for granted the good faith of the researchers who worked in this field, it is likely that the data collected are the result of a series of errors, due in turn to a series of coincidences, and an excessive enthusiasm for the first results reported in literature.

A first glance at the Tables I-IV immediately reveals an excessive spread between results obtained with the same dopant element and with the same technique. More significant are the data that one can take from Table 5, concerning the optical power emitted by ZnO LED based on *p-n* homojunctions: It is surprising that none of the authors have ever measured this essential parameter! Optical emission data were always given in terms of arbitrary units, probably because the emitted power was either too low, or too unstable, to be efficiently measured. Also, surprising is the fact that most of the papers were published in the years between 2006 and 2008 while after these period, research in ZnO emitting devices has gradually moved towards hetero- and nano-structures. Furthermore, the values of mobility and carrier concentration are in most cases either too high to be consistent with theoretical considerations, or too low to be accurately measured without ambiguity. In fact, the high hole concentration and mobility reported in several papers are not in line both with standard electron transport theory of ZnO, and other experimental works published (which report hole concentration and Hall mobility at room temperature ranging from 10^{16} to 10^{17} cm^{-3} and from few to tenths of cm^2/V s, respectively). From an analogy with GaN, Look and Claflin [39] suggested that when the Hall analysis gives ZnO hole concentration higher than 10^{19} cm^{-3} and a mobility higher than 10 cm^2/V s caution should be exercised. Theoretical considerations about these values are reported on [66].

On the contrary, when the values of electrical parameters are too low, ambiguous or unstable results may be measured. Just as an example, with typical values used in a Hall configuration (magnetic flux density B = 0.5 T, current I = 10 mA, ZnO film resistivity ρ =

10^{-3} Ω cm, film thickness d = 500 nm) and for a value of mobility equals to 0.1 $cm^2/(V\ s)$, one obtains a Hall voltage value as low as 1 μV!

Besides the considerations on the values of the electrical parameters, one should also speculate on the technique used to assess these values. The Hall method, commonly adopted in research and industry to evaluate the electrical parameters of a semiconductor wafer, reveals many limits when used in several circumstances. Bierwagen et al. [216] proved that inhomogeneities in carrier concentrations can result in an incorrect assignment of the carrier type. To avoid this wrong assignment of the carrier type, there are some possible remedies, based on an appropriate contact configuration, or a "cloverleaf" contact pattern but they were not generally adopted in ZnO measurements reported in literature.

Given the limits of the Hall method, Macaluso et al. [217] measured the type of conductivity of a ZnO film grown on InP and subsequently annealed at 600°C in air, with three different techniques, including Hall method, *C-V* measurements, and a photochemistry-based characterization technique. Surprisingly, while a *p*-type conductivity was measured by the Hall method, using the two other techniques the ZnO film behaved as *n*-type material. The mechanism responsible of the *p*-type doping measured by Hall measurements was ascribed to the formation of a very thin, high conducting layer at the interface ZnO/substrate during the high-temperature post-growth annealing. This high conductive layer dominated the Hall measurements and instead was invisible to both *C-V* and photocurrent-based methods.

So far, it has been only reviewed the mobility and carrier concentration of ZnO *p*-type films reported in literature, and evidenced some limits of the most common methods used to calculate these values; these considerations have to be taken as the *effects* of some possible mistakes in conductivity-type measurements.

Now, it is worthwhile to investigate about the *causes* of these wrong results in order to establish the real state-of-the-art in ZnO *p*-type doping.

The interest in nitrogen as a possible acceptor of ZnO probably originates from one pioneering paper of 1983 [64] where the authors, studying the energy levels of sp^3-bonded substitutional deep impurities in wurtzite-based semiconductors, predicted that nitrogen impurities would have been good candidates for producing shallow *p*-type dopants. In addition, Park et al. [37] in 2002 calculated that the position of nitrogen energy level in ZnO was 0.40 eV relative to the valence-band maximum. Furthermore, to act as a *p*-dopant, they demonstrated that nitrogen had to occupy an oxygen site in ZnO lattice (N_O). These important outcomes paved the way to an impressive number of studies of N:ZnO doping, with the (not-always-convincing) results reported in Table 3.

From a point of view of the optical properties, PL analysis of *p*-type ZnO films evidenced the presence of one peak at 3.31 eV, which was classified as an acceptor-bound exciton (A^0X) associated with N_O [66, 218]. This was an indirect confirmation of the correctness of the doping-type assignment.

Nitrogen seemed to be the best candidate for ZnO *p*-doping (as well as magnesium for GaN), although a long series of problems jeopardized the efforts to optimize the N-doping process. In particular:

- Limpjumnong et al. [219] showed that defects like $(NC)_O$ and $(N_2)_O$ have sufficiently low energies to convert substitutional N_O acceptors into donors, hindering the efforts of doping ZnO *p*-type.

- In N-doped ZnO films deposited by MOCVD, Keyes et al. [220] revealed a complex growth chemistry, resulting in the presence of C, O, and N-related functional groups; presence of N-H, C-H, and O-H bonds (and a possible metal carbonyl group) were proven to be detrimental to the formation of *p*-type ZnO and explained the relatively low hole concentration or even the creation of *n*-type material that could occur with the incorporation of nitrogen.
- Using ammonia in order to dope ZnO with N could cause several adverse consequences in *p*-type doping. As an example, high ammonia gas flows not only introduce nitrogen in ZnO, but also lattice defects [221]. In addition, hydrogen contained in ammonia could inadvertently passivate nitrogen (notably, N_O^- and H^+ combine to form a neutral defect complex) [222].

Anyway, a huge list of papers dealing with *p*-type ZnO doping with N has been published until the end of 2009 when Lyons et al. wrote a fundamental article, dramatically titled "*Why nitrogen cannot lead to p-type conductivity in ZnO*" [223]. Based on advanced first-principles calculations the authors found that nitrogen has a ionization energy of 1.3 eV (way higher than 0.4 eV!), hence it should be actually considered as a deep acceptor and cannot lead to hole conductivity in ZnO.

In the same year, another serious blow was given to the interpretation of the PL peak at 3.31 eV. Thonke et al. [224] correlated this emission unambiguously with *c*-plane stacking faults, definitively rejecting the assignment of this peak to A^0X emission.

These major discoveries created problems to the model of N_O as a shallow acceptor level. Probably, the researchers were so sure about the nature of nitrogen as a shallow acceptor that did not enough care to carry out other theoretical studies. On the contrary, each experimental result (even if ambiguous or unconvincing) confirmed, as in a chain, the previous experiment.

Indeed, after 2009, the number of papers concerning *p*-type doping with nitrogen considerably decreased. Nevertheless, together with nitrogen, researchers investigated on phosphorus, arsenic, and antimony as possible acceptors of ZnO. Park et al. had well reported on [37] that substitutional P_O and As_O behave as acceptor dopants but, as mentioned in the previous section, they introduce deep levels. Limpjumnong et al. [126] in 2004 demonstrated that As_{Zn}-$2V_{Zn}$ and Sb_{Zn}-$2V_{Zn}$ complexes may have lower formation energy than other defects, introducing shallow acceptor levels. This model, confirmed by Shen et al. [225] and known as the "complex acceptor model", was later (in 2006) confirmed for P, as well [226]. These results promoted the research on VA group elements as shallow acceptors of ZnO. It was recently proven by Wahl et al. [227] that group VA elements do not occupy substitutional oxygen sites, but mostly substitutional Zn sites. Possible fractions of VA elements on oxygen sites – if existing at all – could be a few percent maximum. Although this is a strong argument in favor of the complex acceptor model, to date no experimental evidence regarding the formation of such a complex has been given.

Anyway, doping ZnO with P and As (and to a less extent, Sb) became more and more "popular", although also in this case many researchers had reported on several problems. For example, in ZnO films implanted with P ions, Lin et al. [228] evidenced that below the limit of solubility (~10^{18} ions/cm^3) many defects could be formed in an oxygen atmosphere, whereas above solubility the excess P doping induced the formation of phosphide compounds in ZnO films and seriously deteriorated their cristallinity and optical properties. The formation of phosphide compounds in correspondence of the substrate interface could definitively explain the nature of the misleading interpretation of *p*-type doping in the case of ZnO films doped with phosphorus (and, notably, with VA group elements).

As mentioned in a previous section, ZnO may be *p*-type doped by thermal treatment by inducing out- diffusion of P from InP substrates, or As from GaAs substrates. With respect of ZnO deposited on InP substrate, given the lattice mismatch between film and substrate, the annealing process may lead to the introduction of some defect levels that could compensate the acceptor centers. *Shim* et al. [229] observed that the intensity of the UV peak in PL spectrum of ZnO film deposited on InP substrate, related to the donor-bound exciton (D^0X), decreases after the annealing treatment of the sample. Consequently, it would seem doubtful obtaining UV electroluminescence from a *p-n* ZnO diode, if the *p*-region is obtained by annealing of the substrate. Moreover, Macaluso et al. [217] proved that annealing a ZnO film, grown on an InP substrate, at 600°C in air led to a *p*-type conductivity which may be ascribed to a highly *p*-conductive thin layer formed during the annealing process at the interface ZnO/InP. This circumstance raises many doubts on the effectiveness of the out-diffusion method from a substrate, as a way to get *p*-type-doped ZnO. The problem of the formation of a high *p*-conductive layer is not limited to the above-mentioned out-diffusion method.

When P is used as an acceptor dopant, in order to introduce large concentrations of the impurity into ZnO, the solubility limit may be exceeded. This happens because the impurity species tend to incorporate into an alternate phase rather than on a substitutional lattice site inside the semiconductor, this competing phase appearing as a precipitate in the host material. In the case of VA group impurities in ZnO, the most probable phases to be formed are the Zn_3X_2 compounds where X = N, P, As, or Sb.

When a Zn_3P_2 compound is formed during *p*-doping of a ZnO film, hole accumulation can occur at ZnO/Zn_3P_2 interface due to an unusual valence-band alignment between the two materials, as shown in Figure 6.

Figure 6 clearly shows that a hole accumulation layer is formed at the ZnO/Zn_3P_2 interface on the Zn_3P_2 side. The hole accumulation layer forms within a distance of a few nanometer from the interface, and with a density exceeding 10^{20} cm^{-3} [230].

Such a phenomenon was not only observed in the case of P doping, but also in *p*-type ZnO doped with N. Ye et al. [231] observed the presence of a two-dimensional hole gas (2DHG) between the ZnO film and the Si substrate which led to measure anomalously high values of hole mobility (~150 cm^2/Vs at room temperature).

The presence of these alternative highly-conductive phases may explain why high values of carrier density and/or mobility were often reported in literature in *p*-type ZnO films doped with VA group elements, as shown in Tables III-V.

This phenomenon is analogous to the two-dimensional electron gas (2DEG) that is created in a MOSFET in inversion mode. A similar phenomenon was already well-known for GaN layers grown on sapphire where a highly-degenerate *n*-type 2DEG region, close to the GaN/sapphire interface, affects falsely the mobility and the carrier density measurements [232].

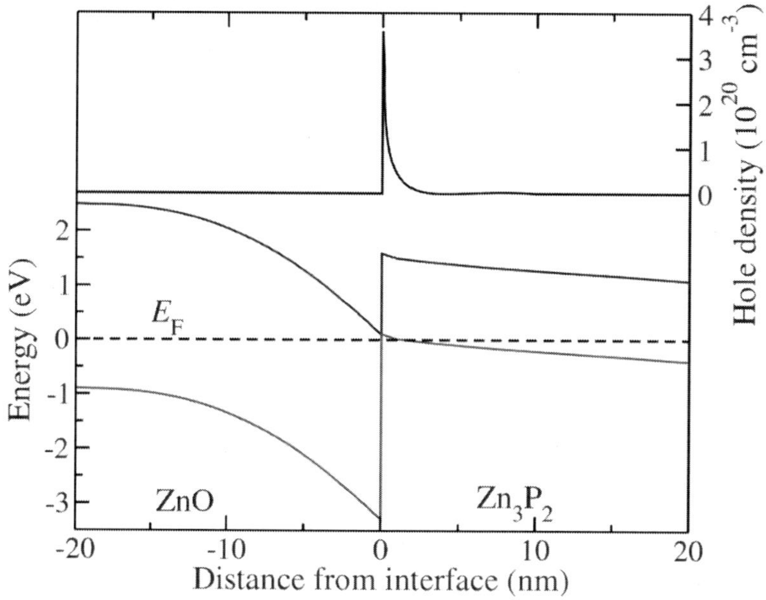

Courtesy of S. Limpijumnong et al. [230].

Figure 6. (Bottom) Band diagram for a ZnO/Zn3P2 interface. (Top) Corresponding density of holes.

Finally, it is worthwhile to point out that ZnO films doped with phosphorus could lead to highly-inhomogeneous distribution of P in the ZnO lattice, as observed by Duclère et al. [233]. These inhomogeneities generally result in a wrong assignment of the carrier type [216], as mentioned before.

After these observations, it seems almost improbable that most of the results reported on Tables 3-5 could be really attributable to *p*-type doping of ZnO. Concerning doping with IA group elements, the high lattice mismatch introduced by these elements when substituting Zn atoms induces a great number of lattice stress points that they are not believed to be very effective in obtaining *p*-type doped ZnO. Moreover, the acceptors Li_{Zn} and Na_{Zn} are compensated by some interstitial donors Li_i and Na_i, respectively [234].

Such compensating defects would need to be eliminated in order to achieve *p*-type conductivity. Also, doping with IB group elements does not seem convincing because they introduce levels in the bandgap too deep to be ionized at room temperature.

The co-doping pathway to *p*-type doping seems to be more promising. The results appear to be more consistent with theoretical considerations, as reported in the previous sections, although not always convincing [235]. Anyway, this is the direction taken by most of researcher currently working on ZnO *p*-doping.

At this point, one could be fascinated by the ZnO *p-n* homojunction-based devices reported in literature (and summarized in Table 6). In fact, is there a better proof of *p*-type doping than the fabrication of a *p-n* rectifying homojunction or, even better, a *p-n* homojunction LED? If all the results reported so far were compromised by a series of mistakes and an effective ZnO *p*-doping has been impossible to obtain, one should not expect to be possible to fabricate ZnO-based homojunction devices. This is particularly true for emitting devices. To sum up, if *p*-type doping results are uncorrected the two main objections that one can raise are:

1 How to explain the rectifying nature of *p-n* ZnO homojunctions?
2 How to explain the electroluminescence from ZnO *p-n* homojunction-based LEDs?

Concerning the ZnO homojunctions, many results were reported on rectifying ZnO *p-n* junctions. Before analyzing these results, it is worthwhile to recall that in a *p-n* junction diode the threshold voltage is approximately given by the energy gap value numerically expressed in V [236]. Although some results are consistent with this assertion, there is a high spread in terms of threshold voltage reported. Further, several papers report on values largely higher than 3 V, as it should be for a ZnO *p-n* junction. This fact could be due to the electrical contacts manufacturing technology that is less mature in ZnO system and in turn causes an additional voltage drop at the ohmic contacts; also, it could be due to the *p*-type conductivity in ZnO which is generally low. If one assumes that *p*-type conductivity is poor, or even null, it may be supposed that a Schottky junction is formed between the metal contact and the *p*-type ZnO. This assumption is also supported by the circumstance that most of the reported values of reverse current are almost high; in other words, the rectifying effect of the junction is far from the ideal, this behavior being typical of Schottky junctions.

In a Schottky diode the barrier height should be numerically equal to the difference between the metal work function and the semiconductor electron affinity; however, it is found experimentally that the barrier height is almost independent of the metal employed because of a distribution of interface states, characterized by a neutral level ϕ_o, which may arise from chemical defects, broken bonds, etc. As a consequence, the barrier height can be numerically expressed as the difference between the values of the energy gap and the neutral level ϕ_o [237]. This means that a threshold voltage numerically equal to the band gap of ZnO does not exclude that the rectifying effect could be due to a parasitic Schottky diode, instead of a *p-n* junction.

Clearly, it is difficult to extrapolate the behavior of the structures from the presented results. Many authors indeed carried out several electrical measurements on the metal contacts of ZnO; the structure generally used for these measurements is reported in Figure 7.

Researchers generally observed a rectifying contact between A and C and ohmic contact between A and B. Nevertheless, one cannot totally exclude that a Schottky junction is formed between the metal contact and the *p*-type ZnO. Let us consider that the characteristic of a Schottky contact is described by two parameters: the effective barrier height and the ideality factor. Due to lateral inhomogeneities of the barrier height, both parameters differ from one diode to another [238]; also, strain-induced piezoelectricity can modify the barrier height [239]. Apart from these local phenomena, it seems that the effective barrier height and the ideality factor are measurably influenced by ZnO surface polarity (which is different on the top and on the bottom part of the semiconductor [240]). For example, in Ag/ZnO junction, fabricated either on Zn-polar, or O-polar wafers, barrier heights varying from 0.77 (O-polar) to 1.02 eV (Zn-polar) were measured [241]. In other experiments performed on Au/ZnO:N junctions, treated at different annealing temperature, a variation of barrier height values close to a factor 2, dependently on ZnO polarity, has been observed [242].

Recently, Ni contacts on single-crystal (0001) ZnO annealed at 800°C. Saw et al. [243] showed an ohmic behaviour in correspondence of the O-polar surface, while a rectifying behaviour in correspondence of the Z-polar surface.

If the conductivity of the *n*-type ZnO is sufficiently high, it is not excluded that a Schottky junction is created between *n*-ZnO (considered as a degenerate semiconductor) and the *p*-ZnO; therefore, a rectifying effect can be measured between A and C contacts of Figure 7. Clearly, specific measurements would be necessary to prove this assumption; anyway, given the controversial data reported in literature one can conclude that the *p*-type region in ZnO *p-n* junction – if really *p*-type – is of poor quality.

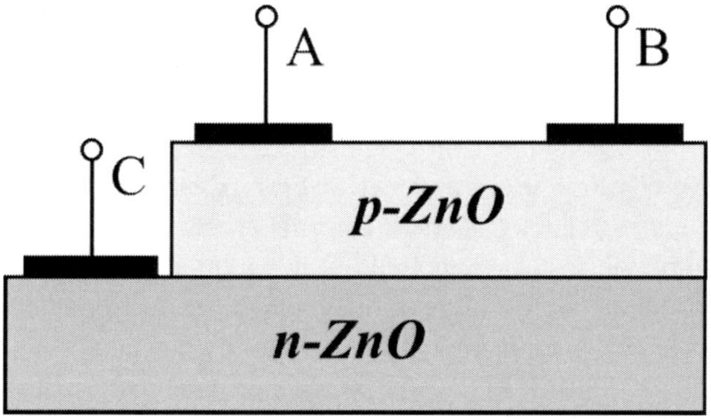

Figure 7. Disposal of the metal contacts on a ZnO-based *p-n* junction. A and B are provided to test the nature of the contacts on *p*-type ZnO, while A and C consisted in standard contacts for the *p-n* homojunction.

If adequately polarized, ZnO *p-n* homojunctions proved electroluminescence, as showed in Table 6. Nevertheless, a careful analysis of the values reveals several data that look unconvincing, e.g. the large data spread concerning the turn-on voltage and the emission wavelengths. Even more surprising is the fact that no paper cited in Table 6 reports on optical power emitted by the LEDs. Probably, the power emitted by the devices was too low or unstable to be easily measured and researchers preferred focusing more on spectral qualitative data rather than quantitative.

Although the electroluminescence could be a convincing proof of the effectiveness of *p*-doping, one should wonder whether the emission be possible in some structures different from a *p-n* junction. A look at some reviews on history of science can immediately give the answer: The first report on light-emitting diodes was published by Henry Joseph Round in 1907 [244]; the light was produced by touching a SiC crystal with two metal electrodes so that a rectifying Schottky contact was formed.

The mechanism of light emission in a Schottky diode is well-described by E. F. Schubert in his famous book "*Light-Emitting Diodes*" [245]. Schottky diodes are usually majority carrier devices but minority carriers can be created even under strong forward-bias conditions, as shown in Figure 8, which displays the band diagram of a metal–semiconductor junction (where the semiconductor is assumed to be of *n*-type conductivity) under (a) equilibrium and (b) strong forward bias conditions.

Under strong forward bias conditions, the band bending allows minority carriers to be injected into the semiconductor by tunneling through the surface potential barrier.

Recombination of the minority carriers with the n-type majority carriers occurs in semiconductor and light is emitted. The voltage required for minority carrier injection in Schottky diodes is larger than typical *p-n* junction LED voltages, because the injection occurs only under strong forward-bias conditions; this could explain why researchers often reported turn-on voltage values higher than expected, as shown in Table 6.

It is worth mentioning here that light can also be generated in a Schottky diode under reverse-bias conditions exploiting the avalanche effect in which high-energy carriers impact-ionize atoms of the semiconductor.

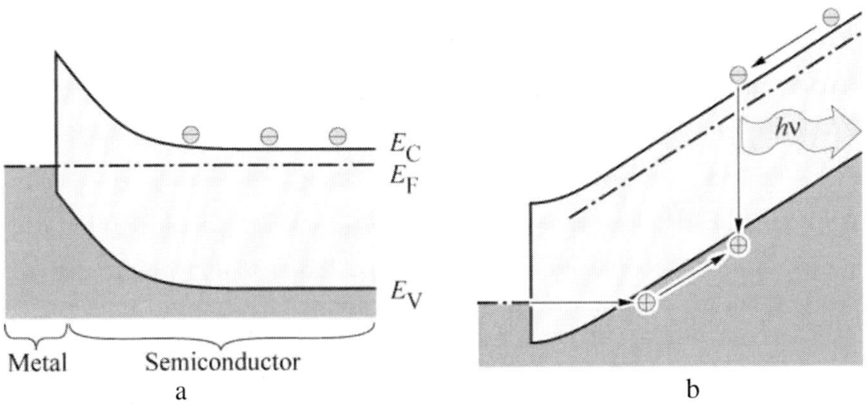

Courtesy of E. F. Schubert [245].

Figure 8. Band diagram of a Schottky diode under (*a*) equilibrium conditions and (*b*) strong forward bias. The figure explains that light emission is achievable even in a metal-semiconductor structure.

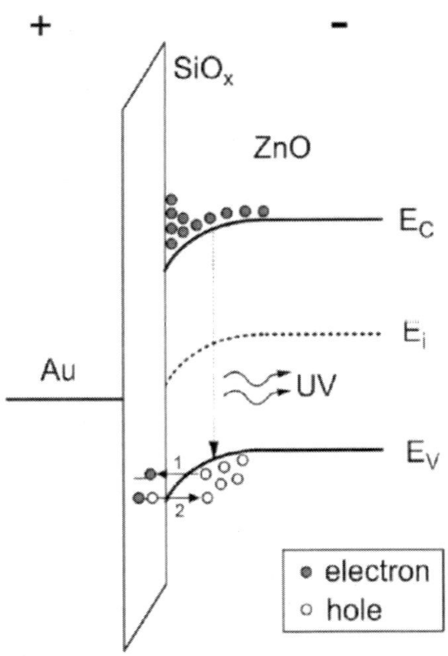

Courtesy of P. Chen et al. [246].

Figure 9. Energy band diagrams of a ZnO/SiO_x/Au structures under forward bias. Arrows 1 and 2 refer to the two possible pathways for the injection of holes into the valence band of ZnO.

The band bending, which causes the minority carriers injection, may also occur in presence of a dielectric interlayer between semiconductor and electrode. Such a layer may be formed especially upon a thermal treatment in air. Figure 9 schematically shows the mechanism of electroluminescence from a ZnO-based metal-insulator-semiconductor structure under forward bias, as reported by Chen et al. [246]. The structure consists of a 100 nm-thick SiO_x ($x \leq 2$) layer deposited on a 300 nm-thick ZnO film over a *n*-type silicon substrate by reactive direct current sputtering; two Au electrodes are finally sputtered, one on the SiO_x film and the other on the backside of the *n*-Si substrate.

Since the SiO_x layer is quite thick, the energy barrier is too wide for the electrons in ZnO to substantially tunnel through the SiO_x; therefore, a large amount of the electrons injected from the silicon substrate can be confined in the downwards-bended region of the conduction band of ZnO. The confinement of electrons in the conduction band of ZnO, beneath the SiO_x barrier, is necessary in order to have recombination with holes of valence band and electroluminescence from ZnO.

The holes can be injected into ZnO through the following mechanism: Under forward bias, some electrons in the valence band of ZnO can wind up into the electron traps located in the band gap of SiO_x, which are inevitably generated during the deposition process; therefore, an equivalent amount of holes is generated in the valence band of ZnO. The electron-hole pairs can be also generated in the insulator layer through impact ionization created by the applied electric field, then swept into the underneath ZnO. This phenomenon was observed in the past in ZnO varistors [247].

Conclusion

ZnO is a promising material for electronic and optoelectronic devices due to its unique properties and some advantages over other wide-gap materials (high exciton binding energies, availability of large-area ZnO substrates, amenability to wet chemical etching, etc.).

If non-intentionally doped, ZnO shows an *n*-type behavior which nowadays is mainly attributed to two causes: strong O – H bonds and complexes consisting of an oxygen vacancy and a hydrogen atom. The latter explains why ZnO growth of films in an oxygen-deficient environment leads to *n*-type conductive films.

A key step to fabricate ZnO-based devices is the achievement of high quality *p*-type ZnO layers. ZnO is expected to be *p*-doped by elements of IA group (such Li, Na, and K), IB group (such as Cu, Ag, Au), or VA group (such as N, P, As, Sb).

Many results have been reported in literature, especially on N-doping because, for a long time, nitrogen was supposed to be the best candidate in terms of energy levels and ionic radii. However, despite many efforts *p*-type ZnO with the requested reproducibility and quality is still lacking. This is in part due to strong self-compensation effects when *p*-type dopants are used. The *p*-type doping of the ZnO can be prevented by many causes, such as the presence of lattice defects, the limited solubility of doping elements, the presence of deep acceptor levels, the instability of *p*-type doping for an unlimited time.

Given the large number of items that may compensate the acceptor dopants, the co-doping pathway to *p*-type doping seems to be more promising because the results appears to be more consistent with theoretical considerations, although not always the results are convincing.

The plethora of unexpected results reported in literature could be explained by several reasons, the mains of which are: failures in electrical parameter measurements; uncorrected assignment of the PL peak previously attributed to A^0X emission; early incomprehension of doping model; presence of an alternative highly-conductive 2DHG path; formation of undesired Schottky junctions between contacts and semiconductors in *p-n* homojunction based devices.

In conclusion, the analysis of the results reported in literature, on the basis of new theoretical and experimental studies, should suggest a bit of prudence for carrier type identification, especially due to erroneous interpretation of Hall measurements.

Therefore, despite the fact that many results evidence that *p*-type doping of ZnO should be considered as a consolidated "reality", a critical analysis of theoretical calculations together with the difficulty to obtain *p-n* junction-based devices should indicate that efficient *p*-type doping of ZnO is still a "mirage", far to be obtained.

Acknowledgments

The authors would like to thank Prof. Monica Santamaria of University of Palermo for taking the SEM images of ZnO films shown in Figure 2.

References

[1] C. W. Bunn, *Proc. Phys. Soc.* 47, 835 (1935).
[2] E. Mollwo, Z. Angew. *Phys.* 6, 257 (1954).
[3] T. C. Damen, S. P. S. Porto and B. Tell, *Phys. Rev.* 142, 570 (1966).
[4] C. A. Mead, *Phys. Lett.* 18, 218 (1965).
[5] I. T. Drapak, *Semiconductors* 2, 624 (1968).
[6] T. Minami, M. Tanigava M. Yamanishi, and M. Kawamura, *Jpn. J. Appl. Phys.* 13, 1475 (1974).
[7] L. J. Brillson, *J. Vac. Sci. Technol.* 15, 1378 (1978).
[8] A. E. Tsurkan, N. D. Fedotova, L. V. Kicherman, and P. G. Pas'ko, *Semiconductors* 6, 1183 (1975).
[9] H. Morkoç and Ü. Özgür, *Zinc Oxide. Fundamentals, Materials and Device Technology*, (Wiley-VCH, Weinheim, Germany, 2009), pp. 387-456.
[10] S. J. Pearton and F. Ren, *Curr. Opin. Chem. Eng.* 3, 51 (2014).
[11] V. A. Karpina, V. I. Lazorenko, C. V. Lashkarev, V. D. Dobrowolski, L. I. Kopylova, V. A. Baturin, S. A. Pustovoytov, A. Ju. Karpenko, S. A. Eremin, P. M. Lytvyn, V. P. Ovsyannikov, and E. A. Mazurenko, *Cryst. Res. Technol.* 39, 980 (2004).
[12] K. Vanheusden, C. H. Seager, W. L. Warren, D. R. Tallant, and J. A. Voigt, *Appl. Phys. Lett.* 68, 403 (1996).
[13] D. C. Look, J. W. Hemsky and J. R. Sizelove, *Phys. Rev. Lett.* 82, 2552 (1999).
[14] C. G. Van de Walle, *Phys. Rev. Lett.* 85, 1012 (2000).
[15] A. F. Kohan, G. Ceder, D. Morgan, and C. G. Van de Walle, *Phys. Rev. B* 61, 15019 (2000).
[16] C. G. Van de Walle, *Physica B* 308–310, 899 (2001).

[17] S. F. J. Cox, E. A. Davis, S. P. Cottrell, P. J. C. King, J. S. Lord, J. M. Gil, H. V. Alberto, R. C. Vilão, J. Piroto Duarte, N. Ayres de Campos, A. Weidinger, R. L. Lichti, and S. J. C. Irvine, *Phys. Rev. Lett.* 86, 2601 (2001).

[18] Y. M. Strzhemechny, H. L. Mosbacker, D. C. Look, D. C. Reynolds, C. W. Litton, N. Y. Garces, N. C. Giles, L. E. Halliburton, S. Niki, and L. J. Brillson, *Appl. Phys. Lett.* 84, 2545 (2004).

[19] M. D. McCluskey and S. J. Jokela, *J. Appl. Phys.* 106, 071101 (2009).

[20] T. M. Barnes, K. Olson and C. A. Wolden, *Appl. Phys. Lett.* 86, 112112 (2005).

[21] S. Majumdar and S. Bhuni, In: *AIP Conf. Proc.* 1447, 667 (2012).

[22] Y. Yan and S.-H. Wei, *Phys. Status Solidi B* 245, 641 (2008).

[23] V. Avrutin, D. J. Silversmith and H. Morkoc, *Proc. IEEE* 98, 1269 (2010).

[24] A. Krtschil, A. Dadgar, N. Oleynik, J. Bläsing, A. Diez, and A. Krost, *Appl. Phys. Lett.* 87, 262105 (2005).[25]A. Dadgar, A. Krtschil, F. Bertram, S. Giemsch, T. Hempel, P. Veit, A. Diez, N. Oleynik, R. Clos, J. Christen, and A. Krost, *Superlattices Microst.* 38, 245 (2005).

[26] B. K. Meyer, H. Alves, D. M. Hofmann, W. Kriegseis, D. Forster, F. Bertram, J. Christen, A. Hoffmann, M. Straßburg, M. Dworzak, U. Haboeck, and A. V. Rodina, *Phys. Status Solidi B* 241, 231 (2004).

[27] H. Kawazoe, M. Yasukawa, H. Hyodo, M. Kurita, H. Yanagi, and H. Hosono, *Nature* 389, 939 (1997).

[28] A. Kudo, H. Yanagi, H. Hosono, and H. Kawazoe, *Appl. Phys. Lett.* 73, 220 (1998).

[29] K. Ueda, S. Inoue, S. Hirose, H. Kawazoe, and H. Hosono, *Appl. Phys. Lett.*, 77, 2701 (2000).

[30] C. F. Windisch Jr., K. M. Ferris, G. J. Exarhos, *J. Vac. Sci. Technol. A* 19, 1647 (2001).

[31] H. Mizoguchi, M. Hirano, S. Fujitsu, T. Takeuchi, K. Ueda, and H. Hosono, *Appl. Phys. Lett.* 80, 1207 (2002).

[32] A. N. Banerjee and K. K. Chattopadhyay, *Prog. Cryst. Growth* 50, 52 (2005).

[33] A. Kudo, H. Yanagi, K. Ueda, H. Hosono, H. Kawazoe, and Y. Yano, *Appl. Phys. Lett.* 75, 2851 (1999).

[34] H. Ohta, K. Kawamura, M. Orita, M. Hirano, N. Sarukura, and H. Hosono, *Appl. Phys. Lett.* 77, 475 (2000).

[35] M.-S. Oh, S.-H. Kim and T.-Y. Seong, *Appl. Phys. Lett.* 87, 122103 (2005).

[36] K. H. Nam, H. Kim, H. Y. Lee, D. H. Han, and J. J. Lee, *Surf. Coat. Tech.* 202, 5463 (2008).

[37] C. H. Park, S. B. Zhang and S.-H. Wei, *Phys. Rev. B* 66, 073202 (2002).

[38] J. J. Lander, *J. Phys. Chem. Solids* 15, 324 (1960).

[39] D. C. Look and B. Claflin, *Phys. Status Solidi B* 241, 624 (2004).

[40] Y.-J. Zeng, Z.-Z. Ye, W.-Z. Xu, L.-L. Chen, D.-Y. Li, L.-P. Zhu, B.-H. Zhao, and Y.-L. Hu, *J. of Cryst. Growth* 283, 180 (2005).

[41] B. Xiao, Z. Z. Ye, Y. Z. Zhang, Y. J. Zeng, L. P. Zhu, and B. H. Zhao, *Appl. Surf. Sci.* 253, 895 (2006).

[42] Y. J. Zeng, Z. Z. Ye, J. G. Lu, W. Z. Xu, L. P. Zhu, and B. H. Zhao, *Appl. Phys. Lett.* 89, 042106 (2006).

[43] J. G. Lu, Y. Z. Zhang, Z. Z. Ye, Y. J. Zeng, H. P. He, L. P. Zhu, J. Y. Huang, L. Wang, J. Yuan, B. H. Zhao, and X. H. Li, *Appl. Phys. Lett.* 89, 112113 (2006).

[44] D. Y. Wang, J. Zhou and G. Z. Liu, *J. Alloy Compd.* 481, 802 (2009).

[45] K. C. Chiu, Y. W. Kao and J. H. Jean, *J. Am. Ceram. Soc.* 93, 1860 (2010).
[46] S. Y. Tsai, M. H. Hon and Y. M. Lu, *J. Cryst. Growth* 326, 85 (2011).
[47] L. L. Yang, Z. Z. Ye, L. P. Zhu, Y. J. Zeng, Y. F. Lu, and B. H. Zhao, *J. Electron. Mater.* 36, 498 (2007).
[48] S. S. Lin, J. G. Lu, Z. Z. Ye, H. P. He, X. Q. Gu, L. X. Chen, J. Y. Huang, and B. H. Zhao, *Solid State Commun.* 148, 25 (2008).
[49] H. Liu, X. Pan, P. Ding, Z. Ye, H. He, and J. Huang, *Mater. Lett.* 80, 175 (2012).
[50] Y. Li, Y. Zhang, H. He, Z. Ye, J. Jiang, and L. Cao, *Mater. Lett.* 76, 81 (2012).
[51] Y. Li, X. H. Pan, J. Jiang, H. P. He, J. Y. Huang, C. L. Ye, and Z. Z. Ye, *J. Alloy Compd.* 584, 466 (2014).
[52] J. Wu and Y. T. Yang, *Mater. Lett.* 62, 1899 (2008).
[53] Y. Huang, W. Zhou and P. Wu, *Solid State Commun.* 183, 31 (2014).
[54] K.-S. Ahn, T. Deutsch, Y. F. Yan, C. S. Jiang, C. L. Perkins, J. Turner, and M. Al-Jassim, *J. Appl. Phys.* 102, 023517 (2007).
[55] H. L. Pan, B. Yao, T. Yang, Y. Xu, B. Y. Zhang, W. W. Liu, and D. Z. Shen, *Appl. Phys. Lett.* 97, 142101 (2010).
[56] B. Marí, M. Sahal, M. A. Mollar, F. M. Cerqueira, and A. P. Samantilleke, *J. Solid State Electrochem.* 16, 2261 (2012).
[57] H. S. Kang, D. Ahn, J. H. Kim, G. H. Kim, S. H. Lim, H. W. Chang, and S. Y. Lee, *Appl. Phys. Lett.* 88, 202108 (2006).
[58] L. Duan, W. Gao, R. Q. Chen, and Z. X. Fu, *Solid State Commun.* 145, 479 (2008).
[59] I. S. Kim, E.-K. Jeong, D. Y. Kim, M. Kumar, and S. Y. Choi, *Appl. Surf. Sci.* 255, 4011 (2009).
[60] L. Duan, C. H. Yu, L. Ni, and Z. Wang, *Appl. Surf. Sci.* 257, 3463 (2011).
[61] M. A. Myers, J. H. Lee, Z. H. Bi, and H. Y. Wang, *J. Phys. Condens. Matter* 24, 145802 (2012).
[62] Y. Xu, B. Yao, Y. F. Li, Z. H. Ding, J. C. Li, H. Z. Wang, Z. Z. Zhang, L. G. Zhang, H. F. Zhao, and D. Z. Shen, *J. Alloy Compd.* 585, 479 (2014).
[63] M. A. Haase, J. Qiu, J. M. DePuydt and H. Cheng, *Appl. Phys. Lett.* 59, 1272 (1991).
[64] A. Kobayashi, O. F. Sankey and J. D. Dow, *Phys. Rev. B* 28, 946 (1983).
[65] X.-L. Guo, H. Tabata and T. Kawai, *J. Cryst. Growth* 223, 135 (2001).
[66] D. C. Look, D. C. Reynolds, C. W. Litton, R. L. Jones, D. B. Eason, and G. Cantwell, *Appl. Phys. Lett.* 81, 1830 (2002).
[67] J. G. Lu, Z. Z. Ye, L. Wang, J. Y. Huang, and B. H. Zhao, *Mater. Sci. Semicond. Process* 5, 491 (2003).
[68] C.-C. Lin, S.-Y. Chen, S.-Y. Cheng, and H.-Y. Lee, *Appl. Phys. Lett.* 84, 5040 (2004).
[69] W. Z. Xu, Y. Y. Ye, T. Zhou, B. H. Zhao, L. P. Zhu, and J. Y. Huang, *J. Cryst. Growth* 265, 133 (2004).
[70] H. W. Liang, Y. M. Lu, D. Z. Shen, Y. C. Liu, J. F. Yan, C. X. Shan, B. H. Li, Z. Z. Zhang, J. Y. Zhang, and X. W. Fan, *Phys. Status Solidi A* 202, 1060 (2005).
[71] E. Kamińska, A. Piotrowska, J. Kossut, A. Barcz, R. Butkute, W. Dobrowolskic, E. Dynowska, R. Jakiela, E. Przeździecka, R. Lukasiewicz, M. Aleszkiewicz, P. Wojnar, and E. Kowalczyk, *Solid State Commun.* 135, 11 (2005).
[72] J.-L. Zhao, X.-M. Li, J.-M. Bian, W.-D. Yua, and C.-Y. Zhang, *J. Cryst. Growth* 280, 495 (2005).

[73] Z. Xiao, Y. Liu, J. Zhang, D. Zhao, Y. Lu, D. Shen, and X. Fan, *Semicond. Sci. Technol.* 20, 796 (2005).
[74] C. Wang, Z. G. Ji, J. H. Xi, J. Du, and Z. Z. Ye, *Mater. Lett.* 60, 912 (2006).
[75] M. L. Tu, Y. K. Su and C. Y. Ma, *J. Appl. Phys.* 100, 053705 (2006).
[76] X. N. Li, S. E. Asher, S. Limpijumnong, B. M. Keyes, C. L. Perkins, T. M. Barnes, H. R. Moutinho, J. M. Luther, S. B. Zhang, S.-H. Wei, and T. J. Coutts, *J. Cryst. Growth* 287, 94 (2006).
[77] J. W. Sun, Y. M. Lu, Y. C. Liu, D. Z. Shen, Z. Z. Zhang, B. H. Li, J. Y. Zhang, B. Yao, D. X. Zhao, and X. W. Fan, *Solid State Commun.* 140, 345 (2006).
[78] C. M. Lee, S. Y. Park, J. M. Lim, and H. W. Kim, *Mater. Lett.* 61, 2495 (2007).
[79] A. R. Poghosyan, X. N. Li, A. L. Manukyan, S. G. Grigoryan, and E. S. Vardanyan, *Proc. SPIE* 6698, 66981D (2007).
[80] G. Epurescu, G. Dinescu, A. Moldovan, R. Birjega, F. Dipietrantonio, E. Verona, P. Verardi, L. C. Nistor, C. Ghica, G. Van Tendeloo, M. Dinescu, *Superlattices Microst.* 42, 79 (2007).
[81] S. Chakrabarti, B. Doggett, R. O'Haire, E. McGlynn, M. O. Henry, A. Meaney, and J.-P. Mosnier, *Superlattices Microst.* 42, 21 (2007).
[82] J. G. Lu, Q. N. Liang, Y. Z. Zhang, Z. Z. Ye, and S. Fujita, *J. Phys. D: Appl. Phys.* 40, 3177 (2007).
[83] Y.-F. Lu, Z. Z. Ye, Y. J. Zeng, H. P. He, L. P. Zhu, and B. H. Zhao, *Chem. Vap. Deposition* 13, 295 (2007).
[84] H. Wang, H. P. Ho, K. C. Lo, and K. W. Cheah, *Mater. Chem. Phys.* 107, 244 (2008).
[85] J. W. Sun, Y. M. Lu, Y. C. Liu, D. Z. Shen, Z. Z. Zhang, B. Yao, B. H. Li, J. Y. Zhang, D. X. Zhao, and X. W. Fan, *J. Appl. Phys.* 102, 043522 (2007).
[86] E. Przeździecka, E. Kamińska, K. P. Korona, E. Dynowska, W. Dobrowolski, R. Jakieła, Ł. Kłopotowski, and J. Kossut, *Semicond. Sci. Technol.* 22, 10 (2007).
[87] S.-Y. Tsai, Y.-M. Lu and M.-H. Hon, *J. Phys. Conf. Ser.* 100, 042037 (2008).
[88] L. Dunlop, A. Kursumovic, J. L. MacManus-Driscoll, *Appl. Phys. Lett.* 93, 172111 (2008).
[89] W. Liu, S. L. Gu, J. D. Ye, S. M. Zhu, Y. X. Wu, Z. P. Shan, R. Zhanga, Y. D. Zhenga, S. F. Choy, G. Q. Lo, and X. W. Sun, *J. Cryst. Growth* 310, 3448 (2008).
[90] P. Cao, D. X. Zhao, J. Y. Zhang, D. Z. Shen, Y. M. Lua, B. Yao, B. H. Li, Y. Bai, and X. W. Fan, *Appl. Surf. Sci.* 254, 2900 (2008).
[91] I. V. Rogozin, *Thin Solid Films* 517, 4318 (2009).
[92] H. Nian, S. H. Hahn, K. K. Koo, E. W. Shin, and E. J. Kim, *Mater. Lett.* 63, 2246 (2009).
[93] E. Gür, S. Tüzemen and S. Doğan, *Phys. Scripta* 79, 035701 (2009).
[94] S. Y. Huang, S. Xu, J. W. Chai, Q. J. Cheng, J. D. Long, and K. Ostrikov, *Mater. Lett.* 63, 972 (2009).
[95] D. Wang, J. Zhang, Y. Peng, Z. Bi, X. Bian, X. Zhang, and X. Hou, *J. Alloy Compd.* 478, 325 (2009).
[96] J. Karamdel, C. F. Dee and B. Y. Majlis, *Appl. Surf. Sci.* 256, 6164 (2010).
[97] S. Majumdar and P. Banerji, *J. Appl. Phys.* 107, 063702 (2010).
[98] M. Ding, D. X. Zhao, B. Yao, B. H. Li, Z. Z. Zhang, and D. R. Shen, *Appl. Phys. Lett.* 98, 062102 (2011).
[99] L. D. Tang, B. Wang and Y. Zhang, *J. Alloy Compd.* 509, 384 (2011).

[100] W. W. Liu, B. Yao, Z. Z. Zhang, Y. F. Li, B. H. Li, C. X. Shan, J. Y. Zhang, D. Z. Shen, and X. W. Fan, *J. Appl. Phys.* 109, 093518 (2011).
[101] X. Chen, Z. Zhang, B. Yao, M. Jiang, S. Wang, B. Li, C. Shan, L. Liu, D. Zhao, and D. Shen, *J. Appl. Phys.* 110, 053305 (2011).
[102] S. Dhara and P. K. Giri, *Thin Solid Films* 520, 5000 (2012).
[103] Z. W. Wang, Y. Yue and Y. Cao, *Superlattices and Microst.* 65, 7 (2014).
[104] K.-K. Kim, H.-S. Kim, D.-K. Hwang, J.-H. Lim, and S.-J. Park, *Appl. Phys. Lett.* 83, 63 (2003).
[105] V. Vaithianathan, B.-T. Lee and S. S. Kim, *J. Appl. Phys.* 98, 043519 (2005).
[106] F. G. Chen, Z. Z. Ye, W. Z. Xu, B. H. Zhao, L. P. Zhu, and J. G. Lv, *J. Cryst. Growth* 281, 458 (2005).
[107] P. Wang P., N. F. Chen and Z. G. Yin, *Appl. Phys. Lett.* 88, 152102 (2006).
[108] Z. G. Yu, P. Wu and H. Gong, *Appl. Phys. Lett.* 88,132114 (2006).
[109] F. X. Xiu, Z. Yang, L. J. Mandalapu, J. L. Liu, and W. P. Beyermann, *Appl. Phys. Lett.* 88, 052106 (2006).
[110] B. J. Kwon, H. S. Kwack, S. K. Lee, Y. H. Cho, D. K. Hwang, and S. J. Park, *Appl. Phys. Lett.* 91,061903 (2007).
[111] A. Allenic, W. Guo, Y. Chen, M. Brandon Katz, G. Zhao, Y. Che, Z. Hu, B. Liu, S. B. Zhang, and X. Pan, *Adv. Mater.* 19, 3333 (2007).
[112] D. K. Hwang, M. S. Oh, Y. S. Choi, and S. J. Park, *Appl. Phys. Lett.* 92,161109 (2008).
[113] R. Q. Ding, H. Q. Zhu and Y. Wang, *Mater. Lett.* 62, 498 (2008).
[114] M. S. Oh, D. K. Hwang, Y. S. Choi, J. W. Kang, and S. J. Park, *Appl. Phys. Lett.* 93, 111905 (2008).
[115] R. Q. Ding, H. Q. Zhu and Q. G. Zeng, *Vacuum* 82, 510 (2008).
[116] J. Jiang, L. P. Zhu, J. R. Wang, X. Q. Gu, X. H. Pan, Y. J. Zeng, and Z. Z. Ye, *Mater. Lett.* 62, 536 (2008).
[117] X. H. Pan, J. Jiang, Y. J. Zeng, H. P. He, L. P. Zhu, Z. Z. Ye, B. H. Zhao, and X. Q. Pan, *J. Appl. Phys.* 103, 023708 (2008).
[118] G. X. Hu, H. Gong, Y. Wang, and H. F. Liu, *J. Appl. Phys.* 106,123522 (2009).
[119] Z.-W. Zhao, L.-Z. Hu, H.-Q. Zhang, J.-C. Sun, J.-M. Bian, H.-W. Liang, B.-Z. Huo, D.-C. Yu, X. Chen, and Q. Fu, *Chin. Phys. Lett.* 26, 057305 (2009).
[120] Y. Y. Kim, W. S. Han and S. K. Cho, *Appl. Surf. Sci.* 256, 4438 (2010).
[121] Y. P. Wang, J. Q. Zhou, Q. Lu, L. L. Liu, X. Zhang, and X. J. Wu, *Jpn. J. Appl. Phys.* 49, 041103 (2010).
[122] N. T. Huong, N. V. Tuyen and N. H. Hong, *Mater. Chem. Phys.* 126, 54 (2011).
[123] Y. Li, Z. Liu and J. Ren, *J. Vac. Sci. Technol. A* 29, 03A109 (2011).
[124] Y. H. Xue, X. D. Zhang, Y. Y. Shen, D. C. Zhang, F. Zhu, L. H. Zhang, and C. L. Liu, *Semicond. Sci. Technol.* 26, 125016 (2011).
[125] H. Q. Zhu, Y. M. Li, J. L. Li, and L. Sun, *Adv. Mater. Res.*, 557-559, 1984 (2012).
[126] S. Limpijumnong, S. B. Zhang, S.-H. Wei, and C. H. Park, *Phys. Rev. Lett.* 92, 155504 (2004).
[127] Y. R. Ryu, S. Zhu, D. C. Look, J. M. Wrobel, H. M. Jeong, and H. W. White, *J. Cryst. Growth* 216, 330 (2000).
[128] D. C. Look, G. M. Renlund, R. H. Burgener II, and J. R. Sizelove, *Appl. Phys. Lett.* 85, 5269 (2004).

[129] N. Xu, Y. Xu, L. Li, Y. Shen, T. Zhang, J. Wu, J. Sun, and Z. Ying, *J. Vac. Sci. Technol. A* 24, 517 (2006).
[130] P. Wang, N. Chen, Z. Yin, F. Yang, C. Peng, R. Dai, and Y. Bai. *J. Appl. Phys.* 100, 043704 (2006).
[131] C. Yuen, S. F. Yu, E. S. P. Leong, S. P. Lau, K. Pita, H. Y. Yang, and T. P. Chen, *J. Appl. Phys.* 101, 094905 (2007).
[132] J. C. Fan and Z. Xie, *Mater. Sci. Eng. B* 150, 61 (2008).
[133] J. C. Fan and Z. Xie, *Appl. Surf. Sci.* 254, 6358 (2008).
[134] M. Kumar and S. Y. Choi, *Appl. Surf. Sci.* 255, 2173 (2008).
[135] H. K. Choi, J. H. Park, S. H. Jeong, and B. T. Lee, *Semicond. Sci. Technol.* 24, 105003 (2009).
[136] S. Nagar and S. Chakrabarti, *Thin Solid Films* 518, 4542 (2010).
[137] A. Kumar, M. Kumar and B. P. Singh, *Appl. Surf. Sci.* 256,7200 (2010).
[138] Y.-C. Cheng, Y.-S. Kuo, W.-H. Li, J.-J. Shyue, and M. J. Chen, *Thin Solid Films* 519, 5558 (2011).
[139] C. O. Kim, D. H. Shin, S. Kim, S.-H. Choi, K. Belay, and R. G. Elliman, *J. Appl. Phys.* 110, 103708 (2011).
[140] T. Aoki, Y. Shimizu, A. Miyake, A. Nakamura, Y. Nakanishi, and Y. Hatanaka, *Phys. Status Solidi B* 229, 911 (2002).
[141] F. X. Xiu, Z. Yang, L. J. Mandalapu, J. L. Liu, and W. P. Beyermann, *Appl. Phys. Lett.* 87, 152101 (2005).
[142] L. J. Mandalapu, Z. Yang, F. X. Xiu, D. T. Zhao, and J. L. Liu, *Appl. Phys. Lett.* 88, 092103 (2006).
[143] L. J. Mandalapu, F. X. Xiu, Z. Yang, D. T. Zhao, and J. L. Liu, *Appl. Phys. Lett.* 88, 112108 (2006).
[144] X. Pan, Z. Ye, J. Li, X. Gua, Y. Zenga, H. He, L. Zhua, and Y. Che, *Appl. Surf. Sci.* 253, 5067 (2007).
[145] W. Guo, A. Allenic, Y. B. Chen, X. Q. Pan, Y. Che, Z. D. Hu, and B. Liu, *Appl. Phys. Lett.* 90, 242108 (2007).
[146] L. J. Mandalapu, Z. Yang, S. Chu, and J. L. Liu, *Appl. Phys. Lett.* 92, 122101 (2008).
[147] X. H. Pan, W. Guo, Z. Z. Ye, B. Liu, Y. Che, H. P. He, and X. Q. Pan, *J. Appl. Phys.* 105, 113516 (2009).
[148] D. H. Kim, N. G. Cho, K. S. Kim, S. H. Han, and H. G. Kim, *J. Electroceram.* 22, 82 (2009).
[149] K. Samanta, P. Bhattacharya and R. S. Katiyar, *J. Appl. Phys.* 108, 113501 (2010).
[150] T. Yang, B. Yao, T. T. Zhao, G. Z. Xing, H. Wang, H. L. Pan, R. Deng, Y. R. Sui, L. L. Gao, H. Z. Wang, T. Wu, D. Z. Shen, *J. Alloy Compd.* 509, 5426 (2011).
[151] Y. Zhang, H.-L.g Lu, Y. Geng, Q.-Q. Sun, S.-J. Ding, and D. W. Zhang, *Vacuum* 103, 1 (2014).
[152] D.-K. Hwang, K.-H. Bang, M.-C. Jeong, and J.-M. Myoung, *J. Cryst. Growth* 254, 449 (2003).
[153] C. K. To, B. Yang, S. C. Su, C. C. Ling, C. D. Beling, and S. Fung, *J. Appl. Phys.* 110, 113521 (2011).
[154] D. Q. Yu, L. Z. Hu, J. Li, H. Hu, H. Q. Zhang, J. M. Bian, J. X. Zhu, S. S. Qiao, X. Chen, and B. Wang, *Appl. Surf. Sci.* 255, 4430 (2009).
[155] A. Mandal and S. Chakrabarti, *Proc. SPIE* 7603, 76031P (2010).

[156] K.-H. Bang, D.-K. Hwang, M.-C. Park, Y.-D. Ko, I. Yun, and J.-M. Myoung, *Appl. Surf. Sci.* 210, 177 (2003).
[157] E.-C. Lee and K. J. Chang, *Phys. Rev. B* 70, 115210 (2004).
[158] K. Minegishi, Y. Koiwai, Y. Kikuchi, K. Yano, M. Kasuga, and A. Shimizu, *Jpn. J. Appl. Phys.* 36 L1453 (1997).
[159] T. Yamamoto, *Thin Solid Films* 420 – 421, 100 (2002).
[160] E. S. Jung, H. S. Kim, H. K. Cho, and J. H. Kim, *Superlattices and Microst.* 42, 62 (2007).
[161] C. Lee, S. Y. Park, J. Lim, and H. W. Kim, *Mater. Lett.* 61, 2495 (2007).
[162] J. G. Lu, S. Fujita, T. Kawaharamura, and H. Nishinaka, *Chem. Phys. Lett.* 441, 68 (2007).
[163] Z.-G. Fei, Z.-Z. Ye, L.-P. Zhu, J.-G. Lü, B.-H. Zhao, J.-Y. Huang, Z.-H. Zhang, L. Wang, and Z.-G. Ji, *J. Cryst. Growth* 268, 163 (2004).
[164] J. G. Lu, Z. Z. Ye, F. Zhuge, Y. J. Zeng, B. H. Zhao, and L. P. Zhu, *Appl. Phys. Lett.* 85, 3134 (2004).
[165] J. G. Lu, L. P. Zhu, Z. Z. Ye, Y. J. Zeng, F. Zhuge, B. H. Zhao, and D. W. Ma, *J. Cryst. Growth* 274, 425 (2005).
[166] F. Zhuge, L. P. Zhu, Z. Z. Ye, D. W. Ma, J. G. Lu, J. Y. Huang, F. Z. Wang, and Z. G. Ji, and S. B. Zhang, *Appl. Phys. Lett.* 87, 092103 (2005).
[167] M. Dutta, T. Ghosh and D. Basak, *J. Electron. Mater.* 38, 2335 (2009).
[168] M. Kumar, J. Prakash Kar, I.-S. Kim, S.-Y. Choi, and J.-M. Myoung, *Curr. Appl. Phys.* 11, 65 (2011).
[169] T.-H. Yang and J.-M. Wu, *Acta Mater.* 60, 3310 (2012).
[170] M. Joseph, H. Tabata, H. Saeki, K. Ueda, and T. Kawai, *Physica B* 302–303, 140 (2001).
[171] K. Nakahara, H. Takasu, P. Fons, A. Yamada, K. Iwata, K. Matsubara, R. Hunger, and S. Niki, *J. Cryst. Growth* 237–239, 503 (2002).
[172] M. Sumiyaa, A. Tsukazaki, S. Fuke, A. Ohtomo, H. Koinuma, and M. Kawasaki, *Appl. Surf. Sci.* 223, 206 (2004).
[173] J. Bian, X. Li, L. Chen, and Q. Yao, *Chem. Phys. Lett.* 393, 256 (2004).
[174] Z. Z. Zhang, Z. P. Wei, Y. M. Lu, D. Z. Shen, B. Yao, B. H. Li, D. X. Zhao, J. Y. Zhang, X. W. Fan, and Z. K. Tang, *J. Cryst. Growth* 301–302, 362 (2007).
[175] K. Kobayashi, T. Koyama, X. Zhang, Y. Kohono, Y. Tomita, Y. Maeda, and S. Matsushima, *J. Phys. Chem. Solids* 74, 80 (2013).
[176] T. Guo, G. Dong, Q. Chen, X. Diao, and F. Gao, *J. Phys. Chem. Solids* 75, 42 (2014).
[177] T. Yamamoto and H. Katayama-Yoshida, *J. Cryst. Growth* 214-215, 552 (2000).
[178] L. Cao, L. Zhu, Y. Li, M. Yang, and Z. Ye, *Mater. Lett.* 86, 34 (2012).
[179] G. Y. Yao, G. H. Fan, F. Zhao, J. H. Ma, J. Chen, S. W. Zheng, S. M. Zeng, L. F He, and T. Zhang, *Physica B* 407, 3539 (2012).
[180] Y. Xu, T. Yang, B. Yao, Y. F. Li, Z. H. Ding, J. C. Li, H. Z. Wang, Z. Z. Zhang, L. G. Zhang, H. F. Zhao, and D. Z. Shen, *Ceram. Int.* 40, 2161 (2014).
[181] L. Balakrishnan, S. Gowrishankar, J. Elanchezhiyan, and N. Gopalakrishnan, *Physica B* 406, 4447 (2011).
[182] S. Gowrishankar, L. Balakrishnan, J. Elanchezhiyan, T. Balasubramanian, and N. Gopalakrishnan, *Physica B* 406, 4085 (2011).
[183] L. G. Wang and A. Zunger, *Phys. Rev. Lett.* 90, 256401 (2003).

[184] T. Yamamoto and H. Katayama-Yoshida, *Jpn. J. Appl. Phys.* 38, L166 (1999).
[185] Y. R. Ryu, W. J. Kim and H. W. White, *J. Cryst. Growth* 219, 419 (2000).
[186] S. Tüzemen, G. Xiong, J. Wilkinson, B. Mischuck, K. B. Ucer, and R. T. Williams, *Physica B* 308–310, 1197 (2001).
[187] Y. R. Ryu, T. S. Lee, J. H. Leem, H. W. White, *Appl. Phys. Lett.* 83, 4032 (2003).
[188] K. Ip, Y. W. Heo, D. P. Norton, S. J. Pearton, J. R. LaRoche and F. Ren, *Appl. Phys. Lett.* 85, 1169 (2004).
[189] J. M. Bian, X. M. Li, C. Y. Zhang, W. D. Yu, and X. D. Gao, *Appl. Phys. Lett.* 85, 4070 (2004).
[190] H. Yang, Y. Li, D. P. Norton, S. J. Pearton, S. Jung, F. Ren, and L. A. Boatner, *Appl. Phys. Lett.* 86, 172103 (2005).
[191] X.-L. Guo, J.-H. Choi, H. Tabata, and T. Kawai, *Jpn. J. Appl. Phys.* 40, L177 (2001).
[192] A. Tsukazaki, M. Kubota, A. Ohtomo, T. Onuma, K. Ohtani, H. Ohno, S. F. Chichibu, and M. Kawasaki, *Jap. J. Appl. Phys.* 44, L643 (2005).
[193] A. Tsukazaki, T. Onuma, M. Ohtani, T. Makino, M. Sumiya, K. Ohtani, S. F. Chichibu, S. Fuke, Y. Segawa, H. Ohno, H. Koinuma, and M. Kawasaki, *Nat. Mater.* 4, 42 (2005).
[194] S. J. Jiao, Z. Z. Zhang, Y. M. Lu, D. Z. Shen, B. Yao, J. Y. Zhang, B. H. Li, D. X. Zhao, X. W. Fan, and Z. K. Tang, *Appl. Phys. Lett.* 88, 031911 (2006).
[195] G. T. Du, W. F. Liu, J. M. Bian, L. Z. Hu, H. W. Liang, X. S. Wang, A. M. Liu, and T. P. Yang, *Appl. Phys. Lett.* 89, 052113 (2006).
[196] W. Liu, S. L. Gu, J. D. Ye, S. M. Zhu, S. M. Liu, X. Zhou, R. Zhang, Y. Shi, Y. D. Zheng, Y. Hang, and C. L. Zhang, *Appl. Phys. Lett.* 88, 092101 (2006).
[197] S. Chu, J. H. Lim, L. J. Mandalapu, Z. Yang, L. Li, and J. L. Liu, *Appl. Phys. Lett.* 92, 152103 (2008).
[198] M. Schirra, R. Schneider, A. Reiser, G. M. Prinz, M. Feneberg, J. Biskupek, U. Kaiser, C. E. Krill, R. Sauer, and K. Thonke, *Physica B* 401, 362 (2007).
[199] W. Zhao, L. Zhao, Z. Shi, X. Xia, X. Li, X. Dong, Y. Chang, B. Zhang, and G. Du, *Appl. Surf. Sci.* 257, 4685 (2011).
[200] Y. I. Alivov, D. C. Look, B. M. Ataev, M. V. Chukichev, V. V. Mamedov, V. I. Zinenko, Y. A. Agafonov, and A. N. Pustovit, *Solid-State Electron.* 48, 2343 (2004).
[201] J.-H. Lim, C.-K. Kang, K.-K. Kim, I.-K. Park, D.-K. Hwang, and S.-J. Park, *Adv. Mater.* 18, 2720 (2006).
[202] W. Z. Xu, Z. Z. Ye, Y. J. Zeng, L. P. Zhu, B. H. Zhao, L. Jiang, J. G. Lu, H. P. He, and S. B. Zhang, *Appl. Phys. Lett.* 88, 173506 (2006).
[203] Y. R. Ryu, T. S. Lee, J. A. Lubguban, H. W. White, B. J. Kim, Y. S. Park, and C. J. Youn, *Appl. Phys. Lett.* 88, 241108 (2006).
[204] M. Pan, R. Rondon, J. Cloud, V. Rengarajan, W. Nemeth, A. Valencia, J. Gomez, N. Spencer, and J. Nause, *Proc. SPIE* 6122, 61220M (2006).
[205] Z. Z. Ye, J. G. Lu, Y. Z. Zhang, Y. J. Zeng, L. L. Chen, F. Zhuge, G. D. Yuan, H. P. He, L. P. Zhu, J. Y. Huang, and B. H. Zhao, *Appl. Phys. Lett.* 91, 113503 (2007).
[206] Z. P. Wei, Y. M. Lu, D. Z. Shen, Z. Z. Zhang, B. Yao, B. H. Li, J. Y. Zhang, D. X. Zhao, X. W. Fan, and Z. K. Tang, *Appl. Phys. Lett.* 90, 042113 (2007).
[207] X. Dong, B. L. Zhang, X. P. Li, W. Zhao, X. C. Xia, H. C. Zhu, and G. T. Du, *J. Phys. D Appl. Phys.* 40, 7298 (2007).

[208] J. C. Sun, J. Z. Zhao, H. W. Liang, J. M. Bian, L. Z. Hu, H. Q. Zhang, X. P. Liang, W. F. Liu, and G. T. Du, *Appl. Phys. Lett.* 90, 121128 (2007).
[209] W. F. Liu, J. M. Bian, L. Z. Hu, H. W. Liang, H. Q. Zang, J. C. Sun, Z. W. Zhao, A. M. Liu, and G. T. Du, *Solid State Commun.* 142, 655 (2007).
[210] J. C. Sun, H. W. Liang, J. Z. Zhao, J. M. Bian, Q. J. Feng, L. Z. Hu, H. Q. Zhang, X. P. Liang, Y. M. Luo, and G. T. Du, *Chem. Phys. Lett.* 460, 548 (2008).
[211] J. Y. Kong, S. Chu, M. Olmedo, L. Li, Z. Yang, and J. L. Liu, *Appl. Phys. Lett.* 93, 132113 (2008).
[212] J. Z. Zhao, H. W. Liang, J. C. Sun, J. M. Bian, Q. J. Feng, L. Z. Hu, H. Q. Zhang, X. P. Liang, Y. M. Luo, and G. T. Du, *J. Phys. D Appl. Phys.* 41, 195110 (2008).
[213] H. S. Kim, F. Lugo, S. J. Pearton, D. P. Norton, and F. Yu-Lin Wang, *Appl. Phys. Lett.* 92, 112108 (2008).
[214] Y. J. Zeng, Z. Z. Ye, Y. F. Lu, W. Z. Xu, L. P. Zhu, J. Y. Huang, H. P. He, and B. H. Zhao, *J. Phys. D Appl. Phys.* 41, 165104 (2008).
[215] X. Xia, Z. Shi, L. Zhao, W. Zhao, X. Dong, B. Zhang, and G. Du, *J. Lumin.* 131, 280 (2011).
[216] O. Bierwagen, T. Ive, C. G. Van de Walle, and J. S. Speck, *Appl. Phys. Lett.* 93, 242108 (2008).
[217] R. Macaluso, M. Mosca, C. Calì, F. Di Franco, M. Santamaria, F. Di Quarto, and J.-L. Reverchon, *J. Appl. Phys.* 113, 164508 (2013).
[218] D. C. Look, *Semicond. Sci. Technol.* 20, S55 (2005).
[219] S. Limpijumnong, X. Li, S.-H. Wei, and S. B. Zhang, *Appl. Phys. Lett.* 86, 211910 (2005).
[220] B. M. Keyes, L. M. Gedvilas, X. Li, and T. J. Coutts, *J. Cryst. Growth* 281, 297 (2005).
[221] A. Dadgar, N. Oleynik, J. Bläsing, S. Deiter, D. Forster, F. Bertram, A. Diez, M. Seip, A. Greiling, J. Christen, and A. Krost, *J. Cryst. Growth* 272, 800 (2004).
[222] X. Li, B. Keyes, S. Asher, S. B. Zhang, S.-H. Wei, T. J. Coutts, S. Limpijumnong, and C. G. Van de Walle, *Appl. Phys. Lett.* 86, 122107 (2005).
[223] J. L. Lyons, A. Janotti and C. G. Van de Walle, *Appl. Phys. Lett.* 95, 252105 (2009).
[224] K. Thonke, M. Schirra, R. Schneider, A. Reiser, G. M. Prinz, M. Feneberg, J. Biskupek, U. Kaiser, and R. Sauer, *Microelectron J.* 40, 210 (2009).
[225] Y. Shen, L. Mi, X. Xu, J. Wu, P. Wang, Z. Ying, and N. Xu, *Solid State Commun.* 148, 301 (2008).[226]W.-J. Lee, J. Kang and K. J. Chang, *Phys. Rev. B* 73, 024117 (2006).
[227] U. Wahl, J. G. Correia, T. Mendonça, and S. Decoster, *Proc. SPIE* 7603, 76030K (2010).
[228] C.-C. Lin, S.-Y. Chen and S.-Y. Cheng, *Appl. Surf. Sci.* 238, 405 (2004).
[229] E. S. Shim, H. S. Kang, S. S. Pang, J. S. Kang, I. Yun, and S. Y. Lee, *Mater. Sci. Eng. B* 102 366 (2003).
[230] S. Limpijumnong, L. Gordon, M. Miao, A. Janotti, and C. G. Van de Walle, *Appl. Phys. Lett.* 97, 072112 (2010).
[231] H. B. Ye, J. F. Kong, W. Pan, W. Z. Shen, and B. Wang, *Solid State Commun.* 149, 1628 (2009).
[232] D. C. Look and R. J. Molnar, *Appl. Phys. Lett.* 70, 3377 (1997).
[233] J.-R. Duclère, R. O'Haire, A. Meaney, K. Johnston, I. Reid, G. Tobin, J.-P. Mosnier, M. Guilloux-Viry, E. Mcglynn, and M. O. Henry, *J. Mater. Sci. Mater. El.* 16, 421 (2005).
[234] M. G. Wardle, J. P. Goss and P. R. Briddon, *Phys. Rev. B* 71, 155205 (2005).

[235] S. B. Zhang, S.-H. Wei and Y. Yan, *Physica B* 302–303, 135 (2001).
[236] E. F. Schubert, *Light-Emitting Diodes*, 2nd Ed. (Cambridge University Press, New York, US, 2006), pp. 59-63.
[237] J. Singh, Semiconductor Optoelectronics. *Physics and Technology*, (McGraw-Hill, Inc., New York, US, 1995), pp. 316-318.
[238] W. Mönch, *J. Vac. Sci. Technol. B* 17, 1867 (1999).
[239] J. Zhou, P. Fei, Y. Gu, W. Mai, Y. Gao, R. Yang, G. Bao, and Z. L. Wang, *Nano Lett.* 8, 3973 (2008).
[240] M. Stutzmann, O. Ambacher, M. Eickhoff, U. Karrer, A. Lima Pimenta, R. Neuberger, J. Schalwig, R. Dimitrov, P. J. Schuck, and R. D. Grober, *Phys. Status Solidi B* 228, 505 (2001).
[241] M. W. Allen, *Schottky Contact Formation to Bulk Zinc Oxide*, *Ph.D. Thesis*, (University of Canterbury, Christchurch, New Zealand, 2008).
[242] D. C. Oh, J. J. Kim, H. Makino, T. Hanada, M. W. Cho, T. Yao, and H. J. Ko, *Appl. Phys. Lett.* 86, 042110 (2005).
[243] K. G. Saw, S. S. Tneh, G. L. Tan, F. K. Yam, S. S. Ng, and Z. Hassan, *PLOS ONE* 9, e86544 (2014).
[244] H. J. Round, *Electr. World* 49, 309 (1907).
[245] E. F. Schubert, *Light-Emitting Diodes*, 2nd Ed. (Cambridge University Press, New York, US, 2006), pp. 1-4.
[246] P. Chen, X. Ma and D. Yang, *Appl. Phys. Lett.* 89, 111112 (2006).
[247] G. E. Pike, S. R. Kurtz, P. L. Gourley, H. R. Philipp, and L. M. Levinson, *J. Appl. Phys.* 57, 5512 (1985).

In: Advances in Semiconductor Research
Editors: D. Persano Adorno and S. Pokutnyi
ISBN: 978-1-63321-755-3

Chapter 13

Thermoelectric and Magneto-Optic Semiconductor Single Crystals and Ceramics - Fabrication and Applications

Mythili Prakasam* and Alain Largeteau
CNRS, Université de Bordeaux, ICMCB, Pessac, France

Abstract

Semiconductor materials have applications in vast fields and are the basis for modern electronics. Semiconductors materials have electrical conductivity intermediate between that of metals and insulators. In this chapter we will discuss about the applications of semiconductors in thermoelectric devices and magneto-optics. Semiconductors are also considered to be ideal for thermoelectric devices because of their band structure and electronic properties at high temperatures. By Wiedemann- Franz law, if the electrical conductivity is high, the thermal conductivity of electron increases and therefore the thermal conductivity of phonon should be minimized. In semiconductors, $\kappa_{electron} < \kappa_{phonon}$, so it is easier to decouple κ and σ in a semiconductor through engineering κ phonon. Semiconductor single crystals of congruently melting compounds are obtained by Czochralski or Bridgman techniques and the peritectic decomposing compounds are prepared by flux-growth. Low defect semiconductor thermoelectric crystals are obtained by vapor transport. Further pressed powder metallurgy and advanced sintering techniques are employed for obtaining semiconductor nanograined ceramics to minimize phonon scattering. Bi_2Te_3 (BT) is one of earliest known semiconductors for thermoelectric device applications. Recently various semiconductor materials are explored and different strategies are adopted [1] to increase the thermoelectric efficiency zT through incorporation of dopants, nano-precipitation and nanograins to scatter the phonon and decrease the thermoelectric conductivity.

The other class of semiconductor's application is in magneto-optic devices, which are widely obtained from the semi magnetic semiconductors (SMSC) or diluted magnetic semiconductors (DMS).

DMS have recently been a major focus of magnetic semiconductor research [2, 3]. These are based on traditional semiconductors, but are doped with transition metals instead of, or in

* E-mail address: prakasam@icmcb-bordeaux.cnrs.fr/ mythili.prakasam@gmail.com. Corresponding author: Mythili Prakasam. CNRS, Université de Bordeaux, ICMCB, 87 avenue du Dr. A. Schweitzer, Pessac, F-33608, France.

addition to, electronically active elements. The most common DMS are II-VI compounds (like CdTe, ZnSe, CdSe, CdS, etc.), with transition metal ions (e.g., Mn, Fe or Co) substituting their original cations. There are also materials based on IV-VI (e.g., PbTe, SnTe) and recently III-V (e.g., GaAs, InSb) crystals. On the other hand, rare earth elements (e.g., Eu, Gd, Er) are also used as magnetic atoms in SMSC. These mixed crystals (semiconductor alloys) may be considered as containing two interacting subsystems. The first of these is the system of delocalised conduction and valence band electrons. The second is the random, diluted system of localised magnetic moments associated with the magnetic atoms. The wide variety of both host crystals and magnetic atoms provides materials which range from wide gap to zero gap semiconductors, and which reveal many different types of magnetic interaction. Several of the properties of these materials may be tuned by changing the concentration of the magnetic ions.

1. Introduction

Semiconductors are a material type with enormous device applications arising from its unique properties which translate into various applications by doping. These belong to the special category of the materials with properties between those of insulators and conductors, with narrow bandgaps of < 2 eV with covalent bonds. There are two types of semiconductor materials, where one of the types is intrinsic/pure semiconductor and the other being extrinsic/ doped semiconductors. Silicon (Si) is the most widely used material in semiconductor devices, followed by Germanium (Ge). But thermal sensitivity of Ge limits its usage in comparison to Si. The other widely used material is Gallium Arsenide (GaAs) usually employed for high-speed devices. In addition other well known materials used for LEDs are Silicon Carbide (SiC), Indium Phosphide (InP), Indium Arsenide (InAs) and Indium Antimonide (InSb). The primary applications of these aforesaid materials are in blue LEDs and in solid state laser diodes. The other class of semiconductor category belongs to organic semiconductors, which was first invented in 1950s. These organic semiconductors consist of charge-transfer complexes and various linear conductive polymers. Similar to their inorganic counterparts, organic semiconductors can be doped. Some of the examples of semiconducting organic compounds are Pentacene, Anthracene, and Ruberene and the organic polymers such as poly (3-hexylthiophene), poly (p-phenylene vinylene), polyacetylene and its derivatives are other few examples. Some of the widely known applications of semiconductors are Thermistors, Hall probes, IR sensors/ optoelectronic devices, diodes, transistors, FETs, LEDs, solar cells, Faraday rotators and thermoelectrics etc.

One of the important components in the semiconductor technology is optimizing the quality and production of the basic semiconductor material used for device construction. Amongst the various forms of semiconductors used for device construction are in the forms of single crystals, thin films and ceramics. Generally the fabricated bulk single crystals which are grown by Czochralski method, Bridgman method and zone melting methods are processed into substrates. Further epitaxial growth techniques are used for depositing thin films of semiconductor materials (in the range of few micrometers) on to the substrates. Chemical Vapor Deposition (CVD) and Physical Vapor Deposition (PVD) are the prevalently employed techniques to obtain the semiconductor ingots.

The other ways of fabrication process for semiconductors is ceramics. The usage of ceramics is currently increasing owing to the facile combinations of various compositions to alter and obtain the desired electrical, mechanical and physical properties. One of the obvious examples is in integrated circuit manufacturing, where the ceramics semiconductor manufacturing leads to faster, desired shaped at a lesser manufacturing price and in short time to achieve the desired device performance. Ceramics related to semiconductor are conventionally fabricated by using traditional forming and densification methods with the high purity raw powders for precisely controlling the electrical properties.

Among the wide range of semiconductor applications, in this chapter we will primarily discuss about the fabrication technologies of single crystals and ceramics for thermoelectric and magneto-optical applications.

2. Thermoelectric Semiconductor Crystals and Ceramics

Thermoelectric (TE) materials are those which exhibit thermoelectric effect. Thermoelectric effect is essentially created by a temperature difference leading to electric potential or vice versa. These phenomenons are commonly referred as Seebeck effect (temperature conversion to electric potential), Peltier effect (Current conversion to temperature difference) and Thomson effect (heating/ cooling conductor). In principle, thermoelectric materials should have the following properties:

- High Seebeck coefficient S - to get the maximum output voltage per degree of temperature difference.
- High electrical conductivity σ - to minimize Joule heating
- Low thermal conductivity λ - to restrict the diffusion of the heat across the device in order to maintain a large temperature gradient.

In order to judge the efficiency of thermoelectric materials as an energy conversion, a factor named as "Figure of Merit, Z" is used. Where Z is defined as,

$$Z = \frac{\sigma S^2}{\lambda} \dashrightarrow \quad (1)$$

Conventionally thermocouples which were fabricated from metallic conductors including various other metals restricted its efficiency only ~ 3%. Since the discovery of semiconductors in 1950s the efficiency of thermoelectric generators [4, 5] was increased significantly. Materials with high thermoelectric figures of merit are typically heavily doped semiconductors and for many years the best materials had a figure of merit of around 1. Recent advances in materials science have increased this number to around 4. In practical applications to obtain a thermoelectric material with large voltage while holding a temperature gradient, its thermal conductivity must be low. This ensures that when one side is made hot, the other side stays cold. The best known examples of low thermal conductivity and high power factor were Bismuth Telluride (Bi_2Te_3) (BT), Lead Telluride (PbTe) (PT), and Silicon Germanium (SiGe) (SG) the three semiconductor expensive compounds with rare elements. Today, low thermal conductivity can be achieved by creating nanoscale features such as particles, wires, or interfaces in bulk semiconductor materials. These nanoscale features lower the thermal conductivity [6] of the semiconductor and do not affect their strong electrical properties.

Before studying the properties of semiconductor TE materials, an introduction on TE semiconductor materials is discussed in the following section. Ernest search for alternative energy sources has now recently revived interest for research of TE materials. High thermoelectric figures of merit [7-9] are exhibited by the heavily doped semiconductor materials. In 1954, H Julian Goldsmid [4] was the one who demonstrated 0°C cooling by using the thermoelectric quality factor based on BT. This led to the important identification of significance to have high mobility and effective mass combination and low lattice thermal conductivity in semiconductors up on good choice of dopants.

Henceforth researchers started to look for materials with high zT in small band gap semiconductors made of heavy elements. Later Glen Slack [5] generalized the material requirement with the concept of phonon behavior as in glass and electron mobility behavior similar to crystalline semiconductors came in to existence. Currently the materials engineering in the nanoscale lead to the improvement of thermoelectric efficiency.

Thermoelectric device efficiency is closely related to the choice of semiconductor. Semiconductors could either be p-type or n-type, in either of these types the fundamental principle is based on charge carrier movement. These charge carriers in the presence of a temperature gradients leads to the ability of charge diffusion from hot to cold zone until equilibrium is attained between the potentials of diffusion and electrostatic repulsion, resulting in a charge carrier accumulation (Seebeck effect). TE devices are constructed by alternating p and n type semiconductors thermally connected in parallel and electrically in series (Figure 1).

The material based on the common semiconductor lead telluride, is the most efficient thermoelectric material known. It exhibits a thermoelectric figure of merit (so-called "ZT") of 2.2, the highest reported to date. Figure 2 shows the figure of merit of various semiconductor TE materials. Efficiency of the current TE materials inhibits its commercialization. If in future upon successful achievement of TE parameters, there are various potential applications for domestic and industrial applications.

Some of the mention-worthy are automobile industries, cooking stations and large combustion engines, where the generated heat energy can be converted into electricity.

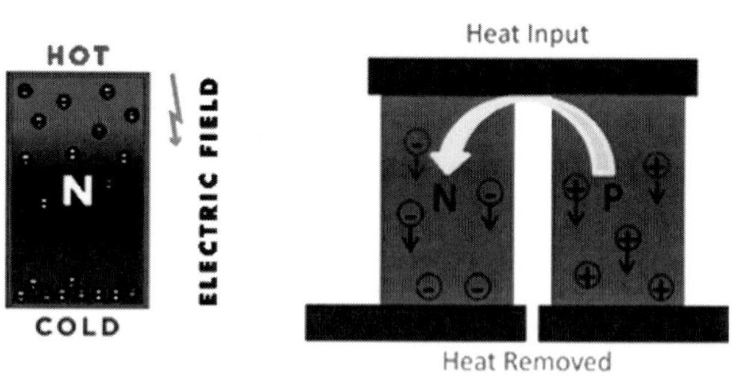

a b

Figure 1. (a) Mechanism showing Seebeck effect (b) Conventional thermoelectric device.

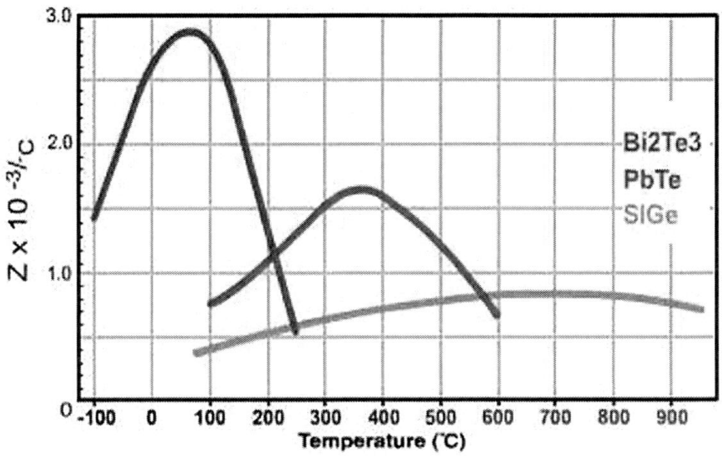

Figure 2. Figure merit of some well known thermoelectric materials.

TE materials are available both in the form of the crystals and ceramics. Initially in years of 70's and 80's semiconductor single crystals were streamlined only for Silicon (Si), III-V and II-VI compounds. The main problem experienced in these materials was the high vapor pressure and toxicity of the materials for TE applications. Later, V-VI materials have also been explored to study their thermoelectric properties. The bandgaps of several of the pure materials fall in the long wavelength region (Bi_2Te_3, 0.13 eV, 9.5 μm; Bi_2Se_3, 0.16 eV, 7.8 μm and Sb_2Te_3, 0.21 eV, 5.9 μm). Growth of these TE materials has been attempted by Czochralski method, Vertical Bridgman method, Floating zone and flux methods. However, there is also usage of ceramics for TE applications. Dresselhaus [10] in 1990s suggested that the low-dimension and nanostructure can significantly enhance the thermoelectric performance. Following which many research groups around the world have successfully demonstrated the TE efficiency of materials obtained through low-dimensional/ nanoparticles and nanocomposites [10, 11]. In the low dimensional structures, phonons contribute significantly to improve the "z". A phonon is a quantum of vibrational energy with different wavelength. When heat flows through a material, a spectrum of phonons needs to be scattered at different wavelengths (short, intermediate and long). Many of these nanoparticles are produced by using the ball milling and subsequently employing hot pressing has yielded TE ceramic [12] with average grain size of 20 nm. The aforesaid method is cost-efficient which can be translated to industrial scale production in short time. However, there are other factors that have to be taken in to consideration during conventional ball milling, because there is a large amount of heat causing lattice distortion and destruction/ disturbance of chemical bonds leading to amorphous phase and decomposition.

Ceramics are fabricated by forming the alloy either by induction or arc melting (Figure 3) followed by milling and densifying. Further it has to be noted that lower thermal conductivity and improved thermoelectric performance can be obtained when average grain sizes are made smaller than 100 nm.

Also the densification stage could lead to grain growth, which can be avoided by the usage of advanced rapid sintering methods such as Spark Plasma Sintering (SPS). In SPS due to the rapid heating and cooling rate, the grain growth can be limited to stay in the nano domain region. The rapid heating and Joule heating help in reducing the densification temperature and sintering time.

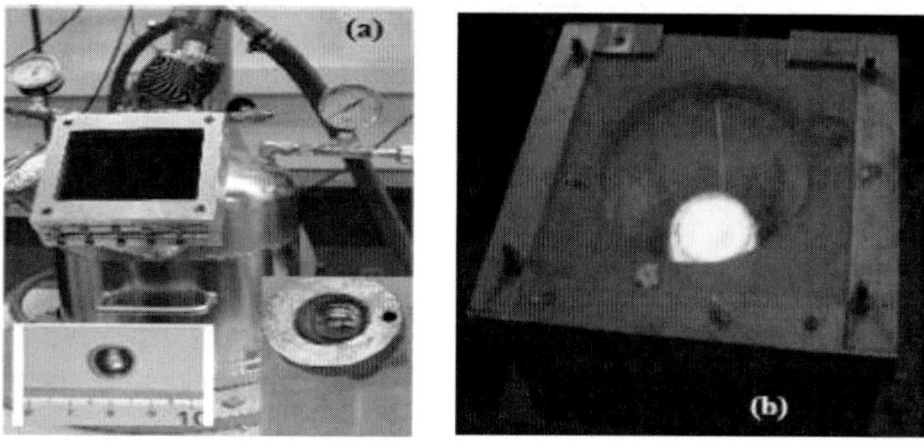

Figure 3. (a) Arc Melting furnace and (b) Induction melting furnace for alloy preparation.

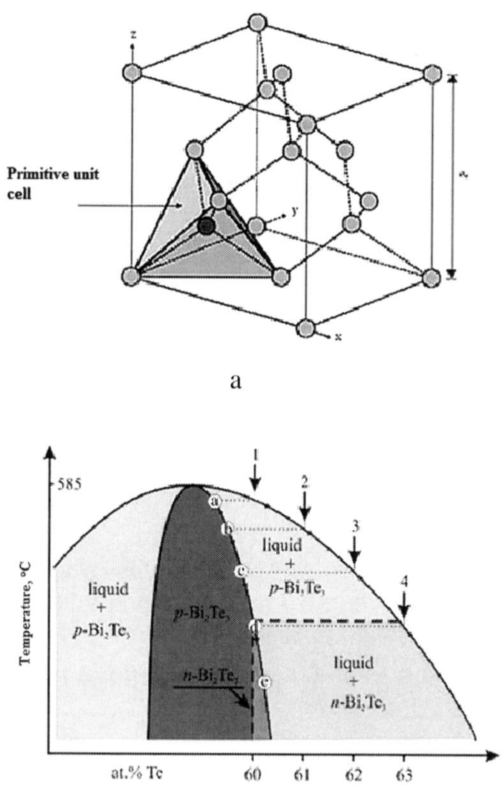

Figure 4. (a) Crystal structure and (b) Phase diagram of Bi_2Te_3.

In the following section, Bi_2Te_3 (BT) single crystal growth and ceramics will be discussed. To start with, a brief discussion about the structure is presented. BT belongs to the class of V_2VI_3 binary chalcogenide semiconductors with rhombohedral unit cell (Figure 4 (a)) symmetry space group R3⁻ m with 2 and 3 fold aced of symmetry C2 and C3.

BT has the capability to produce significant temperature gradient with the applied current, which has potential applications for power supply and in cooling systems. However these materials are prone to dislocations and cleavage structures due to its layered structure. Layered structures results from anisotropy of crystal bonding energy in different lattice direction, pronounced, where Vander Waals forces are involved along one principal crystallographic axis. Hexagonal symmetries are predominant in such types of structures. Generally, deformation stacking faults produce peak shifts, twin faults produce peak asymmetries and both types of faulting contribute to the peak broadening.

The phase diagram of Bi_2Te_3 has been discussed by Satterthwaite and Ure Jrr. Figure 4 (b) shows a part of the phase diagram of Bi–Te binary system close to the melting point of Bi_2Te_3. It can be observed that the composition with congruent melting is not a stoichiometric compound but is shifted to the Bi-rich side, which implies that the equilibrium melt for electrically neutral stoichiometric Bi_2Te_3 should be enriched with tellurium. The "boundary" melt composition, which separates primary crystallization of p- and n-type regions of Bi_2Te_3, was reported to be ~63 mol% Te + 37 mol% Bi. It was evidenced that the change of composition of bulk BT can be obtained by annealing at the same time with the possibility to incorporate the doping ions. Theoretically it can be verified the surface state consists of a single nondegenerate Dirac cone, while the bulk state remains insulating. Due to the nonstoichiometry, all the grown BT crystals exhibit either p- or n- type and hence behaves as a metallic compound in the bulk form. Much of the defects and nonstoichiometry arises from Te segregation. But the electrical properties of the BT can be tuned to the desired composition by annealing the crystal in the Te-vapour. The non-stoichiometry in BT is said to have lead to the various domains of different carrier types.

Ainsworth employed vertical Bridgman (VB) (Figure 5 (a)), Chemical Vapor Transport (CVT) (Figure 5 (b)), and Czochralski (Cz) methods for the growth of BT single crystals. Segregation of Te in the last to freeze material was a general problem in VB and CVT methods at slow growth rates and polycrystals growth was prevalent in these techniques at rapid heating rates. However Cz method yielded the best crystals upon stirring the melt at frequent intervals.

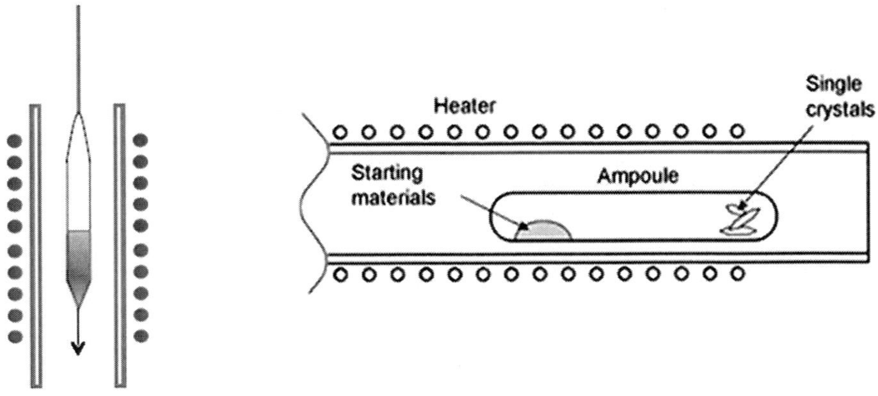

a b

Figure 5. Schematic diagram of (a) Vertical Bridgman technique and (b) Chemical Vapor transport.

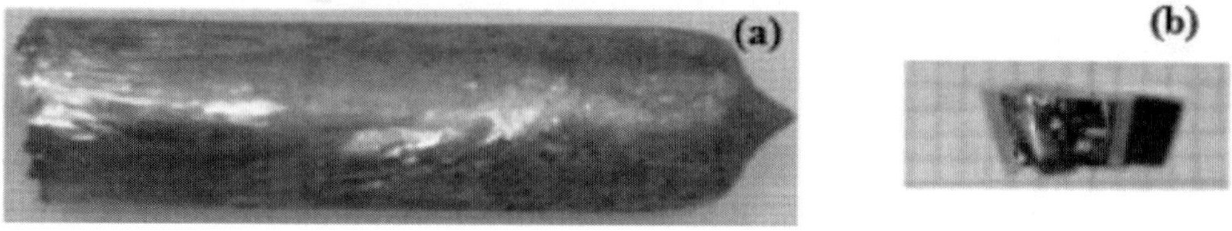

Figure 6. (a) Single crystal of a layered structure semiconductor (b) cleaved single crystal of Bi_2Te_3.

Recently various authors have successfully employed VB by conducting the crystal growth in fused quartz tube. Seeded growth and slow heating and cooling rates avoided the cleavage planes. Figure 6 shows one of grown layered structure crystal and the cleaved crystal of BT. Currently VB crystal growth technique is widely used for the growth of BT single crystals.

From the crystal growth of BT, it can be inferred that to obtain the crystals of BT is not easy and yet to be optimized. Nano-engineering of the bulk material is found to be an efficient way to improve figure of merit.

Various reports show that by using the ball milling followed by hot pressing resulted in significant enhancement of figure of merit. Currently hot-pressing, hot-extrusion have been extensively applied to prepare polycrystalline BT-based alloy bulks with preferred grain orientation, but their figure of merit was not enhanced due to the difficulty in controlling the grain growth. Mechanical alloying (MA) has several advantages over conventional melting or grinding techniques [13], and has been recently applied to fabricate nano-sized alloys. Spark plasma sintering (SPS) is suitable for fabricating TE materials [14] since this technique can restrict grain growth by sintering at a low sintering temperature and in a short time. Researchers have confirmed that MA and SPS are simple but effective methods to prepare TE materials even with complex compositions. BT-based materials with fine grains in the nanometric order were fabricated by MA and SPS. The aforesaid results indicate that the thermal conductivity is reduced by refining grain size, but the electrical transport properties are not as high as those of single crystal, which is due to the grain orientation of the crystals. By using SPS with multi step sintering and hot forging, it is possible to obtain the grain orientation and to yet retain the nano grain form.

It is believed that with further improvement in sintering conditions, it is possible to obtain BT ceramics with improved thermoelectric properties. The problem of segregation can be avoided in ceramic form of BT. These type of research is in the initial stage and hence cannot be converted yet in to a commercial process.

3. Magneto-Optic Semiconductor Crystals and Ceramics

Magneto-optics is the study of the effects of a magnetic field on specific properties of light, such as polarization. Magneto-optics [15, 16] enables us to see spins of carriers and magnetic ions in bulk semiconductors and quantum structures. A magneto-optic crystal is a photonic crystal that is comprised of magneto-optical material such that the optical response of the device depends on the magnetization of the magneto-optical material. The magneto-optical effect can be enhanced by means of optical resonances. Magneto-optics (MO) can be observed in medium and large gap semiconductors, and by interband transitions.

By doping and by exchange of *sp-d*, we could convert a semiconductor magnetic. Three types of semiconductors: (A) a magnetic semiconductor, in which a periodic array of magnetic element is present; (B) a diluted magnetic semiconductor, an alloy between nonmagnetic semiconductor and magnetic element; and (C) a nonmagnetic semiconductor, which contains no magnetic ions. Figure 7 shows the structure of semiconductors and different types of semiconductors. Diluted magnetic semiconductor (DMS) is one of the key elements in spintronics. Spintronics, or spin electronics, involves the study of active control and manipulation of spin degrees of freedom in solid-state systems. It employs integrated use both of charge and spin of electrons. The idea of a DMS, where a small (<10%) concentration of magnetically active atoms (most often Mn although Co and Cr have also been used occasionally) is distributed at the cation sites of the host semiconductor, is favorable because such a system may have both semiconducting and ferromagnetic properties.

However, fabrication of such materials has been difficult due to the magnetic transition metal (i.e., Mn) which is not thermodynamically stable in the semiconductor host (i.e., GaN) and tends to segregate. The Faraday effect or Faraday rotation involves an interaction between light and magnetic field in a medium. The Faraday effect causes a rotation of the plane of polarization which is linearly proportional to the component of the magnetic field in the direction of propagation. In an istropic medium all the polarization represent eigen modes of the electromagnetic wave. In Faraday configuration only σ^+ and σ^- represent eigen modes. Decomposition of linearly polarized light in two circularly polarized waves, causing the phase shift and rotation of the polarization plane. Figure 8 shows the effect of Faraday rotation in the semiconductor materials. In the following sections, we will discuss about the one of the well known DMS magneto-optical material of Cadmium manganese telluride ($Cd_{1-x}Mn_xTe$) (CMT) [17].

This magneto-optical material has a giant Verdet constant. Potential applications of CMT include Faraday rotation devices (optical isolators in the range 630 nm to 1100 nm), LEDs, Spintronics, electro-magnetic interference free devices, field tunable phase shifters, small coupled solar cells for fly-by-light applications, phased array radar with completely fiber-fed antennae, bomb detonators and harsh environment sensors.

$Cd_{1-x}Mn_xTe$ is a wide-gap diluted magnetic semiconductor known to form in the zinc blende structure in the Mn composition range of $x = 0$–0.77. In the absence of a magnetic field, the CMT crystal has an electronic structure of a typical II–VI semiconductor. In the presence of a magnetic field, many interesting magnetic, optical and magneto-optical properties have been observed and these have been widely investigated.

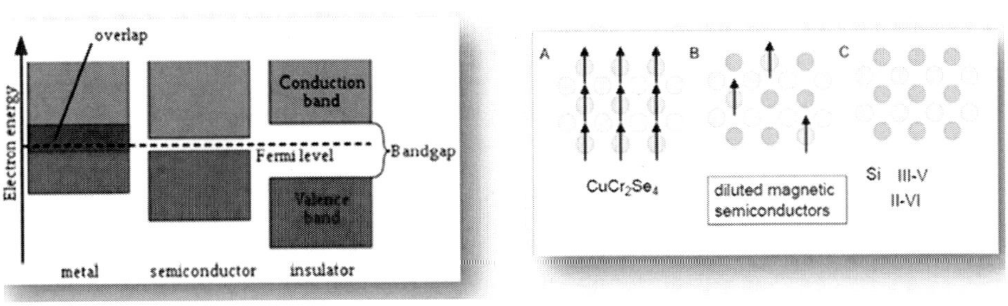

a b

Figure 7. (a) Band gap energy structure and (b) types of semiconductors.

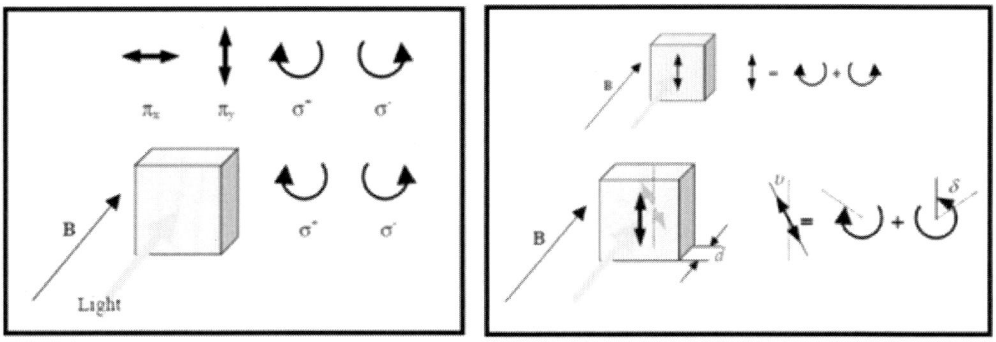

a b

Figure 8. Effect of (a) Faraday rotation in the semiconductors and (b) Decomposition of polarized light in to two circularly polarized waves.

The material has been identified as a candidate for many practical applications, such as magnetic field sensors, solar cells, optical isolators and lasers.

Recently, increased attention has been paid to CMT because it is revealed that CMT is a good candidate to compete with $Cd_{1-x}Zn_xTe$ in X-ray and γ–ray nuclear detector applications. Some of the well-known heavy atom semiconductor detectors of gamma radiation, operating at room temperature, such as Hg_2I_2, CdTe, CdZnTe, as well as GaAs were investigated earlier by many researchers. A common problem observed with these currently used materials is the low mobility and short lifetime of the holes, and their tendency to get trapped in deep levels within the semiconductor forbidden band. Compared to the widely used detector material CdZnTe, CMT has several advantages in terms of segregation co-efficient of Mn and homogeneity in the melt grown crystal. Though CMT crystals are reported to exhibit detector properties, there are several practical difficulties hindering it to be used for detector applications and other devices due to: The presence of a solid-state phase transition at high temperature and low stacking fault energy make it extremely difficult to grow single phase CMT bulk crystal without twin. Diluted Magnetic semiconductors (DMS) [18-20] based on II-VI compounds have been realized for its application in optical isolators utilizing large Verdet constant in the region of optical absorption edge. The bulk crystals of $Cd_{1-x}Mn_xTe$ have been grown from melt with different growth methods; however there are no reports on the CMT with high crystalline perfection until to date, as it requires more work. Figure 9 (a) and (b) are the crystals of CMT with different composition grown by us [21] at ICMCB-CNRS, France.

The procedure of the single crystal growth of CMT done at ICMCB-CNRS, France by our research team is given below.

High purity raw materials of Cd, Mn and Te were used as starting materials for synthesis. In order to avoid chemical reaction of Mn ion with ampoule, the ampoules were prepared by coating with graphite. The graphitized quartz tube containing the starting material was sealed under pressure of 10^{-6} Torr. The synthesizing process was controlled in a self-designed rocking furnace for 48 h and the crystal growth was carried out by vertical Bridgman technique in a two zone furnace with the pulling rate of 1 mm/h and the temperature gradient of 13 K/cm was maintained to avoid constitutional supercooling. CMT crystals were grown below 1150°C to avoid the peritectic decomposition of MnTe. The temperature profile was optimized to make an ideal condition for minimizing the occurrence of twins in the crystal.

The crystal growth experiments were carried out by varying various growth parameters such as growth rates and temperatures. The liquidus temperature does not alter with composition for x=0.1 to 0.7. This has been ascertained by the homogeneous crystals obtained from the growth cycle. However, when x>0.7, MnTe melts peritectically and decomposes in presence of Cd at temperatures below 1150°C.

With stoichiometric quantities of the starting compounds sealed in the quartz ampoule, the temperature of the materials enclosed in the sealed ampoule was slowly increased to reach 50°C above the melting point. The top and bottom zones were subjected to an inverse temperature gradient for 24 h in order to forcibly mix and homogenize the material.

After the growth cycle, to avoid the solidification of excess element, the temperature gradient was inverted. The stability of the temperature at the growth interface was controlled and varied to be less than 1°C. After the growth, the ingot was in-situ annealed and then slowly cooled down to room temperature to minimize the thermal stress. Twins and polycrystals were prevalent throughout the last to freeze region.

The occurrence of twins can be attributed mainly to Zincblende-wurtzite phase transition. CMT crystals grown by Bridgman method are found to contain higher concentration of twins The twins occurring at the last to freeze region in the CMT crystal could be due to solid state phase transition. The as-grown p-type CMT crystals were annealed with Cd vapor to compensate the native defects of Cd vacancies caused due to the high vapor pressure of Cd at high temperatures. Figure 9 shows the example of CMT crystals with different compositions obtained at ICMCB-CNRS, France.

From the aforesaid it can be seen that fabrication of magneto-optical single crystals is not yet matured, so an alternative fabrication technique in the form of Ceramic is in progress. It has been reported that MO single crystals have a figure-of-merit of an order of magnitude higher than magneto-optical glasses and a large thermal shock parameter, which allows their usage in laser systems with kilowatt average power.

An alternative way is to use magneto-optical ceramics that possesses figure-of merit comparable with single crystals and a larger thermal shock parameter. One of the widely used MO materials to date is terbium gallium garnet (TGG) which has high value of Verdet constant and thermal conductivity and low absorption losses in the 500-1100 nm wavelength range. High opti'cal quality TGG ceramics with the temperature dependence of the Verdet constant have been reported very recently.

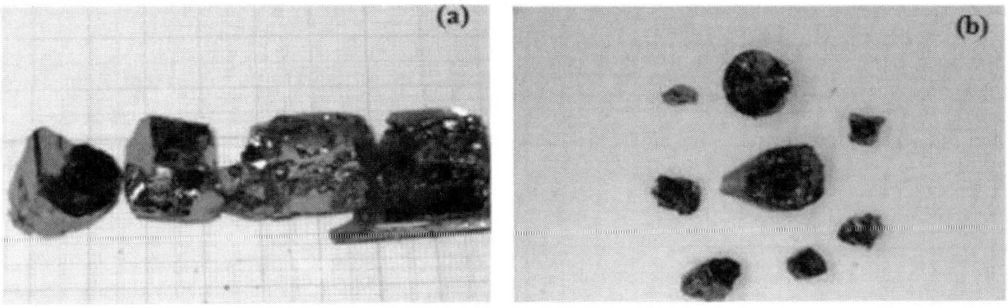

Figure 9. Cleaved semiconductor crystals of CMT with different compositions in (a) and (b).

Conclusion

Semiconductors are multifunctional and have usage in various devices, it is essential to know about the various existing fabrication techniques. In this chapter we have discussed about the applications of semiconductors in thermoelectric devices and magneto-optics. Earlier the only form that was used for the thermoelectric and magneto-optical devices were single crystals. But currently due to the advancement of sintering technology, it is possible to realize the thermoelectric and magneto-optics in the form of ceramics. Currently the most commonly and successfully established technique for the elaboration of single crystals of single crystals of semiconductor thermoelectric and magneto-optical devices is Bridgman technique. In order to obtain a dense form with nanograin and with equivalent properties as that of single crystals is densified nanostructured thermoelectric and transparent magneto-optical ceramics. The aforesaid type of ceramics is realized by Spark Plasma Sintering. This chapter could help in understanding the processing of the semiconductor fabrication materials for device applications.

References

[1] Timothy, D. Sands (2005), *Designing Nanocomposite Thermoelectric Materials.*

[2] Tomasz Dietl, (2010) *Nature Materials* 9, 965-974.

[3] Hideo Ohno, (2010) *Nature Materials* 9, 952-954.

[4] Goldsmid, H. J., Douglas, R. W. (1954) *British Journal of Applied Physics*, Volume 5, Issue 11, pp. 386-390.

[5] Glen Slack, (1995) *CRC Handbook of Thermoelectrics*; Rowe, D. M., Ed.; CRC Press: Boca Raton, FL, pp. 407-440.

[6] Jeffrey Snyder, G. and Eric Toberer, S. (2008) *Nature Materials* 7, 105-114.

[7] Ure, R. W. and Heikes, R. R. (1961) *Thermoelectricity: Science and Engineering*, Interscience Publishers: New York Ch. 11, 339-388.

[8] Cadoff, I. B. and Reinhold, E. (1960) *Thermoelectric Materials and Devices,* Reinhold Publishing Corporation, New York, 143-162.

[9] Egli, P. H., (1960) *Thermoelectricity*. John Wiley and Sons., Inc.: Washington DC, pp. 26 - 207.

[10] Dresselhaus, M. S., Chen, G., Tang, M. Y., Yang, R. G., Lee, H., Wang, D. Z., Ren, Z. F., Fleurial, J.-P., and Gogna, P. (2007) *Adv. Mater.* 19, 1043-1053.

[11] Bell, L. (2008) *Science*, 321, 1457-1461.

[12] Kanatzidis, M. G. (2010) *Chem. Mater.* 22, 648-659.

[13] Xiao Yan, Giri Joshi, Weishu Liu, Yucheng Lan, Hui Wang, Sangyeop Lee, J. W. Simonson, S. J. Poon, T. M. Tritt, Gang Chen, Z. F. Ren, (2011) *Nano Letters*, 11 (2), 556-560.

[14] Rowe, D. M. and Bhandari, C. M., (1983) *Modern thermoelectrics* (Holt-Saunders Ltd, London.

[15] Gaj, J A., Galazka, R. R. and Nawrocki, M., (1978) *Solid State Communication* 25, 193-195.

[16] Bartholomew, D. U., Furdyna, J. K. and Ramdas, A. K., (1986) *Phys. Rev. B* 34, 6943-6950.

[17] Mycielski, A., Burger, A., Sowinska, M., Groza, M., Szadkowski, A., Wojnar, P., Witkowska, B., Kaliszek, W., and Siffert, P. (2005). *Phys. Status Solidi C* 2, 1578-1585.

[18] Furdyna, J. K. (1988) *J. Appl. Phys*. 64, 29-64.

[19] Kossut, J., Beer, A. C. and Furdyna, K. *Diluted magnetic semiconductors, semiconductors and semimetals 25*, Academic, 1988.

[20] Burger, A. (1999) *Journal of Crystal Growth* 198, 872 - 876.

[21] Mythili Prakasam, Oudomsack Viraphong, Lionel Teule-Gay, Rodolphe Decourt, Philippe Veber, Encarnacion G. Villora, Kiyoshi Shimamura (2011). *Journal of Crystal Growth* 318, 533 - 538.

INDEX

A

B

C

D

E

F

G

H

M

N

O

P

Q

R

S

T

U